The Art of Rice

The Art of Rice

Spirit and Sustenance in Asia

Roy W. Hamilton

UCLA Fowler
Museum
of Cultural History
Los Angeles

With contributions by

Aurora Ammayao
Mutua Bahadur
Francesca Bray
Francine Brinkgreve
Eric Crystal
Pamela Deuel
Mariko Fujita
Ruth Gerson
Stephen C. Headley
Rens Heringa
Gene Hettel
Garrett Kam
Thitipol Kanteewong

Pattana Kitiarsa
Gisèle Krauskopff
Nanditha Krishna
Kwang-Kyu Lee
Kurt W. Meyer
Sohini Ray
Michael Reinschmidt
Toshiyuki Sano
Suriya Smutkupt
Kik Soleh Adi Pramono
David J. Stuart-Fox
Vi Văn An
Vũ Hồng Thuật

The Art of Rice was made possible by major funding from

National Endowment for the Humanities,
dedicated to promoting excellence in the humanities
Rockefeller Foundation
Getty Grant Program
Henry Luce Foundation
UC Pacific Rim Research Program
UCLA School of the Arts and Architecture
Nikkei Bruin Committee
Yvonne Lenart Public Prorams Fund

NATIONAL ENDOWMENT FOR THE
HUMANITIES

Any views, findings, conclusions, or recommendations expressed in this exhibition do not necessarily represent those of the National Endowment for the Humanities.

Lynne Kostman, *Managing Editor*
Diane Mark-Walker, Michelle Ghaffari, Dinah Berland, *Editors*
Gassia Armenian, *Editorial Assistant*

Lausten and Cossutta Design, *Designers*
Danny Brauer, *Production Manager*
Don Cole, *Principal Photographer*
David L. Fuller, *Cartographer*
Robert Swanson, *Indexer*

UCLA Fowler Museum of Cultural History
Box 951549
Los Angeles, California 90095-1549

Requests for permission to reproduce material from this volume should be sent to the UCLA Fowler Museum Publications Department at the above address.

Printed and bound in Hong Kong by
South Sea International Press, Ltd.

Library of Congress Cataloging-in-Publication Data

Hamilton, Roy W.
    The art of rice : spirit and sustenance in Asia /Roy W. Hamilton ; with contributions by Aurora Ammayao...
    [et al.]. p. cm.
    ISBN 0-930741-98-6
    1. Rice—Social aspects—Asia. 2. Rice—Asia—Folklore. 3. Rice planting rites—Asia. 4. Rice gods—Asia. 5. Asia—Social life and customs. 6. Asia—Religious life and customs. I. Ammayao, Aurora. II. University of California, Los Angeles. Fowler Museum of Cultural History. III. Title.

GR265.H35 2003
398'.36849'095—dc22

2003058024

Cover: Detail of figure 1.7. *Page 2:* Figure 18.16. *Page 10:* Detail of figure 33.1. *Pages 18–19:* Women give trays of rice a final cleaning before cooking (photograph by Stephen P. Huyler, Uttar Pradesh, India). Back cover: Figure 18.13.

Contents

Foreword

The title of this volume, *The Art of Rice: Spirit and Sustenance in Asia*, unmistakably declares its scope and ambition. Its sheer size, with thirty-five essays and additional entries written by twenty-seven authors from ten different countries, is further testimony to its impressive aims. This is one of the UCLA Fowler Museum's largest and most complex undertakings, and one with goals that are equally far-reaching. Unlike our usual investigations, which have dealt in depth with single ethnic groups, particular artistic genres, or systems of visual culture that can be traced from a point of origin to its diaspora, this effort encompasses the vast geographical construct called "Asia" and the innumerable ways that the growing and eating of rice have become intimately bound to aspects of personal identity; notions of family, community, and state; and systems of religious belief and ritual activity. More than three billion people, most of them in Asia, consider rice the staple of their diet. Taking on a topic of this scope was no small task for a museum of our size!

First discussed by staff in 1995, the project was formally initiated the following year when the first planning grant was received to develop institutional partnerships with scholars in six regions of East, Southeast, and South Asia. Roy Hamilton, Curator of Asian and Pacific Collections at the Fowler, has served as the project director, and from the outset he has had the ambitious goal of forging highly productive and lasting links among international scholars, artists, museums, and universities across Asia, America, and Europe. The initial seed money for the endeavor was provided by the University of California's Pacific Rim Research Program; it allowed for the establishment of six regional teams. In 1997, a second planning grant received from UCLA's School of the Arts and Architecture allowed for project development in India, China, and Japan. Over the next few years *The Art of Rice* garnered two research grants from the Getty Grant Program and again from the UC Pacific Rim Research Program for work in Vietnam, Thailand, the Philippines, Indonesia, China, India, Japan, and Korea. Three planning meetings were held to convene participating scholars from around the world, one in the Netherlands, one in Thailand, and one at UCLA. Hamilton also began the enormous task of identifying objects for the exhibition by surveying museums in the United States. We are indebted to the aforementioned grantors for recognizing the significance of the project's goals, one of which was the creation of an international, interdisciplinary collaborative research team. On the basis of these early years of intensive investigation, *The Art of Rice* received additional major support from the National Endowment for the Humanities, the Rockefeller Foundation, and the Henry Luce Foundation to enable several significant outcomes: a major traveling exhibition with many loans from museums in America and abroad, as well as newly commissioned works from Asia; a substantial multi-author scholarly publication; and a program of innovative rice-related educational activities for audiences of all ages. The latter programs were also supported by funds from the Nikkei Bruin Committee and the Fowler's Yvonne Lenart Public Programs Fund.

The subject matter of this project is deceptive—rice, after all, is just a food. However, as the book and the exhibition it accompanies attest emphatically, it is more than just food. It is indeed "sustenance" and of a type that functions on many levels for different Asian nations and peoples. For example, there are many beloved rice deities, especially goddesses, who create and protect the sacred grain. Rice deities are associated directly with a bountiful crop as well as with prosperity more generally. Indeed, rice is so fundamental to the many Asian peoples who grow and eat it that it has become synonymous with life itself.

As a cultural history museum the Fowler's particular interest in the subject has been the intersection of rice's significance with a diverse and fascinating array of cultural practices and artistic expression. The visual culture of rice in Asia dates back thousands of years yet continues to evolve in response to modernization and globalization. The great accomplishment of this project is that it surveys the full range of visual expression, from the urban to the rural, from the courtly to the popular, from the past to the present.

Hamilton is to be congratulated for the successful realization of this impressive enterprise. It is his remarkable energy, penetrating focus, deep commitment, and wide-ranging intellectual curiosity that have propelled *The Art of Rice* forward. His engagement with the subject matter has been infectious, inspiring those who have partnered with him, and he has been the glue that has held this complex global effort firmly together. On behalf of the entire Fowler Museum staff, I wish to extend our profound thanks to Roy Hamilton. We also extend our gratitude to all his worldwide collaborators on the exhibition and book; the contributors to this volume are too numerous to cite here but some of their many accomplishments are briefly described in the list of contributors at the back of the book. We thank the many lenders to the exhibition, both private and institutional, who are recognized in the front matter. Two museums are partnering with us on the exhibition's national tour, and we gratefully acknowledge at COPIA: The American Center for Wine, Food and the Arts, Betty Loar, Executive Director, and Betty Teller, Assistant Director for Exhibitions, and at the Honolulu Academy of Arts, Stephen Little, Director.

It is not an exaggeration to say that projects of this scale and scope are only possible in an internal context of cooperation and collective enthusiasm. Members of the Fowler Museum staff are exemplary in their commitment to bringing complex projects like this one to fruition. *The Art of Rice* presented unique challenges by virtue of its internationalism and the number of participants involved in the publication and the exhibition. My thanks, first of all, go to my predecessor and friend, Doran Ross, who spearheaded the project in its early years and worked closely with Hamilton to establish its parameters and goals. This made it easier for me to step in at a later stage and provide support and assistance as the project took final shape. Polly Roberts, Deputy Director and Chief Curator, has been a generous and helpful colleague. The impressive funding *The Art of Rice* has received is due in no small part to the superb efforts of our development staff, Lynne Brodhead, Director of Development, and Leslie Denk, Associate Director of Development. The complexity of the exhibition—its preparation and implementation—has required the concerted efforts of many key staff and departments. Sarah Kennington, Registrar, and Farida Sunada, Associate Registrar, have done a masterful job of requesting and tracking worldwide loans. Rachel Raynor, Curatorial Assistant, provided essential support in this lengthy and often cumbersome process. One of the project's biggest concerns involved the ways certain objects, many made of unusual organic and ephemeral materials, were to be responsibly conserved for the purposes of exhibition. Jo Hill, Director of Conservation, handled these challenges with her usual exactitude. Her greatest triumph was the reconstruction and treatment of a wood, interior house partition from Java made of 109 pieces and forty feet in length. She was assisted in these efforts by Patricia Measures, Assistant Conservator, and a number of interns and volunteers. Fran Krystock, Collections Manager, and Jason DeBlock, Assistant Collections Manager, also solved some thorny problems regarding the storage of certain items, such as this "wall" from Java, and did their usual meticulous job of ensuring the safety of the objects as they made their way into the galleries. David Mayo, Director of

Exhibitions, has proven yet again his enormous talent in transforming a large open gallery into a unique exhibition experience. In this instance, the elegant choices of color and the skillful manipulation of space to create moments of high drama have provided the perfect stage for an exhibition of this thematic and artistic diversity. As the installation morphed over time, he has responded with resilience to the need for redesign. Members of his team are to be congratulated for their execution, especially Martha Crawford for her highly sympathetic graphic package, Mike Sessa for his skillful coordination, and Wendy Phillips for her remarkable resourcefulness. The myriad details of the exhibition's tour have been diligently structured and overseen by Karyn Zarubica, Traveling Exhibition Coordinator.

Our education staff, led with a very sure hand and keen eye by Betsy Quick, Director of Education, was integral to the development of the exhibition. Quick worked closely with Hamilton to develop the exhibition's various strategies for interpretation, from didactic panels to several kinds of labels and interactive media. A "Rice Box" has been developed for self-guided family tours and a curriculum resource unit for K–12 students. Lyn Avins has provided critical support in the development of the curriculum guide, and Alicia Katano, Assistant Director of Education, has assisted with the school-based visitor programs. Quick was also responsible for developing a diverse menu of outreach activities to complement the exhibition. The obvious culinary possibilities this project offers have been creatively exploited with a "Rice Fest" family day and a restaurant tour to several of the many Asian ethnic communities in the greater Los Angeles area. Ilana Gatti, Program and Events Coordinator, and Jonathan Ritter, Campus Outreach Coordinator, have been indispensable to the success of these complex undertakings. Stacey Ravel Abarbanel, Director of Communications, has used an effective combination of tenacity and innovation in her efforts to bring this project the publicity and promotion it deserves. David Blair, Assistant Director, has done his usual magic in keeping the project on budget and under control given the multiple sources of revenue and many participants and vendors to be compensated. He has been ably assisted by Jennifer Garbee, Accounting Administrator. Additional recognition goes to Marylene Foreman, Human Resources Manager, Betsy Escandor, Executive Assistant, and Gassia Armenian, Curatorial Assistant, for their generous help with whatever needed to be done.

Given the size and scope of this publication, it has presented its own set of challenges. The tasks of overseeing the editing and production of the volume were managed with great competence and equanimity by Danny Brauer, Director of Publications, and Lynne Kostman, Managing Editor. Editing the many essays and entries was shared among several fine editors: Diane Mark-Walker, Michelle Ghaffari, and Dinah Berland. Gassia Armenian also provided critical editorial assistance. Kostman coordinated all of their efforts and did a large share of the editing herself, ensuring stylistic and semantic consistency across chapters written by many authors for whom English is not a first language. Her work on this publication has been nothing less than amazing. The book was designed handsomely and sensitively by Lausten and Cossutta Design; Brauer managed the complex details of production and printing with the ease and assurance that can only come from tremendous experience. The beautiful studio photography was done by Don Cole, whose talents are evident in images of objects as small as sake cups or as large as the Java wall.

I have the sense that after reading this book—and, for those who can, seeing the companion exhibition—that eating rice will never be the same! We will begin to recognize that the grain we consume with various Asian-inspired cuisines in this country is not the same as the minute rice of our childhoods. Moreover, it will also become clear that for many people across Asia it is not enough to expect fully stocked supermarket shelves; rituals and offerings must still be made to insure a successful crop and the beneficence of a pantheon of rice deities. Likewise, the rich and various arts produced today—like those created for generations before—continue to express the essential role and importance of rice to Asian peoples worldwide.

Marla C. Berns
Director

Preface

Culture's beginnings:
rice-planting songs from the heart
of the country.
         —Bashō (Japanese, 1644–1694)

None of the keywords in the subtitle of this book—*spirit, sustenance, Asia*—is as straightforward as it might seem. *Asia*, outwardly the most concrete, is actually a problematic and highly artificial construct. Nothing—not language, not religion, not politics, not even geography—unites this enormous region or makes the name a useful construct. In this volume it is used as a rather crude but necessary short-hand, for the book does not deal with all parts of the continent equally. Rather, our subject is specifically those parts of South, Southeast, and East Asia (home, to be sure, to the vast majority of the population) where rice is not only the staple food but also the focal point of a pervasive set of interre-lated beliefs and practices. This book explores those beliefs and practices, especially as they can be traced through the arts and material cultures of the peoples involved.

      Ironically, rice is perhaps the one thing that comes closest to uniting at least a large part of what we know as Asia. In the course of my research, I came across authors from almost every Asian nation (and Madagascar as well) who claimed that rice cultivation originated in their respective coun-tries. Archaeologists have, for the moment, settled on China as the likely place of first domestication, but more than national pride is behind the claims of these authors. If we pause to think that Asian rice farmers have developed an astonishing 120,000 varieties of rice over untold generations, we can see that the people of each community were in fact the originators of *their* rice as *they* know it and love it. Everybody in "Asia," it seems, has a stake in rice.

      Sustenance too has special meaning here. Rice indeed sustains those who eat it, in a purely metabolic sense, whether in Asia or elsewhere. But many in Asia feel that rice is a divinely given food that nourishes them in a way that no other food can. Their bodies, they feel, are essentially com-posed of this sacred grain. *Sustenance*, then, is not used merely in a nutritional sense but may also be understood as constitutional and even spiritual.

      This leads us to the most complicated component of the subtitle, *spirit*. Many of the under-lying ideas about rice in Asia have common roots in spirit beliefs, no matter how much they may have evolved subsequently under the influence of diverse religious systems. By spirit beliefs I mean the attri-bution of conscious life to natural objects and the concept that living things have spirits or souls that are separable from their physical forms. Both of these principles underlie Asian rice beliefs. Rice has a spirit (singular), and spirits (plural) of the rice exist as well. Phrased another way, the rice plant pos-sesses a living spirit, and that spirit lives from year to year in the form of the rice spirits, sometimes envisioned as a rice "mother" or rice deity. Despite the range and depth of these complex ideas, it was a rather different meaning of spirit that caught the attention of the Asian-born community advisors from around Southern California who gathered to discuss their feelings about rice. It is noteworthy in this regard that *spirit* can also mean "a special attitude or frame of mind; the feeling, quality, or disposition characterizing something" (*Merriam Webster's Collegiate Dictionary* 1997, 1134). To our advisors, rice was a quintessential feature of human life. Without it, they would not be human. Life, in short, would be unimaginable without rice.

I first became aware of many of the ideas that are discussed in this book as resident in Indonesia in 1971. The little I knew about other Asian countries at the time suggested that similar, and possibly related, ideas might be found in varying combinations in many different cultures. This supposition was reinforced as I later on, as a researcher, began digging into the vast but widely scattered body of literature dealing with cultural aspects of rice in Asia. I initiated this project as an exploration of those connections, and it was purposefully planned from the beginning as an international collaborative effort. As a result, this volume draws together the work of twenty-seven authors representing many different cultures and writing from many different perspectives. The essays have been organized according to the main themes of the book, but not otherwise heavily moderated. Most of the objects in the exhibition are discussed in the chapters that I have written to lead off each part of the book, while the contributing authors of the chapters that follow have been encouraged to use their own voices to explore diverse byways. We are proud of the many international ties among artists, scholars, and institutions that have emerged as a result of this collaboration, and we hope readers will find this deliberate polyphony stimulating.

Each object in the exhibition was chosen for the story it has to tell and positioned according to its place in the unfolding of the various themes addressed. In the exhibition and the book, we have purposefully juxtaposed items from various cultures and time periods, which we hope will encourage visitors and readers to make cross-cultural comparisons. It must not be assumed, however, that a particular belief found in one location necessarily holds sway everywhere. For this reason we have endeavored to provide as much information as possible about the specific origin of each object and the context in which it was made or used. Some may be nonplussed by our mixing of priceless works of fine art with agricultural tools battered through years of hard use, popular religious depictions intended as souvenirs, ephemeral objects made of straw, and ritual rice-dough offerings, but that is to miss the point. Each has its place in our story, and each speaks with equal eloquence, if we let it, for the cultural history of the community that produced it and invested it with meaning.

Curating an exhibition of this scope is in some ways an unenviable job, as it is impossible for any one person to have a truly authoritative understanding of so many diverse objects. Like most curators, I would be more comfortable writing about a tiny range of objects that I know well. I can only apologize in advance for the shortcomings that undoubtedly have resulted from such a wide-ranging and syncretic project.

This book comes at a time of unprecedented familiarity with, and appreciation for, rice in the United States. In the 1950s, my mother purchased rice in one-pound plastic bags, in a supermarket where it occupied a couple of feet of shelf space surrounded by dried beans and macaroni. Her choices were limited to long grain, short grain, or Uncle Ben's. She would cook half the bag and put the rest away in a jar in the cupboard, where it would stay—for weeks or months—until we ate rice again. Today I buy my rice in an Asian-owned supermarket in Los Angeles, where an entire aisle is filled with rice in sacks weighing up to fifty pounds, and the choices have expanded to include jasmine, basmati, glutinous or "sticky" rice, brown rice, black rice, red rice, California-grown "Japanese" rice, pricey real Japanese rice in fancy packets, parboiled rice from India, popped rice, broken rice (once a necessity for the frugal, now turned into a culinary triumph by Vietnamese cooks), and the Philippine specialty *pinipig*, immature rice cooked and then flattened to make a snack. The next aisle over has rice noodles, and another rice wine. Across the street at the local Thai restaurant, on Sunday afternoons Mexican and Armenian families jostle for table space. It seems that in the twenty-first century, all of us are increasingly made of rice.

Roy W. Hamilton
Curator of Asian and Pacific Collections

Acknowledgments

From the beginning this project was planned as an international collaborative effort, and this volume has been immeasurably enriched by the participation of twenty-six contributing authors (listed on the contents page) from ten countries. I thank all of those who contributed for their dedication and for sharing their knowledge and friendship. In addition, we received support and assistance from many institutions abroad; in particular I would like to thank Gene Hettel and Aurora Ammayao at the International Rice Research Institute in the Philippines; Nguyễn Văn Huy, Vũ Hồng Thuật, Vi Văn An, Trần Trung Hiếu, Nguyễn Thị Thu Hương, and Trần Thị Thu Thủy at the Vietnam Museum of Ethnology; Toshiyuki Sano at Nara Women's University; Vithi Phanichphant and Thitipol Kanteewong at Chiang Mai University; Suriya Smutkupt and Pattana Kitiarsa at Suranaree Institute of Technology in Nakhon Ratchasima; Kik Soleh Adi Pramono and Karen Elizabeth Sekar Jaya at the Mangun Dharma Arts Center in Java; Dr. Nanditha Krishna and M. Amirthalingam at the C.P.R. Environmental Education Centre in Madras; Soojung Kang at the National Museum of Contemporary Art (Seoul); and Hyun-Mee Oh at the Seoul Museum of Art. For additional assistance abroad, I thank Rens Heringa, Nguyễn Anh Hiếu, Garret Kam, and Dwi Sutaryantha.

My search for objects for the exhibition that accompanies this book took me to many museums in the United States, and I sincerely appreciate the efforts of all those who assisted me—unfortunately, they are too numerous to mention individually. For help well beyond the call of duty, however, I would like to express my gratitude to Patricia Graham at the Nelson-Atkins Museum, Andrew Maske at the Peabody Essex Museum, and Meher McArthur at the Pacific Asia Museum. For my cross-town colleagues at the Los Angeles County Museum of Art, who were no doubt called upon more frequently than others, I want to express my particularly warm regard and thanks: Hollis Goodall, Keith Wilson, Robert T. Singer, Robert L. Brown, Steven Markel, and June Li.

We have borrowed objects from twenty-six institutions and twenty-one private collectors on four continents, and we gratefully acknowledge the many staff members who handled the often complex institutional loan arrangements and truly recognize their vital contribution to the exhibition. The many private lenders, who are listed in the front of the book, not only loaned their objects but also graciously and willingly shared their expertise.

A number of works were commissioned from abroad for the exhibition and have helped us to present Asian rice cultures as living traditions. For these I thank Gourishankar Bandopadhaya in Calcutta, Kik Soleh Adi Pramono and Daniel Mulyana in Java, Saito Shigeya and the Hirose Rice

Straw Craft Association (Hirose Wara Zaiku no Kai) in Japan, Diego Gerry Villanueva in the Philippines, and Jero Ni Madé Rénten in Bali (whose beautiful offering unfortunately could not ultimately be included in the exhibition).

I am grateful to Garrett Solyom, Harold C. Conklin, Karen Smyers, Sylvia Fraser-Lu, and Hirokazu Kosaka for sharing scholarly information from experienced perspectives. I feel especially privileged to have been able to include in the book and the exhibition images by two outstanding photographers who have previously presented their work in museum exhibitions, Stephen P. Huyler and Eric Crystal.

In her foreword Marla Berns has expressed our shared appreciation of the many Fowler Museum staff members who worked so diligently on this project. How hard they worked, I think, only they and I know. It remains for me, therefore, to express my gratitude to Dr. Berns herself, former Fowler Museum director Doran Ross, Deputy Director and Chief Curator Polly Roberts, and Assistant Director David Blair for their steady support for this long and challenging project. I was also assisted by a number of research assistants, translators, volunteers, and others over the six years that I have worked on this project. For this I thank Juliana Wilson, Sohini Ray, Jean Concoff, Dustin Leavitt, Zlata Zukanovic, Khoi Ta, Bokyung Kim, Sam Bartels, and most consistently, Rachel Raynor, without whose assistance I could not have managed. Finally I thank Lillis Ó Laoire for his unflagging support.

Roy W. Hamilton
Curator of Asian and Pacific Collections

Notes to the Reader

In a multi-author book spanning such a large, linguistically and culturally diverse area of the world, problems of stylistic treatment must necessarily arise. While our goal has been to strive for as much consistency as possible, multiple systems of transliteration exist for many of the languages encountered in this volume. Regional variation also accounts for differing orthography. For example, the book contains material covering several closely-related groups speaking languages from the Tai-Kadai language family, including not only the Thai of central Thailand, but also the Khon Muang and Tai Yong of northern Thailand, the Isan and Thai Khorat of northeastern Thailand, and the Black Tai and White Tai of Vietnam; no single standard exists for all of these languages. Our authors have, therefore, been permitted some flexibility in their choice of spellings, and endnotes and cross-references explain these variations.

Sanskrit-derived terms present a special case, as there are not only many transliteration systems occurring in the literature, but the terms themselves were adopted into other Asian languages, exposing them to further variability. Thus what is Śrī Devī according to one reckoning has become Dewi Sri in others.

We have followed the preferences of our colleagues and contributing authors, many of whom have previously published in English, for the presentation of their own names. Otherwise, we have attempted to follow the appropriate convention for the country in question, unless this runs contrary to the way in which an individual is best known to an English-speaking audience.

Finally, Madras and Calcutta have been retained in lieu of the more current Chennai and Kolkata, which may as yet be unfamiliar to many readers, and diacritical marks have been omitted from well-known Japanese place names, e.g., Tokyo, Honshu, Kyushu.

•

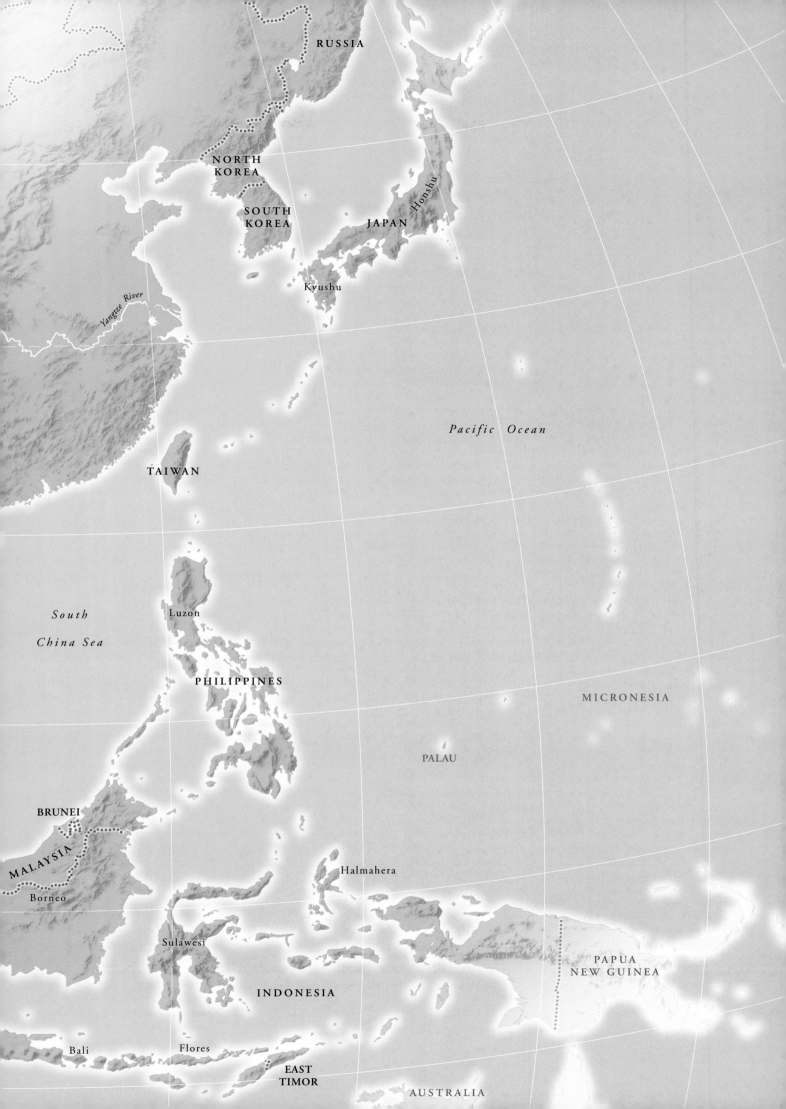

I

Introduction

Roy W. Hamilton

Oh sacred padi,
You the opulent, you the distinguished,
Our padi of highest rank;
Oh sacred padi,
Here I am planting you:
Keep watch o'er your children,
Keep watch o'er your people,
Over the little ones, over the young ones,
Oh do not be laggard, do not be lazy,
Lest there be sickness, lest there be ailing;
You must visit your people, visit your children.
You who have been treated by Pulang Gana;
Oh do not neglect to give succour,
Oh do not tire, do not fail in your duty.

—Iban prayer, quoted in Freeman (1970, 154–55)

With these words of prayer addressed to the rice (*padi*) spirits residing in a special sacred clump of rice plants, an Iban farmer in Borneo circa 1950 replanted the shoots in his field. They had been uprooted the day before and treated in an all-night curing ceremony invoking the god of fertility, Pulang Gana, to restore their health. This special intervention was undertaken because the community's rice crop was threatened by disease and failing to thrive. In returning the sacred plants to the field, the farmer urged their spirits to watch over the other rice spirits in the field and spur them on to produce a bountiful supply of grain. This prayer was recorded and translated by anthropologist Derek Freeman and published in his pioneering study of Iban rice agriculture. Another anthropologist, Harold C. Conklin, conducted similar fieldwork among the Hanunóo on the Philippine island of Mindoro from 1952 to 1954. Both men were struck by the importance of rituals conducted to please the rice spirits:

> In Iban eyes *padi* is by far the most precious thing which they possess; it is the main source from which wealth flows, and upon its successful cultivation all well-being depends. To the Iban, however, the cultivation of *padi* is not so much a problem in agricultural method, as a problem in ritual knowledge and

1.1   Mae Phosop. Thailand. Circa 1980. Commercial print. H: 27 cm. FMCH X98.14.14.

This popular religious poster depicts Mae Phosop, the Thai goddess of rice, with incense and rice cakes offered to her for her pregnancy ritual. Posters of this type are sold in religious shops and at rural festivals. The inscription is a lunar counting device farmers can use for timing the proper ritual tasks.

skill. *Padi* is a spirit, and a farmer's success depends pre-eminently on his ability to order his dealings with the *p'adi* spirits in such a way as to win their approval, and to attract to himself the bountiful crops which all men desire. [Freeman 1970, 50]

Ritual observations for the rice plant spirits include the most obligatory taboos as well as the most sacred rites practiced by the Hanunóo. It is universally felt that the welfare of every individual, as well as that of the entire region, depends on the nature of the intimate relationship between the swidden farmers and these hypersensitive rice "people." [Conklin 1975, 88]

Rituals involving rice spirits or deities are not limited to swidden (or "slash and burn") agriculture, or to the islands of Southeast Asia. In the Shinto tradition of Japan, for example, the mountain spirit (*yama-no-kami*) is ritually invited in the springtime to descend to the irrigated rice fields and take up residence as the rice field spirit (*ta-no-kami*). There she presides until the harvest is gathered, when another ritual is held to release her back to the mountain. As in many other Asian traditions, this deity for rice is female (Earhart 1970, 15). In the Malay state of Negeri Sembilan, the rice spirit is sometimes personified in legends as a little girl and called by affectionate nicknames such as Flower Princess (Si Dang Sari) or Crystal Princess (Si Dang Gembala). Before the harvest, a Malay shaman traditionally selected the first seven stalks to be cut. This was done by inspecting the rice for several consecutive evenings to determine which heads of grain embodied the rice spirit (Zainal 1985, 140–41, 149). In central Thailand, when the rice grains begin to swell in the fields, the rice plants are said to be pregnant and a pregnancy ritual may be held for the Rice Goddess, Mae Phosop (fig. 1.1). Phya Anuman Rajadhon wrote about this ritual in 1948, describing how an orange, a banana, slices of sugarcane, powder, perfume, and a comb were placed for the Rice Goddess at a shrine in the rice field. The powder and perfume were sprinkled onto the growing rice plants and their leaves were gently combed. This was an act of "dressing" the goddess, while the orange was intended to prevent her morning sickness. At the conclusion of the ritual, a bamboo marker was set at the boundary of the field, to warn people not to disturb the pregnant goddess (Anuman Rajadhon 1961, 23). Every grain of rice is considered to be a part of the body of Mae Phosop and to contain a tiny bit of her spirit, or *khwan* (J. Hanks 1960, 299).

All of these rituals are expressions of the belief that a vital spirit or soul animates all living organisms and sometimes inert objects as well. They were all recorded in the mid-twentieth century, well before the beginning of the sweeping changes in world agriculture that have come to be known as the green revolution (see chapter 30). It cannot be assumed that these same rituals continue unchanged today, yet, as the chapters of this book demonstrate, many traditional beliefs and practices related to rice still have profound meaning in contemporary Asian societies. As Richard O'Connor has written about Southeast Asian societies, "agriculture is a locus of meaning, not just a means to subsist. As these societies arise performatively, farming's technical practices easily become ritual acts" (1995, 969). Rice rituals comprise what French anthropologist Georges Condominias has called "ritual technology" (1986, 28–29). To the farmer who performs them, the rituals are no less necessary than planting the seed, weeding the field, or guarding the ripening crop from damage by birds or wild animals. Both are "work" that must be successfully completed in order to meet the end goal of a bountiful harvest. Even from an outsider's point of view, the rituals can be seen to regulate the agricultural activities, both in their timing and in the way they are performed. In many cases the functional effects are plainly evident. Throughout this book, rituals are treated as an integral part of the agricultural process.

Rice rituals attest to the extraordinary importance that Asian farmers have invested in their primary crop, for rice is in fact the world's most important food. Over three billion humans, almost half of the total world population, depend on it as their staple. It accounts for more human calories consumed than any other food. Eleven percent of all the land farmed by humans is devoted to it, more than any other crop. The origins of domesticated rice have long been debated by archaeologists, but the most recent evidence suggests that it was first cultivated in the middle Yangtze River Valley about eight thousand years ago (Higham and Lu 1998, 870). Almost all cultivated rice belongs to a single species, *Oryza sativa*. A separate species, *O. glaberrina*, was domesticated in Africa and is still grown there but represents only a tiny fraction of the worldwide harvest. It is estimated that there are now an astonishing 120,000 varieties of *O. sativa* (Bray 1986, 16). This diversity represents the heritage of generations upon generations of primarily Asian farmers, who annually selected seed to nurture their favorite varieties or develop new desirable characteristics.

Rice yields can dramatically increase when agricultural practices are intensified. Some obvious ways to augment the harvest include the building of irrigation systems or the construction of terraces on mountain slopes. The more labor that is invested, the greater the population that can be supported from a given area of land, which in turn further increases the pool of labor. This spiraling process, described by Clifford Geertz (1963, 32) as "agricultural involution," has created the most densely populated rural districts anywhere in the world in prime rice-growing regions like Java, Bengal, and the Yangtze River Valley.

Although rice farming has now been spread around the globe by humans, rice still occupies a special place in Asia. This book explores the cultural aspects of rice as they are expressed through the visual arts and material culture of South, Southeast, and East Asia. This vast and culturally diverse region can in fact be characterized by the role rice plays for its people, as the focus of a rich complex of interrelated cultural practices and beliefs that might be called "rice culture."

A key tenet of rice culture is that rice is a sacred food divinely given to humans that uniquely sustains the human body in a way no other food can. A woman from Borneo summed up this belief in the 1980s when she told anthropologist Christine Helliwell: "If I lived in the West…I would surely die. Because Westerners can't eat rice. There isn't any rice there. But if I couldn't eat rice I'd certainly die. I wouldn't want any other foods. They wouldn't be right for me" (Helliwell 2001, 45). Since humans live by eating rice, the human body and soul are regarded as being made from rice. Therefore, it is by eating rice that humans are defined. Many rice cultures also hold that rice plants and rice grains have a special spirit or soul, comparable to the human soul, that must be nurtured with rituals in order to procure a bountiful harvest. Often seed is saved with special ritual procedures to keep the spirit of the rice alive from year to year, which at the same time maintains sacred varieties inherited from ancestors. The box on page 30 provides a fuller enumeration of many basic tenets of rice culture that are widespread in Asia.

In general terms, rice culture holds sway over a vast part of Asia stretching from India and Sri Lanka in the southwest, encompassing all of Southeast Asia, and extending through China and on to Korea and Japan in the northeast. In some places it is stronger than in others. A large proportion of the tenets will be found, for example, in almost any rural society in Southeast Asia or Japan. The situations in China and India, the two most populous countries, are more complicated. The Yangtze basin, as well as the rest of southern China, remains indisputable rice country, befitting its status as the cradle of rice cultivation. In northern China, however, millet has historically been the most important crop. Centuries of centralized rule from the north appear to have weakened many rituals and

1.2   Irrigated lowland rice fields in Bali surround an irrigation temple (*pura bedugul*). Each section of fields watered from a single main source has a temple of this type, where the religious observances of the irrigation society (*subak*) are held. Holy water, taken from higher up in the irrigation system, is stored in a jar in the temple for use in rituals conducted in the fields. On the post in the foreground is a field shrine (*sanggah*) where offerings were placed a few weeks before this photo was taken, when the grain-bearing panicles first began to emerge. Photograph by Roy W. Hamilton, Petulu Gunung, 2001.

beliefs related to rice in the south, although from the mid–Song dynasty (960–1279) onward, the northern capital was in fact dependent on imported southern rice (see chapter 29). Many of the ritual practices that are part of rice culture in Southeast Asia appear to be confined in modern times in China primarily to the minority peoples of the deep south.

In India the states of the south and the east are very strong rice culture areas, including not only major population centers such as Bengal and Tamil Nadu but also midsized states with distinctive cultural traditions such as Kerala and Orissa. In much of the rest of India, both rice and wheat are grown. Rice fields can even be found up to moderate elevations in India's Himalayan states (as well as in Nepal and Bhutan), but wheat is more important in India's northwestern states and barley in the higher Himalayas. As in China, the north in India has been politically dominant, meaning that many of the strongest rice traditions are regional rather than national.

If this book can be said then to cover those parts of South, Southeast, and East Asia where rice culture predominates, it may be equally important to say what it does not cover. There are indeed many places in the world where rice is economically important and even the staple food, but the extensive system of interrelated beliefs and practices that make up rice culture in Asia, as discussed in this book, are not found. Southwest Asia, West Africa, the Caribbean, and Madagascar[1] are among the main examples, and even Italy, Spain, the United States, and Australia have become important producers of rice in modern times.

Where Asian rice culture prevails, rice agriculture characterizes the landscape. Perhaps the most famous example is the terraced mountain slopes of northern Luzon, in the Philippines, but many other landscapes have been shaped by rice farming as well (figs. 1.2, 1.3). Along the coastal canals and lagoons of Kerala, mile after mile of fertile rice fields lie below water level. This can only have been accomplished by constructing dikes and draining vast areas of marsh, an effort that has been ongoing since ancient times. In neighboring Tamil Nadu, the landscape is divided according to classical Tamil philosophy into five ecological zones: seacoast, desert, mountains, pasture, and agricultural lands called *marutham*. In former times, any temple would have corresponded with the type of landscape in which it was located. Indra, the deity who is responsible for the origin of rice according to some Hindu texts, was considered the appropriate deity for temples in the *marutham* agricultural lands. That Indra is also the god of thunder and rain is perhaps not coincidental considering the importance of precipitation for Tamil Nadu's rainfed rice fields.

In China, Korea, and Japan, there is a well-established concept of an ideal landscape, imagined as rice fields spread out over a valley floor at the base of a mountain. Villages are situated just where the mountain slope meets the rice fields (fig. 1.4). This makes sense ecologically, as the forested mountain slope helps to protect the water supply for the fields, while the village does not take up valuable bottom land. It also corresponds perfectly with the Shinto concept of the *yama-no-kami* and *ta-no-kami* described above. Additionally, in Japanese art there is a strong tradition of depicting rice plants or rice agriculture as a part of an idealized natural environment (fig. 1.5).

Traditional settlement patterns are usually shaped by agricultural needs. The villages in Vietnam's Red River Delta are compact and densely built, saving the surrounding fertile land for agriculture. In central Thailand, villages formerly were strung out along canals or located on high ground, leaving the best lowland for the rice fields, while the canals provided transportation (today Thai villages tend to be strung out along roads, and the harvest is transported by truck rather than barge). In the rugged interior of Borneo, longhouses were built along the rivers, also used for transportation, and the fields were scattered around the surrounding mountain slopes. In the terraced landscape of northern

1.3   Hmong farmers in northern Vietnam grow transplanted rice in these irrigated terraces constructed along the Kim River. The seeds are first planted in hillside plots that have been cleared by burning, like the field visible on the slope above the river, rather than in irrigated nursery beds. This unusual procedure seems to represent the adoption of irrigating and transplanting by a population previously more accustomed to farming hillside swiddens. Photograph by Eric Crystal, Mù Cang Chải District, Yên Bái Province, 2000.

1.4   A village in the Duliu River Valley, Congjian District, Guizhou Province, China, closely matches the rice-agriculture environment ideal, with forested slopes providing water for fields on the floor of the valley. Photograph by Ted Dardis, 2001.

Luzon, the Ifugao people live in small hamlets consisting of a few houses each, surrounded by their terraced rice fields. In many cultures where people live some distance from their fields, small field huts are constructed for temporary residence when the rice demands it, most often when the ripening crop must be guarded from damage by birds or wild animals. In some cases the year-round village may be nearly deserted except for children and the elderly, while the able-bodied workers are away at the field huts.

This wealth of varying landscapes has contributed to the creation of the phenomenal number of varieties of rice. Scientists divide *O. sativa* varieties into two main groups, tropical *indica* and temperate *japonica*. Within each type are varieties that are grown under different ecological conditions. The most basic distinction is between wet- and dry-rice varieties. Dry varieties are typically grown in swidden systems, often on hillsides (fig. 1.6). The natural vegetation is felled and burned before the annual rains set in. The seeds are sown directly in the ground, usually in holes made with a dibble stick. The plants grow much like wheat, without ever being submerged in water, and are dependent on adequate rainfall. In typical swidden fields, the rice is interplanted with other food crops. Ideally, new fields are cleared each year, which means that swidden farming is only suitable for relatively low population densities. Swidden agriculture has gained a bad reputation primarily because it damages the environment when the population density grows too high. Under ideal conditions it has been shown to be both sustainable and efficient, but those conditions have become the exception as the populations of Asian nations have

1.5   One of a pair of six-panel
          screens. By Maruyama
          Ōshin (1790–1838). Japan.
          Ink, color, and gold on
          paper. L: 378 cm. Los
          Angeles County Museum of
          Art L.83.45.6b; Etsuko and
          Joe Price Collection.

          This magnificent gilded
          screen depicting rice plants
          heavy with grain in the
          autumn is part of a pair; the
          other screen shows verdant
          shoots of barley in the
          spring. The screens epito-
          mize the concept of season-
          ality in Japanese art and the
          Japanese reverence for rice as
          an element in an idealized
          landscape.

spiraled. Today swidden rice farming is found primarily in relatively remote mountain areas of Southeast Asia, both on the mainland and in the islands. It is the norm, for example, in upland Laos and in the interior of Borneo. Overall, however, it accounts for only a relatively small part of the total rice harvest.

Wet-rice varieties can be divided into four ecological types: (1) irrigated, (2) rainfed lowland, (3) floodplain, and (4) deep water. Irrigated rice is the most familiar type and is typical of the most highly productive systems (fig. 1.7). The irrigation may be provided naturally by gravity; some examples include series of terraces with water running through them, tanks that drain into fields, or weirs and canals that divert river water. Tremendous effort sometimes goes into the engineering of these systems, as in the Luzon terraces, or the systems of tunnels and canals that deliver water to some Balinese fields miles from where it was diverted out of streams. Irrigation water can also be delivered against gravity if necessary by the use of pumps, waterwheels, or even hand-operated scoops. In irrigated systems the rice is usually sown in nursery beds and later transplanted into the flooded fields. Transplanting gives the rice plants a vital head start over weeds, ultimately increases the strength of the plants, and in some cases helps to make double or

Twenty Tenets of Rice Cultures

Each rice-growing society has its own unique set of values and beliefs associated with rice. Listed below are twenty of the most common tenets that are widespread in many parts of South, Southeast, and East Asia. The list is a composite. No single society holds to every tenet, and the wording below is generalized rather than specific to any particular group of people. Many societies, however, do follow a large number of the tenets, and taken together, they can be said to posit a creed of rice culture.[2]

1. *Rice is a special sacred food, divinely given to humans.*

2. *The rice plant has a living spirit or soul comparable to that of humans, and the life cycle of the rice plant is equated with the human life cycle. The rice spirits must be honored and nurtured through rituals in order to assure a bountiful harvest.*

3. *The stages of rice agriculture determine the annual cycle of human activity, including the conducting of the proper rituals at each phase of the rice crop's growth process.*

4. *The work involved in growing rice is the ideal form of human labor, reflecting a well-ordered, moral society.*

5. *The mythological origin of rice is attributed to a Rice Mother or Rice Goddess; in many versions of the story the goddess is killed and the first rice grows from her body.*

6. *The fertility of the rice crop is metaphorically equated with the fertility of the Rice Goddess and with the fertility of human females; therefore rice is often regarded as female and in exchange systems it functions as a categorically female good.*

7. *Rice must always be treated with respect in order to avoid offending the rice spirits or Rice Goddess; at harvesttime rice may be cut with a special type of knife to avoid harming them.*

8. *The granary is the home of the rice spirits and is often built to resemble a small human house. After the harvest, the grain is ritually installed in its home.*

9. *Special objects may be placed in the granary to accompany the rice; these include anthropomorphic figures made of rice stalks symbolizing the Rice Goddess, or in other cases carved wooden figures or even copies of religious texts.*

10. *The spirit of the rice remains alive at least until the rice is milled; thus the rice that is set aside before milling to serve as seed rice perpetuates the rice spirit, keeping it alive until the rice is planted again in the following agricultural cycle.*

11. *The maintenance of special ancestral genetic strains of rice is a primary link between living humans and their ancestors.*

12. *The daily milling of rice by pounding it in a mortar is traditionally one of the most characteristic activities of village life. Only after it is milled can the rice be brought into the house.*

13. *The daily milling, cooking, and eating of rice determine the daily schedule of human activity.*

14. *Language reflects the special nature of rice as the primary food of humans; often there is no general word for "food" other than the word for rice and an invitation to "eat" implies the eating of rice.*

15. *The household or family unit is defined as those who eat rice together, especially the rice that is produced through the joint efforts of the family members.*

16. *Cooked rice is the ultimate human food and only rice is capable of properly nourishing humans; other foods are regarded as condiments to accompany the rice or as snacks and do not constitute a meal if rice is not served.*

17. *Because humans live by eating rice, their bodies and souls are made from rice.*

18. *Rice and rice alcohol are quintessential offerings made to spirits, deities, and ancestors.*

19. *The offering of a portion of the daily cooked rice to spirits, deities, or ancestors sanctifies the remainder of the rice, which becomes the sacred daily food to be eaten by humans. The living humans, the ancestors, and the deities are united through the daily sharing of this sacred food.*

20. *Because rice ties humans to their ancestors, defines the family unit, and provides the ultimate human nourishment, the growing and eating of rice define what it means to be human.*

1.6   A woman plants rice seed in a hillside field that has been cleared by burning. Men pass through the field first, poking holes into the earth with dibble sticks. The women follow, dropping seeds into the holes. Photograph by Christine Helliwell, Gerai, West Kalimantan, Indonesia, 1986.

triple cropping possible by shortening the time the plants spend in the field. In irrigated systems the fields are usually allowed to dry before harvest, but there are some exceptions (the Luzon terraces, for example, have water in them year-round). Often a different plant crop is grown in rotation during the off-season on the dried fields. In parts of China and India, for example, rice is grown in the fields during the rainy summer and wheat during the cooler and drier winter. More often, though, the plants grown in rotation are root or vegetable crops.

Rainfed lowland systems simply rely on the heavy monsoon rains of Asia to fill the fields with water. Usually the fields are bunded, meaning that they have a dike or bund surrounding them to keep the water in the field. Being at the mercy of rainfall, these systems are typically somewhat less productive than irrigated systems, but large populations in relatively poor areas often depend on them. They are the norm in many parts of India and in northeast Thailand. The seed may be sown in a nursery bed for transplanting later, or it may be broadcast directly into the field, usually just as the rains are beginning. Rainfed lowland landscapes are remarkable for their seasonal variation. During the rainy season the fields present a sea of brilliant green; six months later there may be nothing but a dusty, barren plain, with only the telltale bunds to suggest that the land is green and productive at another time of the year.

Floodplain systems involve seeding or transplanting into low-lying areas that are naturally flooded with water during the rainy season. These systems are found along the edges of shallow lakes or in fluvial floodplains, and they sometimes involve techniques that are distinct from those used in rainfed or irrigated systems (fig. 1.8). Finally, deepwater rice is the rarest and most unusual ecological type, occurring where annual flooding deeply inundates the land. The rice is seeded as the water is just beginning its rise and the growth of the plants keeps up with the rising water level. The plants bear their grain above the surface of the water, sometimes in depths of up to twenty feet. This remarkable system is found in some parts of Bangladesh and Cambodia.

Rice varieties are even more variable in their culinary qualities than in their agricultural requirements. Long grain or short grain; white, brown, red, or black; glutinous, basmati, or jasmine: these are some of the choices that face shoppers even in some American grocery stores. Every part of Asia has its own favorite varieties and characteristics. In most cases these were bred locally and people feel that they go best with the local style of cooking. Even American diners can sense that the nutty taste of India's famous basmati rice doesn't quite seem to go with Chinese food, and the delicious fragrance of jasmine rice, so appreciated with Thai food, is not right for sushi. Japanese rice is appreciated particularly for its delicate texture. Glutinous or "sticky" rice, sometimes known as "sweet" rice, differs from ordinary rice in that it contains a high level of amylopectin (Nguyễn Xuân Hiên 2001, 50), which is a recessive genetic trait. In most countries it is used only for making sweets or rice wine, including sake in Japan. In the Ifugao region, a family that grows a lot of sticky rice is plainly identified as a wealthy family, because they can afford to grow a luxury product that is not eaten but made into rice wine for use in rituals. For many of the Tai and Lao groups of Southeast Asia, however, glutinous rice is the beloved daily staple.

As the characteristics of the local rice differ from place to place in Asia, so too do the cultural and religious traditions to which the rice farmers adhere. The region covered by this book encompasses extraordinary religious diversity, with every major world religion represented, and many minor religious traditions as well. Hinduism is the majority religion in India, Islam in Indonesia, Buddhism in Thailand, Roman Catholicism in the Philippines, and so on. Many countries have mixed traditions, such as Shinto and Buddhism in Japan, or Taoism and Confucianism in China. Every country has minorities

1.7   Balinese rice farming. Bali, Indonesia. 1930s. Paint on cloth. W: 86 cm. FMCH LX74.289; Anonymous Loan.

This painting was collected by the Mexican modernist painter Miguel Covarrubias when he lived in Bali in the 1930s and is an early example of the naturalistic style of painting that was just becoming established in Bali at that time. One of the conventions of Balinese art is that a single painting may include a sequence of scenes, as in this portrayal of several stages of Balinese irrigated rice farming. The activities depicted include plowing, repairing the dikes, transplanting, and weeding. The farmer standing on a flat board pulled by draft animals is leveling the surface of the mud so that the field will retain water at an even depth.

that follow different faiths from the majority. Some places, most notably Java and Vietnam, are famous for their religious syncretism. One of the most interesting things about rice culture is that it is expressed in all of these diverse religious environments, like an underlying stratum bubbling to the surface. This suggests that many aspects of rice culture, especially those that are based in rice spirit beliefs, were broadly established prior to the development and spread of the major world religions. With rice agriculture dating back eight thousand years, this is not surprising. Shinto is in some ways a special case, as it essentially represents a formalization of spirit belief; as might be expected, rice spirits play a prominent role in Shinto, whereas in other religious traditions they may be partly sublimated, or recognized as standing outside of the predominant religion while at the same time being embraced by its practitioners. Even where people currently adhere to monotheistic religions, elements of spirit belief often survive and find expression in rice culture.

The chapters of this volume were developed to give some sense of the tremendous power and diversity of rice culture in Asia at the opening of the twenty-first century. Both the book and the exhibition it accompanies have been organized thematically into eight parts. Each part of the book begins with an introductory essay that lays out its main themes and highlights many of the objects from the corresponding section of the exhibition. Each introduction is followed by chapters that explore those themes more fully. As in the exhibition, disparate objects and divergent practices have been purposefully juxtaposed. It is hoped that this allows for cross-cultural insights while at the same time fostering an appreciation of the breadth and diversity of rice culture in the home continent of humankind's most important food.

•

1.8 Digging stick (*tantajuk*). Banjar peoples. South Kalimantan, Borneo, Indonesia. Probably early 1900s. Carved and painted wood L: 49 cm. FMCH X82.635; Gift of Mr. and Mrs. Theodore Lowenstein.

Banjar farmers grow rice in floodplains along the banks of rivers, where the water rises during the rainy season to cover ground that has hardened during the previous dry season. The soil remains firm even when flooded and is not worked into a soupy consistency as is normally the practice in irrigated rice agriculture. When it is time to transplant the seedlings, digging sticks are used to open holes in the flooded soil. No such tool is necessary in irrigated rice farming because the seedlings can simply be pushed by hand into the softer mud. The elaborately decorated and somewhat fragile form of this digging stick suggests that it may have been used for the ceremonial opening of the transplanting season rather than everyday labor. The carving is in the style of Javanese shadow puppetry (*wayang*) imagery (see chapters 6 and 33), reflecting the strong influence of the Javanese courts in the Banjar sultanate.

Part One

Labor, Ritual,
and the Cycle of Time

2

Labor, Ritual, and the Cycle of Time

Roy W. Hamilton

> Cane goad in hand,
> his feet on the plough,
> he drives the well-bred oxen
> and puddles the earth
> for the nursery plot,
> while the kingfisher
> high in the sky
> tells of rain.
> —Burmese plowing song[1]

> Some folks transplant rice for wages,
> but I have other reasons.
> I watch the sky, the earth, the clouds,
> observe the rain, the nights, the days,
> keep track, stand guard till my legs
> are stone, till the stone melts,
> till the sky is clear and the sea calm.
> Then I feel at peace.
> —Vietnamese folk song, quoted in Nguyễn Ngọc Bình 1985

These two verses present rather idealized images of the grueling work of growing rice, yet they are not outsiders' views. They are songs sung by rice farmers themselves, presenting their own views of their labor. The songs lighten the burden of the work no doubt, but they also show an abiding love and respect for rice farming despite its hardships. The Burmese song is sung when the nursery bed is plowed in preparation for sowing the seed, while the Vietnamese song depicts the transplanting of the seedlings from the nursery into the field where they will grow to maturity. Both songs show a keen awareness of the rice landscape as part of a natural environment, where kingfishers sing before the nurturing monsoon rains set in, or peace comes through the observation of the passage of the season.

In Asia's rice-growing communities, it is this seasonal agricultural cycle that largely determines the timing and organization of human labor and ritual (fig. 2.1). Most traditional varieties of rice take from five to seven months to mature. Before the

2.1 An Indian woman beats a sheaf of rice against the ground to thresh the grain from the stalks. Photograph by Stephen P. Huyler, Uttar Pradesh.

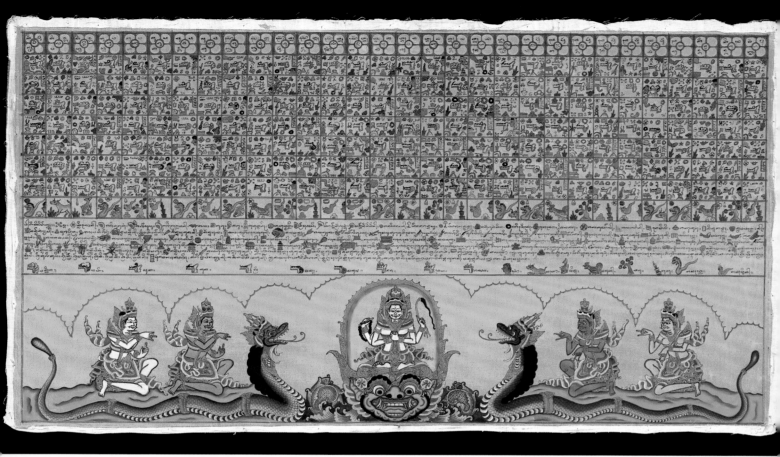

The Balinese Tika

Garrett Kam

The *tika* is a visual representation of the Balinese 210-
day calendar (*pawukon*). The boxes are read from top
to bottom and left to right, with each vertical column
indicating the days of the week from Sunday to
Saturday. Each box is filled with symbols that indicate
whether the day is auspicious or inauspicious for conducting certain activities. A key to the symbols is
located beneath the boxes. At the bottom of the painting are the Hindu gods of the cosmic directions:
Ishvara (white, east), Vishnu (green, north), Brahma
(red, south), and Mahadeva (yellow, west). Shiva (center) sits on the cosmic turtle Badavang Nala, between
the *naga* (serpents) Vasuki (green) and Anantaboga
(red). These figures have no calendar function and are
purely decorative.

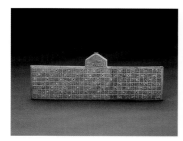

2.3, 2.4 210-day calendars (*tika*).
Bali, Indonesia. 1930s or
earlier. Carved wood.
Top: W: 35.5 cm. FMCH
LX74.213; Anonymous
Loan. Bottom: W: 32.5 cm.
FMCH LX74.214;
Anonymous Loan.

The front side of one *tika*
(top) is decorated with
carved motifs of Vasuki and
Anantaboga. The pointer is
used for counting on the
grid of 210 days marked on
the back, similar to that of
the second *tika* (bottom).
The symbols are denoted
with various combinations
of holes, beads, and carved
markings.

2.5 Gift cover (*fukusa*). Japan.
Late Edo period, late eighteenth or early nineteenth
century. Silk and gold thread
embroidery on blue silk
satin; red-orange silk crepe
lining. H: 91 cm. Los
Angeles County Museum of
Art AC1997.125.1; Costume
Council Fund.

The entire agricultural cycle
is depicted on this cloth,
beginning with the sowing
of the seed at the top and
continuing on down the
cloth with transplanting,
harvesting, threshing (with
flails), milling, and storing
the grain in the granary.

2.6   Votive plaque (ema). Okayama, Japan. Late Edo period (1603–1868). Paint on wood. W: 152 cm. Collection of Lynn S. Gibor.

Ema, the painted wooden plaques placed by donors at Buddhist temples or Shinto shrines, may serve either as supplications by a petitioner or as expressions of gratitude for blessings already bestowed (Okada et al. 1989, 53). This unusually large and detailed ema featuring scenes of the various stages of rice agriculture must surely have expressed a wish for a successful harvest. The scenes include standardized images drawn closely from the Chinese illustrations known as the Gengzhi tu (see chapter 29). Ema are normally personalized with a message or the name of the donor, which must once have appeared in the whitewashed area in the upper right.

development of fast-maturing modern varieties, only one rice crop per year was possible in most places. In temperate climates, such as in China, Korea, and Japan, the colder winter weather was the limiting factor. Nearly all of the rice-growing regions of Asia also have pronounced wet and dry seasons due to the monsoon cycle, and this too often limited farmers to a single rice crop per year even in tropical climates. Only in the most favored tropical areas with ample rainfall or developed irrigation systems, some constructed in ancient times, were farmers able to produce more than one crop per year. In the Burmese kingdom of Pegan (ninth to the thirteenth century), for example, irrigation and the planting of varieties that needed different lengths of time to mature allowed for year-round production (Aung-Thwin 1990, 10). The Khmer kingdom of Angkor (ninth to the fifteenth century) produced up to four crops per year (Allen 1997, 82).

The advent of fast-maturing modern varieties of rice beginning in 1966 has allowed double and triple cropping in many places where previously only a single crop was possible. This has led to widespread changes in the traditional annual cycles of labor and ritual. Work may now go on year-round as one rice crop follows the next in a constant series of rotations. Neighboring fields may no longer be worked in a synchronized manner, sweeping away any sense of organized seasonal activity. Currently only swidden farmers or farmers in places with relatively extreme climates are limited to a single crop per year. This includes areas with long cold winters, such as the snowy parts of northern Japan, or with pronounced dry periods and no irrigation, such as in the rainfed systems of India or northeast Thailand.

The pervasive changes that have engulfed rice agriculture in the past thirty-five years present a problem for any author writing about the agricultural cycle. If patterns of labor and ritual are described in the past tense, it suggests that agricultural lifestyles are entirely a thing of the past. This is certainly not the case. Using the present tense is equally problematic, as it assumes that practices recorded and described long ago have continued unchanged into the present. I have therefore adopted a strategy of shifting

2.7 Water puppet. By Đặng Văn Thiết. Hanoi, Vietnam. 2000. Carved and painted wood, rubber, nails. H: 37 cm. FMCH X2000.24.4a-g.

Water puppetry is a unique Vietnamese form of theater originally performed in village ponds. The puppets float on the water, manipulated by long rods that run below the surface. The puppeteers stand in hip-deep water hidden behind a screen. Themes drawn from the daily routines of agricultural life are the most popular, such as this farmer pulling a harrow behind a water buffalo.

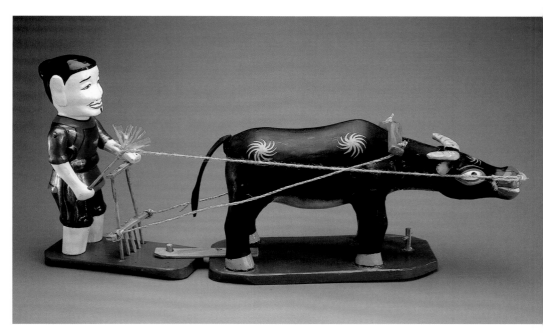

2.8 Water puppet. By Đặng Văn Thiết. Hanoi, Vietnam. 2000. Carved and painted wood. H: 39 cm. FMCH X2000.24.5a,b.

The rice fields in Vietnam's Red River Delta are watered from irrigation canals that lie below the level of the fields. Motorized pumps are often available to raise the water from the main canals into smaller feeder channels, but from there it must often be raised again by hand into the fields. This water puppet is equipped with one of the tools developed for this task, a basketry water scoop (gầu sòng).

2.9 Water scoop (gầu sòng). Vietnam. Bamboo. L: 135 cm. FMCH X2000.34.11.

The scoop is normally suspended from a tripod, which allows the worker to swing it back and forth with rapid scooping motions.

2.10   An Ifugao stonemason
reconstructs a terrace that
has been damaged by flooding. Photograph by Harold
C. Conklin, Baynīnan,
Ifugao Province,
Philippines, 1962.

2.11   Waterwheels raise river water to irrigate rice fields in a Tai community in Vietnam. Photograph by Eric Crystal, Nghệ An Province, 2001.

tenses as necessary on a case-by-case basis, depending on the circumstances. Where there are no clear indications, I have defaulted to the "ethnographic" present tense in honor of those Asian communities and individuals who have struggled to adjust their cultural traditions and preserve them in the face of changing agricultural realities. It is now quite common that harvest rituals, for example, are held after the harvest of a "main" crop ripening roughly at the time of year when the traditional harvest was once held, while harvests of a "second" crop may be conducted without any accompanying ritual.

Because agricultural labor must be timed in accordance with the rice-growing cycle, special means are sometimes needed to mobilize labor at the times of highest demand. Many different forms of communal labor organization have evolved in Asian communities, ranging from individual labor exchanges to community-wide assemblies. The agricultural cycle thus tends to structure social interactions as well as technological ones, both to meet the need for labor organization and also for the purposes of community-wide festivals or celebrations. The majority of Asian festivals are in fact tied to the rice agricultural cycle. As John Stephen Lansing has put it, "To the extent that the agricultural cycle of rites becomes the master calendar of social life, the analysis of one is equivalent to the analysis of the other" (1991, 6). The post-harvest period, when there is relatively less work to do and the supply of food is abundant, is in many Asian communities the most social part of year—not only for agricultural festivals but also for weddings and other community celebrations. The celebration of New Year, which marks the turning point from the old agricultural cycle to the new, typically takes place during this time period as well, and many New Year celebrations also include a ritual offering of the recently harvested crop.

2.12   A Hmong woman trans-
plants rice. Photograph by
Eric Crystal, Yên Bái
Province, Vietnam.

2.13   (Opposite) *Distant View of
Oyama near Ono* (*Ono
ōyama embō*). By Andō
Hiroshige (1797–1858).
Japan. Woodblock print. H:
34 cm. San Diego Museum
of Art 1954:42a; Gift of Mrs.
Clark Cavenee.

Hiroshige's scene of women
transplanting rice in a late
spring rain is from his series
of landscapes *Famous Places
in the Sixty-Odd Provinces*
(*Rokujūyoshū meisho zue*).

The rhythm of the rice agricultural cycle is so ingrained that often it is used
to mark the passage of time. In many cultures, the months are named, or at least characterized, according to activity in the rice fields. Often there is a series of monthly festivals,
with many of them having at their core rituals for the rice crop. Chapter 3, which discusses the complete annual cycle of agricultural rituals and the related Buddhist festivals
in northeast Thailand, provides a comprehensive example. Chapter 4 explores the meanings of the cycle of ceremonial dances that demarcate the agricultural year for the Tharu
peoples of lowland Nepal. Chapter 5 presents previously unpublished information about
the rice agricultural cycle in the northeastern Indian state of Manipur.

In some cases, the coordination of the agricultural cycle and the calendar went
even further. In tropical, irrigated Bali, the "year" is actually calculated according to the
growth cycle of the rice crop. Traditional Balinese varieties of rice take about 210 days to
ripen, and the Balinese annual calendar consists of six months of thirty-five days each.

2.14   Deities overseeing the
transplanting of rice. Japan.
Late nineteenth century.
Hanging scroll; paint and
ink on paper. L: 150 cm.
Peabody Essex Museum,
Salem, Massachusetts,
E11,689. Photograph by
Jeffrey Dykes.

Amaterasu Ōmikami (top),
Japan's Sun Goddess and
deity of the imperial family,
oversees a transplanting ceremony (*taue*), assisted by two
of her immediate relatives,
Tsukiyomi-no-mikoto
(right) and Toyoukehime-
no-kami (left). Directly
below Amaterasu Ōmikami
is Ōkuninushi-no-kami,
depicted atop two rice bales
(*tawara*); this is a formal
name for Daikoku, one
of the Shichifukujin, the
Chinese-derived Seven Gods
of Good Fortune. The deity
riding the fox, with freshly
harvested sheaves of rice
carried over his shoulder,
is a version of Inari, the
Japanese deity for rice (see
chapter 18).

2.15   *Than Nong.* Sanchay peoples. Vietnam. Mid-twentieth century(?). Paint on paper. H: 57 cm. FMCH X2002.15.1.

Than Nong, the Taoist god of agriculture, oversees the planting, with scenes from both wet- and dry-rice agriculture. In the upper half, a woman drops seed into a dibbled hole in a dry field, while just below workers transplant seedlings into an irrigated field.

This means that the major holidays occur once in every 210-day period. Balinese priests use a *tika* or 210-day divination calendar (fig. 2.2) to calculate the auspicious days for conducting activities, including farming tasks. Farmers use simpler *tika*, in the form of wooden reckoning devices (figs. 2.3, 2.4). This system allows priests and farmers together to coordinate agricultural activities.

Japanese culture in particular is known for the importance it places on seasonality. One expression of this is the frequent use of the passing seasons as a theme in the visual arts. Not surprisingly, rice often plays a part, and formalized depictions of rice agriculture in the four seasons are found in many different forms (figs. 2.5, 2.6). Much of the art and material culture of rice agriculture, on the other hand, is tied to the activities of a single part of the agricultural cycle. In the discussion that follows, the seasonal activities of rice agriculture are presented from the beginning of the cycle to the end, starting with the preparation of the land and proceeding through to the harvest and storing of the crop.

2.16   Rice seed container.
Northern Thailand. Bamboo,
lacquer. H: 26.5 cm. FMCH
X2002.17.1a,b.

Large seed boxes are used to
store seed and transport it to
the fields for sowing.

2.17   Rice-sowing basket
(*pemoneh*). Dayak peoples.
Gerai, West Kalimantan,
Borneo, Indonesia. Plaited
rattan, bark cloth. H: 21 cm.
Collection of Christine
Helliwell.

Bark-cloth straps allow Iban
women to carry their
sowing baskets through the
fields while leaving their
hands free to drop the
rice seeds into holes poked
in the ground by men using
dibble sticks.

2.18   Writing box (*suzuribako*)
with rice-transplanting
design. Japan. Seventeenth
century. Lacquer with
mother-of-pearl inlay and
maki-e (sprinkled powder)
decoration over a wooden
core. L: 41 cm. Los Angeles
County Museum of Art
M88.83; Gift of the 1988
Collectors Committee.

2.19   Stirrups with rice-transplanting design. Japan. Edo period, eighteenth century. Lacquer with *maki-e* (sprinkled powder) decoration. Museum of fine Arts, Boston 11.5833 Charles Goddard Weld Collection.

Japanese stirrups provided samurai with a platform for brandishing their swords from atop their horses. This extremely elegant pair was part of the ceremonial armor of a feudal lord.

Land and Water

Much of the hardest work in rice agriculture comes in the preparation of the land. This is particularly true for swidden farmers who have to fell and burn the existing vegetation (once primary forest, now often secondary scrub) before planting can begin. In irrigated areas, terraces, ditches, or weirs often need to be readied for the new season (fig. 2.10). Where transplanting is practiced, the nursery beds need to be worked over and over until the mud in which the seeds will be planted resembles a smooth pudding. The fields into which the seedlings will be transplanted need scarcely less tending, as the surface of the mud must be perfectly level to assure an even depth of water once the plants are growing in the field.

The opening of the fields for the new season's work is typically accompanied by important ceremonial events. Among the Iban of Borneo, for example, elaborate rites are performed to procure auspicious omens for the new agricultural season and propitiate the spirits of the earth and forest where the rice will soon grow (Freeman 1970, 173). The social context in which such rites are conducted varies according to the prevailing system of land ownership. In Iban communities, which are known for their rather egalitarian

character, unused forest land could be claimed by any family. In the societies of eastern Indonesia, where there is typically a more clearly defined system of lineage rights and obligations, a "lord of the land" from a particular founding lineage may parcel out land use rights for each new season. Even royal courts developed procedures for the ritual opening of the fields, and royal plowing ceremonies, in which the king performed the first ceremonial plowing, were once widespread in Asia.

In irrigated rice fields, farmers turn the soil with a plow or, if they are working on narrow terraces, sometimes with a spade. Plows are often pulled behind water buffalo, though small hand-operated tractors that can function in flooded fields are now common in all but the poorest regions. The fields may be dry for the first working, but afterward they are flooded and the labor takes place in deep mud. A harrow is often used to work the mud more finely (fig. 2.7).

The supply of water is as important in wet-rice farming as the land itself. In the drier parts of Sri Lanka, for example, thousands of small artificial ponds or "tanks" (*wewa*) were constructed in ancient times to water small patches of rice fields. Some of the tanks were drained via sluices laid in their bottoms to convey the water out under the tank. Many were built in relatively open terrain by creating bunds, or embankments, up to twenty-four feet high and a mile and a half in length, making the "tank" actually an artificial lake (Brohier 1934, 4). In Japan, where small ponds were also constructed to water rice fields, Hyōgo Prefecture alone has fifty thousand of them (Grist 1965, 42).

Where possible, gravity is used to conduct the water to the fields, often through elaborate systems of weirs, channels, and sluices. Shrines are sometimes constructed for irrigation rituals at critical places in the system of waterworks. This idea is perhaps the most developed in Bali, where special temples serve the irrigation system (see fig. 1.2). Membership in Balinese irrigation societies (*subak*), which maintain the irrigation temples and conduct irrigation rituals at them, is distinct from membership in other types of village temples or community organizations. Elsewhere, in places where gravity will not suffice because the source of water is lower than the field, the water must somehow be raised. The backbreaking use of hand-operated buckets and scoops is still found in some regions (figs. 2.8, 2.9), but many ingenious devices have also been developed to make this task somewhat easier (fig. 2.11).

Planting

There are many variations in the planting process, depending on the type of rice farming. For swidden systems, the seeds are sown in holes poked into the ground with a pointed dibble stick. Labor exchange arrangements are typically used to assemble a large group of workers for this task. Typically the men pass through the field first making the holes, while the women follow, carefully dropping in the seeds. For rainfed lowland fields the seed is often broadcast directly into the field after the rain begins, but sometimes transplanting is practiced.

Irrigated fields are almost always transplanted with seedlings raised in a nursery. Transplanting is perhaps the most unique and characteristic aspect of rice agriculture, and nothing is quite so evocative of rural life in Asia as a row of workers bent over the flooded field pushing the seedlings into deep rich mud (figs. 2.12, 2.13). The seedlings are specially raised in nursery beds planted several weeks before the transplanting takes place. The seed is usually soaked until it germinates, then thickly sown over the surface of the nursery bed. In regions where the seed rice is stored still attached to the stalk, the whole stalks may be laid over the surface of the nursery bed. After a few weeks, the seedlings are pulled from the nursery, trimmed, bound into bundles, and taken to the prepared fields for transplanting.

2.20   In the rice fields surrounding Ubud, Bali, a ritual is held when the panicle is just emerging and the grains are about to swell. The name of the ritual, *mebiu kukuing*, refers to the cravings of the Rice Goddess, Dewi Sri, in the first days of her pregnancy. For this ritual an *ublagablig* (more commonly just called *ubagabig*), made from intricately folded strips of coconut palm leaf, is hung in the fields. Fertility symbolism is expressed in the hermaphrodite form for the *ublagablig*, which has both female breasts and male genitalia. Photograph by Roy W. Hamilton, 2001.

2.21   (Opposite) *A Farmer's Wife in Harvest Season*, from the series *Notable Occupations in the Country of Plenty*. By Yoshimasu Utagawa (active 1830s–1854). Japan. Circa 1851. Woodblock print. H: 37 cm. Sweet Briar College Collection, Sweet Briar, Virginia; Gift of Ruth W. Smith.

In most Asian farming communities there is a strong tradition of providing food and drink to harvest workers. Where there are communal or shared labor arrangements, feeding the workers is usually one of the main obligations of the field owner. Here the wife of a Japanese farmer brings refreshments to the fields, with a sickle tucked into her clothing to signify the season.

2.22   A Balinese farmer passes time by playing a musical instrument, while at the same time keeping watch over the ripening fields. Photograph by Roy W. Hamilton, Nangasepaha, north Bali, 1998.

2.23   A scene from the painted ceiling of the Kerta Ghosa, or Palace of Justice, in Klungkung, Bali, shows rats overrunning the rice fields. The painting is part of a series depicting the evils foretold by earthquakes. Photograph by Roy W. Hamilton, 1998.

2.24   (Opposite) A Balinese woman decorates the first cut stalks of rice, transforming them into representations of the Rice Mother (Nini Pantun). Photograph by Roy W. Hamilton, 2001.

Both the planting and transplanting processes are extremely important occasions for rice ritual (figs. 2.14, 2.15). In Southeast Asia planting rituals are often conducted at a sacred place in or beside the field, which may be marked by a special sacred strain of rice or other sacred plants. Karo Batak farmers in Sumatra, for example, create a "navel" at the center of the field where special plants form an attractive refuge for the rice spirit (van der Goes 1997, 385). On the island of Halmahera in eastern Indonesia, Sahu swidden farmers formerly erected miniature granaries and conducted planting rituals at the center of their field where the felled forest trees had been burned (Visser 1989, 65). There they offered rice colored yellow with turmeric, an egg, and palm wine to the spirits of the ancestors before the planting began (as the Sahu have converted to Christianity, today these rituals have been replaced by Christian prayers). Often the planting or transplanting are specifically regarded as women's work. In the traditional Thai concept of the annual cycle, the entire period when the seeding and transplanting occur is regarded as a "female" time; this is Phansa, the Buddhist "rains retreat," when male monks withdraw to the monasteries and temples (Trankell 1995, 52).

In Japan communal transplanting festivals (Ōtaue) were organized for important fields owned by wealthy landowners, shrines, or temples. Young men prepared the fields and then provided music while the young women transplanted the seedlings. The dress and comportment of the young people were keenly observed by parents sizing up prospective marriage partners for their children. Many of the verses sung to accompany the work had veiled or overt sexual themes (Hoff 1971, 12), which indicates that human fertility and the wish for a bountiful crop were linked. Transplanting festivals are still held today in Japan, but they now function more as public spectacles and less as practical means of organizing labor.

Special tools such as decorated digging sticks (see fig. 1.8) or seed baskets (figs. 2.16, 2.17) are often used for planting. Both planting and transplanting are so important that stylized scenes of these activities are widespread in Asian art. These are more than purely decorative, for they have become icons with power to evoke fecundity and promise prosperity. In Japan, for example, standard scenes of transplanting decorate objects that have nothing to do with agriculture and would have belonged to persons of higher status than ordinary farmers (figs. 2.18, 2.19).

Guarding the Ripening Crop

Once the rice is growing in the field, the work lets up a bit, but there is still regular weeding to be done. This is often the work of women, and many anthropologists have noted the pride that women take in well-maintained fields. Sometimes special rituals are conducted to protect the crop from harm. These may be held in response to a specific disease or pest threat, or they may be held routinely to protect the crop from attack by pest spirits. Chapter 6 describes Javanese rituals that formerly were conducted to expel evil spirits from the rice fields. As the season progresses, farmers await the emergence of the panicle, which will soon open to bear its tiny flowers. The flowers of rice plants, like those of most types of grasses, are so tiny that they easily go unnoticed by city people. For farmers, however, their appearance is full of portent. In vast tracts planted with rice, even the sweet fragrance of the flowers can be detected. When the grains begin to form and swell, special rituals may be held in the fields, such as the pregnancy rituals for the Rice Goddess performed in Bali (fig. 2.20), Java (see chapter 32), and Thailand (see fig. 1.1).

As the crop ripens, the fields must be guarded to prevent birds from carrying off the grain. Ingenious forms of scarecrows and noisemakers are widely used for this purpose, but often a human presence must be maintained in the fields around the clock (fig. 2.22). Children may be engaged to operate noisemakers, or entire families may take up

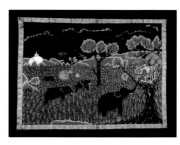

2.25   Elephants raiding a rice
field. Gampaha, Sri Lanka.
1999. Batik on cotton cloth.
W: 117 cm. FMCH
X99.36.9.

Elephants are a threat to the
rice fields in Sri Lanka, and
farmers often spend the
night in field shelters to protect their crop. This scene
depicts a tragic case in
which a farmer was killed by
a raiding elephant. This
cloth was commissioned by
Benille Emmanuel.

residence temporarily in field huts. Rats are another pest, so common that they are closely associated with rice farming (see fig. 8.7). In Bali, plagues of rats are considered to be one of the evils foretold by earthquakes (fig. 2.23), and ritual cremations are held for them when they multiply out of control. fields may also have to be guarded from depredations by larger animals, including wild pigs, monkeys, and even elephants (fig. 2.25).

Harvest

The harvest brings another period of intense labor (fig. 2.21). This is the most momen-tous time in the agricultural cycle, when the promise of abundance is fulfilled if all has gone well. Special rituals are often held for the cutting of the first grain. Harvesttime rituals that celebrate the fertility of the Rice Mother in two Tai communities in northern Vietnam, for example, are described in chapter 7. In Bali, the female head of household pays a special visit to the rice fields shortly before the harvest is to get underway. Beginning at the most sacred corner of the plot, where the irrigation water enters, she cuts the first stalks of grain. She forms these into two bundles, which are bound side by side and decorated with fresh flowers and palm leaf ornaments (fig. 2.24). This construc-tion is called the Nini Pantun, or Rice Mother. Although it is thus feminine in its entirety, the two component bundles are also widely held to represent the Rice Goddess and her consort. The symbolism of the Nini Pantun therefore has to do with the fertile union of male and female elements. When the offerings and prayers are concluded, the Nini Pantun is placed on a shelf at the edge of the field, or sometimes in the crotch of a tree growing within the field, from whence it presides over the harvest. When the entire field has been cut and the grain carried to the granary, the Nini Pantun is taken along in pro-cession and installed in the rafters. There it represents the vital force or spirit of the rice.

Another important task that takes place at the beginning of the harvest is the selection of the seed rice that will be saved for the following crop cycle. This is very often a task for skilled senior women. Senior Ifugao women, for example, make their way through the fields at dawn on the day selected for the first harvest, carefully select-ing the best heads of rice to be kept for seed. The women who perform this task must be familiar with the desired characteristics of perhaps several dozen different varieties of rice. The genetic heritage of the community's rice varieties is literally in the hands of these women, and this weighty responsibility speaks eloquently of the association of women, rice, and fertility.

Archaeological sites thousands of years old have yielded stone knives thought to have been used to cut rice stalks at harvesttime. In some areas, most notably in Southeast Asia, delicate knives held in the fingers are still used for this task (fig. 2.26). These unique tools are perfectly suited to cutting stalks of rice one at a time, a practice that may have evolved because many traditional varieties of rice ripen unevenly. Rice stalks cut in this manner are normally bound into sheaves and transported to the granary for storage still attached to the stalk. The rice keeps longer in this form than if it is threshed prior to storage. Inevitably, these practices came to be reflected in the lore associ-ated with rice, and typically the reason given for the use of the hand knife is that it does not "hurt" the rice and is therefore not offensive to the spirits of the rice or to the Rice Goddess. In Indonesia, where the finger knives are known as *ani-ani*, they are made in a myriad of fanciful forms (figs. 2.27–2.32). Where hand knives are not used, the normal practice is to cut the rice with sickles. These larger tools are more efficient for cutting quantities of grain. They too are often made in highly decorative forms (fig. 2.33).

How the rice is treated after it is cut depends on the storage method. If it is to be stored in sheaves, paddle-like tools may be used to even the bottoms of the stalks so that the sheaves will stand upright (figs. 2.34, 2.35). If it is to be stored in the form of

2.26   (Opposite) A Black Tai
woman uses a finger knife to
harvest rice. The stalks are
cut by pressing them against
the blade. Photograph by
Eric Crystal, Bản Đốc, Nghệ
An Province, Vietnam.

2.27   Rice knife (*ani-ani*) in the form of a boar. Lombok, Indonesia. Carved bone, iron, paint, metal. L: 15 cm. Collection of Clare and Joseph Fischer.

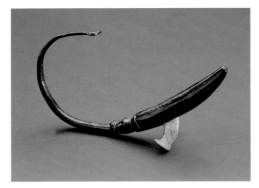

2.28   Rice knife (*gamulang*). Ifugao peoples. Northern Luzon, Philippines. Carved horn, iron. L: 22 cm. Collection of Margaret and Dan Sullivan, Alexandria, Virginia.

2.29   Rice knife (*ani-ani*). Java, Indonesia. Carved wood, paint, iron, mirror. L: 29 cm. Collection of Clare and Joseph Fischer.

2.30   Rice knife (*ani-ani*) with heads in the style of Javanese shadow puppets. Java, Indonesia. Carved wood, iron, batik cloth. W: 13 cm. Collection of Ross G. Kreamer and Christine Mullen Kreamer, Washington, D.C.

2.31   Rice knife carved in the form of the Borneo dragon (*aso*, lit., "dog") motif. Borneo. Carved wood, iron. L: 15 cm. Collection of Ross G. Kreamer and Christine Mullen Kreamer, Washington, D.C.

2.32   Rice knife and sheath. Minangkabau peoples. West Sumatra, Indonesia. Carved wood, ivory, iron. L: 33 cm. Private Collection.

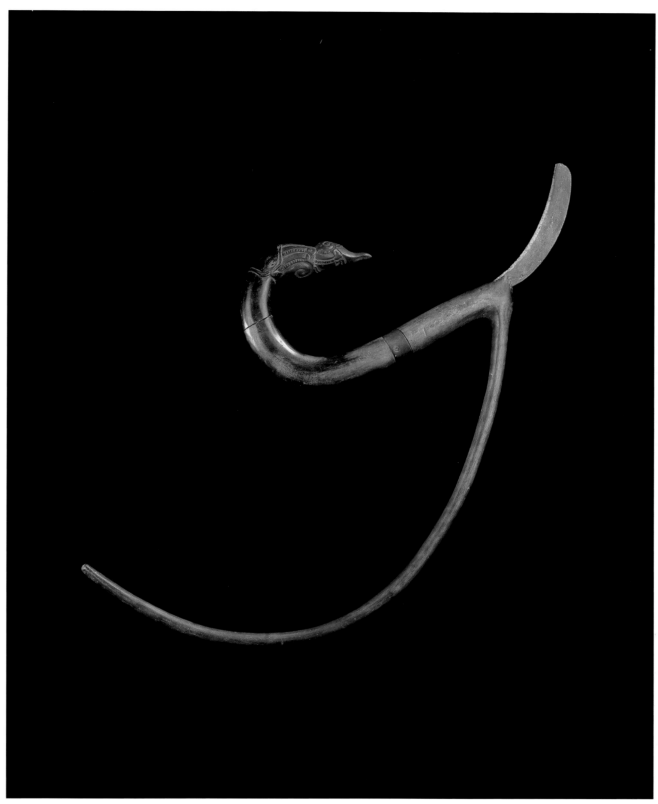

2.33   Sickle in the form of a
            dragon or serpent (*naga*).
            Khmer peoples. Cambodia.
            Carved wood, metal. L: 54
            cm. Collection of Ross
            G. Kreamer and Christine
            Mullen Kreamer,
            Washington, D.C.

2.34   Rice sheaf paddle. Bali,
Indonesia. Carved wood. H:
36 cm. Collection of Steven
G. Alpert, Dallas.

The flat back of the paddle
is used to beat the bottom of
the sheaf to even all the ends
of the stalks so that the sheaf
will stand securely.

2.35   Rice sheaf paddle in the
form of the head of the Rice
Goddess, Dewi Sri. Bali,
Indonesia. Carved and
painted wood. L: 32 cm.
Collection of Ross G.
Kreamer and Christine
Mullen Kreamer,
Washington, D.C.

Rice sheaf paddles evolved
into items of folk art; it
is doubtful that a sheaf with
delicate features such as
this would have withstood
frequent use in the fields.
Many other paddles,
intricately but more solidly
carved, show signs of
heavy use.

2.36   In northern Thailand, rice is threshed by beating the stalks on the sloping inner sides of a huge basket (*ku*) placed in the field. The grain falls directly into the basket. Photograph by Vithi Phanichphant.

2.37   A Hmong harvester pours threshed grain from a height to allow the breeze to separate the grain from the chaff. Photograph by Eric Crystal, Yên Bái Province, Vietnam, 2000.

2.38   Winnowing fan (*wee*). Thailand. Bamboo, nails, string. L: 73 cm. FMCH X98.14.9.

loose grain, it is normally threshed in the field. By eliminating the stalks, less total weight has to be transported to the granary. Threshing is often accomplished by trampling the grain stalks, either with human feet or by driving animals over them. Grinding wheels pulled by livestock can also be used. Another method is to simply beat the stalks of grain against a board or other firm surface (fig. 2.36). After threshing, the grain is winnowed to blow away the chaff. Often this is accomplished by tossing it in a winnowing tray or simply pouring it out from a height (fig. 2.37). Other methods include actively fanning the grain to blow off the chaff (fig. 2.38), or using a hand-cranked winnowing machine with an internal rotating fan.

Transporting the harvested grain to the granary is in many cases one of the hardest tasks of the season (fig. 2.41). This is especially true in swidden agriculture, where the heavy grain often has to be carried long distances over rough terrain in a burden basket. Various types of carrying frames and poles can also be used (see fig. 35.6). Water transport allows for the moving of larger quantities of grain more easily (fig. 2.39). Today, of course, much of the rice grown in Asia is moved by tractor or truck.

The storage of the rice in the granary concludes the cycle of labor. From that point onward, the way in which the grain will be processed and prepared for eating assumes countless forms in households throughout the rice-growing regions of Asia. One major ritual task typically remains—the giving of thanks to the spirits or deities for the success of the harvest (fig. 2.40). This ritual is of such importance that it often becomes the focus of an entire festival, whether celebrated directly in the form of a harvest festival, or more obliquely as a New Year celebration. Either way, the scene is then set for the turning of the cycle and the beginning of the new agricultural season to come.

•

2.39   Vietnamese water puppet depicting a harvest barge. By Đặng Văn Thiết. Hanoi, Vietnam. Carved and painted wood, bamboo, dried rice plants, plant fiber. L: 63.5 cm. FMCH X2001.6.1a–h.

The vast network of interconnected waterways in Vietnam's Red River Delta provides the easiest means of moving grain from the fields to the village. The depiction of such commonplace vignettes from rural life is the mainstay of Vietnamese water puppetry. This harvest barge would literally float across the "stage."

2.40   *Giving Thanks to Ancestors for the Crops.* Final print in the agriculture series *Agriculture and Sericulture Illustrated* (*Gengzhi tu*). China. Qing dynasty, reign of the Yongzheng emperor (r. 1723–1735). Woodblock print on paper. L: 32 cm. © The Seattle Art Museum, Eugene Fuller Memorial Collection 40.83.23.

The final scene from the rice agriculture sequence in the *Gengzhi tu* shows the offering of thanks for the harvest at an ancestral shrine. The entire series, which emphasizes the regularity of the routines of rice agriculture and in many of the scenes depicts the social relationships that are a part of the agricultural process, became a metaphor for the well-ordered Confucian society. This explains in part why these famous scenes, first painted in the Song dynasty by Lou Chou (1090–1162), were still being widely distributed by the imperial printing house six hundred years later (see chapter 29).

2.41   (Opposite) A Black Hmong harvest worker carries a sheaf of grain from the field. Photograph by Eric Crystal, Lào Cai Province, Vietnam, 1999.

3.1   Farmers carry homemade
        rockets (*bung fai*) and
        a wooden female puppet in
        preparation for the annual
        Skyrocket Festival (Bun
        Bung Fai). Photograph by
        Suriya Smutkupt, Sisaket,
        1980.

3

Rice Festivals in Northeast Thailand

Suriya Smutkupt and Pattana Kitiarsa

3.2   A parched rice field during the dry season. Photograph by Suriya Smutkupt, Khon Kaen, 1987.

3.3   Piles of manure can be seen in the background of a vacant rice field during the third lunar month (around mid-February). This is an integral part of the ritual of "Taking Manure to the field" (Phithii Aw Fun Sai Naa). Photograph by Suriya Smutkupt, Khon Kaen, 1987.

Isan, the northeast region of Thailand, constitutes one-third of the country's total area and is home to a population of over twenty million. Known for its harsh climate, arid soil, and severe droughts, this poverty-stricken area is the most underdeveloped in the nation. Because irrigated land is scarce and most arable land is located in the infertile rolling plateau (figs. 3.2–3.4), the inhabitants of Isan have practiced rainfed rice farming for generations (Lovelace et al. 1988, 9–16). Despite conditions highly unfavorable to its propagation, rice has remained integral to Isan culture and society. Like other Tai ethnic groups,[1] the people of Isan believe that rice is the most important substance in the human diet, a belief reflected in a popular Isan poetic genre known as *phayaa*: "Suffering from having no clothes, / One can hide oneself in the house, / Suffering from having no rice, / One can hide nowhere" (Phimworamathakul 1996, 6).[2]

Because Isan is a rainfed rice region, its rice festivals have traditionally been shaped by its unpredictable climate and its geophysical characteristics.[3] A large number of contemporary rice festivals in northeast Thailand, however, have come to reflect the influence of local and international socioeconomic and political forces as well. It is not uncommon at present to witness a modern Isan rice festival run by a politician or local bureaucrat seeking personal advancement. Changes of this sort reflect the rapid agroeconomic and cultural transformation of Isan in particular, and of Thailand in general, and will be explored in the conclusion of this essay.

The Isan rice festivals are part of a larger twelve-month ritual cycle (Hiid Sib Song) that is observed in most rural villages and market towns of northeast Thailand.[4] This traditional cycle consists of cultural practices based on a syncretic religious system characterized by a popular form of Theravada Buddhism, folk Brahminism, and traditional spirit beliefs (see Kirsch 1977). The cycle is central to Isan life and society, forming the shared religious and cultural values observed for generations throughout the region. The heart and soul of this traditional cycle is rice cultivation.

The twelve-month cycle, begins in the first lunar month (Dyan Aai or Dyan Jiang), which falls in December of the Western calendar,[5] with rituals occurring throughout the succeeding lunar months. Seven of these twelve communal rituals pertain directly to rice cultivation and are discussed in detail below.

3.4   The rice crop as it appears in November, not long before the harvest. Photograph by Suriya Smutkupt, Sisaket, 2000.

3.5   Isan women prepare food, fruit, water, flowers, and cooked and uncooked rice for the ritual of Bun Chok Law, which entails opening the granary and the first consumption of freshly harvested rice. Photograph by Suriya Smutkupt, Nakhon Ratchasima, 1999.

The Rice Festival Cycle in Northeast Thailand

As a staple food and an important commodity, rice is vital to the survival of the people of Isan and essential to the larger economy as well. As a ritual object, it resides at the center of belief and practice. It may be the only plant for which a series of rituals accompanies each stage of growth and cultivation—from seedling to harvesting, from the paddy field to the dinner table.

The following brief excerpt concisely describes the climate of the northeast and the related geophysical characteristics that affect the annual rice cultivation calendar:

> Northeast Thailand has a monsoon climate similar to other parts of Southeast Asia, but the region's geophysical characteristics create special conditions. The Northeast is the most continental of the four regions of Thailand and the western mountain ranges of the region create a rain shadow effect over much of the western part of the plateau during the period of the southwest wet monsoon. The overall climate is divided into three basic seasons: the cool season (November through February); the dry or hot season (March through May); and the wet season (June through October).... The actual amount and pattern of rainfall can be extremely erratic and unpredictable, however, not only from year to year, but also within the wet season itself. It is not uncommon for farmers to experience too little, and then too much, rain. This creates considerable risks for agricultural production, 80 percent of which involves rainfed cultivation. [Lovelace et al. 1988, 13]

These features of climate and terrain should be kept in mind while reading the detailed discussion of Isan rice cultivation rituals that follows.

The Ceremonies of Sampling the first Rice and Fertilizing the Paddy field

The third day of the full moon of the third lunar month (Dyan Saam Khyn Saam Kham), around mid-February, is held to be auspicious in the Isan agricultural calendar. It is believed that no evil will take place on this day and that all human endeavor will bring forth fruit. The proverb used to describe this fortuitous moment in time suggests that nothing will be lost or wasted and that metaphorically no rice will escape from the granary: "On the third full moon day of the third lunar month, the frog has no mouth, the otter [*naag*] has no anus" (Kettes 1995, 13).

On this day Isan farmers perform the ritual of the first consumption of freshly harvested rice from the granary (Bun Chok Law; fig. 3.5). The owner of the rice field ritually informs the granary guardian "Today is an auspicious day. We would like to open the granary and gather rice for [the first] consuming."[6] Villagers recite a brief Buddhist chant, followed by expressions such as "Endlessly multiply! May the Rice Mother remain plentiful in this granary! Never recede! Never recede!"[7] The first unhusked rice taken is usually a minimal amount that can easily be contained in a bowl or basket. More actual consuming will soon follow this ritualized act, as from this moment on the granary is considered open for use.

This is also the day of the ritual and actual fertilization of the paddy field and thus the beginning of the new farming season (fig. 3.6). Although since the 1960s chemical fertilizers have been used increasingly in Isan rice farming, farmers continue to use cow or buffalo manure, especially as part of the ritual of "Taking Manure to the field" (Phithii Aw Fun Sai Naa). Early in the morning before the full onslaught of the tropical heat, male and female members of the household throughout the region head to their harvested rice fields with baskets of dry dung.

3.6   An Isan man pushes a cart of cattle manure to be spread in the vacant rice field for the ritual of "Taking Manure to the field" (Phithii Aw Fun Sai Naa). Photograph by Suriya Smutkupt, Khon Kaen, 1987.

One resident of Nakhon Ratchasima described the ritual thus:

On the same day as the first sampling of rice, an amount of dry manure is brought to the field. Manure should be spread throughout the field. Some [farmers] apply the manure primarily to the plots where young rice shoots will be seeded and nursed. While applying the manure, one should inform the rice field's guardian by saying "Today is an auspicious day. I take the manure to the field." In the old days, manure was shared among relatives and neighbors. Nowadays one has to buy it, if you do not keep buffalo or oxen.[8]

In some Isan areas, such as the Soong Noen and Muang Districts in Nakhon Ratchasima, farmers also bring seven dry chili peppers and a few bowls of cooking salt to enhance the soil's fertility. They believe that these ingredients prevent the soil from being "flat in taste" (*cyyd*)[9] and provide nutrients for the rice plants. Farmers catch a handful of manure, salt, and dried chilis and sow the mixture in each paddy plot. In addition to this symbolic fertilization, they carry larger loads of manure to every plot of the field.

Calling Down the Rain: The Bathing Ritual, the Skyrocket Festival, and Other Rain-Calling Rites

In Isan rain-making deities must be propitiated to avert crop failure. The Bathing Ritual (Bun Songkran or Bun Song Naam) and the Skyrocket Festivals (Bun Bung Fai) are two means employed to insure abundant rain and a healthy rice harvest (see fig. 3.1).

Organized in mid-April—the height of Thailand's dry season—Songkran coincides with the celebration of the traditional Thai New Year and with it the wish for an auspicious rice cultivation cycle. In the past, Isan villagers invited their grandparents or respected village elders to bathe in water they brought to bless them on this occasion. In some areas, this ritual was performed as a means to elevate village monks to a higher rank within the traditional religious system, and this aspect has become increasingly prevalent.

Many Isan villagers, however, organized Songkran at a local pond, canal, or forest, where a guardian spirit or rain-making deity was thought to reside. They invited a monk to lead the ritual in this sacred spot. Alms were given and sacred chants relating to rain-making deities were recited. This ritual usually ended with communal bathing and water play aimed at appeasing the spirit (see Smutkupt et al. 1993). These traditional variants of the Songkran ritual existed in Isan long before the popular tourist-oriented versions that may be witnessed in contemporary Thailand.

The Skyrocket Festival has its origin in the *Phaa Daeng Nang Ai* (*Legend of Prince Phaa Daeng and Princess Ai*), which tells of a large city called Thii Taa Nakhon that had experienced a long, severe drought. Phya Khom, the ruler, decided to arrange a skyrocket contest—a means of worshiping the rain-making deity—and invited young princes from other cities to participate. He promised the hand of his beautiful daughter, Princess Ai, to the winner. The charming young Prince Phaa Daeng won the competition and became eligible to marry the princess. Prince Phang Kii, however, the son of Naga (the serpent or dragon king of the underworld), longed for the princess, who had been his lover in a past life. Prince Phang Kii transformed himself into a white squirrel in order to stay close to the princess. When the princess saw the little squirrel, she fell in love with it and ordered her servants to catch it alive. One of them, however, shot it with an arrow. Before its death, the squirrel made a vow that its meat would be abundant and that whoever ate it would perish in an earthquake and Naga's subsequent invasion. Everyone in the city was killed, and the city itself sank to the bottom of Nong Han, a large freshwater lake located in present-day Sakhon Nakhon Province. According to one account Prince Phaa

Daeng was reincarnated as Phayaa Thaan, the Isan/Laotian god of creation. Today Isan farmers fire skyrockets to worship Phayaa Thaan and beg him to release the seasonal rainfalls that are so vital to their rice crops (see Phinthong 1991, 86–111).

The Skyrocket Festival is held on selected days during the sixth lunar month (May–June). It is a communal rite in which leaders and villagers join to build rockets, run the contest, arrange a temple fair, and earn religious merit from Buddhist monks (fig. 3.7). Men are responsible for building, decorating, and firing the rockets. They also prepare wooden phallic symbols (linga) and other symbols of a sexual nature that are played with during the parade from the village temple and center to the firing range in an open field. Villagers chant songs calling on the rain to fall. The association of human fertility and an abundant harvest is further emphasized by pantomines of sexual behavior including the coupling of puppets (fig. 3.8) and male cross-dressing. Within a few weeks of this festival, the rice cultivation season will begin, and Isan farmers will engage in the backbreaking work of tilling their land (Smutkupt and Kitiarsa 1998).

In the 1980s the Skyrocket Festival was modified and made an annual fair used to attract tourists in some provinces, among them Yasothorn, Roi-et, and Sisaket. Although the form and content of the festival have been altered to conform to the needs of the tourist industry and to appeal to more urban audiences, its original purpose of calling forth rain has been preserved. Isan farmers also perform a number of other rain-calling rituals, especially in times of severe drought. These include "Parading the Female Cat" (Hae Naang Maew), "Reciting the Snake-Head fish King's/ Toad King's Prayers" (Suad Khaathaa Plaa Khoo/Phayaa Khaang Khog), and the "Bamboo Basket's Female Spirit Possession" (Phii Naang Doong).[10]

The first Plowing, The first Transplanting, and the first Harvesting: Offering Sacrifices to the Guardian of the Paddy field

The "first Plowing Ritual" (Raeg Thai) resembles the "Royal Plowing Ceremony" (Raeg Naa) held annually in Bangkok during the sixth lunar month to mark the start of the growing season (Gerson 1996, 21–25). As with the royal ceremony, Isan farmers perform the "first Plowing Ritual" to commemorate the beginning of rice-planting season. Unlike its royal counterpart, however, the folk ritual does not deal with Brahministic tradition and Thai state symbolism. The farmers instead worship their rice field guardian (Taa Haeg). The name *Taa Haeg* has a double meaning. On the one hand, it refers to the spirit of the "maternal grandfather" (*taa*), with *haeg* meaning "the first" or "the original." Since most of the rice fields in the Isan region are inherited through the female line, the guardian represents a maternal ancestral spirit. On the other hand, the term also means the "first plot" of the paddy field, as *taa* can also refer to a quadrilateral plot of land.

Sacrifices are offered to Taa Haeg on three occasions. The "first Plowing Ritual" takes place after there is sufficient rainfall in the field, around June or July. The guardian, however, is fed and informed of the first plowing prior to its occurrence. Then a makeshift spirit shrine is set up in the corner of the selected plot, which the owner traditionally regards as the first or original plot in a particular paddy field. A male member of the family usually officiates with assistance from female members. Offerings include a boiled chicken, a set of male and female loincloths (*sarong* and *pha siin*), a sweet treat made of glutinous rice, and gold and silver coins made from rice flour. Homemade tobacco and areca nut are also included, since the guardian is thought to be male. These offerings are arranged in a set of seven tiny banana-leaf cones, and the ritual is performed after breakfast on a selected day. Isan farmers believe that any day except Friday is auspicious for the offering of sacrifices to the rice field guardian.[11]

3.7   Spectators watch a rocket launching at the "Skyrocket Festival" (Bun Bung Fai) in Yasothorn. Photograph by Suriya Smutkupt, 2000.

3.8   An Isan man carries copulating puppets that will be used during the "Skyrocket Festival" (Bun Bung Fai). Photograph by Suriya Smutkupt, Sisaket, 1980.

3.9   An elderly Isan man transplants seven young rice shoots, a symbolic act known as Pag Taa Haeg or Haeg Dam. Photograph by Suriya Smutkupt, Sisaket, 1980.

Prior to this "first Plowing Ritual," farmers will have prepared their rice seedlings. This entails soaking the rice seeds in water overnight. The seeds are brought to the well-plowed and fertilized seeding plots where they are broadcast across the muddy soil. Within a few days, the seeds grow into young rice shoots that will soon be ready to transplant.

Farmers make a second set of offerings to the rice field guardian when rice transplanting begins (fig. 3.9); this is known as the "first Transplanting" (Raeg Dam). The same procedure used in the first plowing ceremony is followed with one difference. The male owner of the rice field must prepare a small, ready-to-transplant plot of land, usually a corner of the plot where the first ceremony was carried out. Seven young rice shoots are transplanted here as a symbolic act (Pag Taa Haeg). Most farmers offer only a homemade cigar and a mouthful of areca nut at this time, placing them inside the previously established shrine. The guardian is ritually invited to "smoke the cigar" and "chew the areca nut." The farmer requests protection and a fertile rice crop from the guardian while transplanting these symbolic rice shoots in the sacred plot. Toward the end of the second feast, a bamboo fence is erected to prevent cattle and other wild animals from entering.[12]

A final feast called the "first Harvesting" (Raeg Kiaw) is offered to the rice field guardian. Before mature rice stalks are cut from the first plot, an offering (again usually a homemade cigar and a mouthful of areca nut) is placed in the shrine with the recitation of sacred phrases. The ceremony is intended to give thanks to the rice field guardian for the protecting and nurturing of the grains. The rest of rice stalks will be harvested soon afterward.

Celebrating the Survival of Young Rice Shoots

The ritual of "Celebrating the Survival of Young Rice Shoots" (Bun Khaaw Pradabdin) is scheduled on the fifteenth full-moon day of the ninth lunar month (September). *Khaaw pradabdin* may be literally translated as "rice crops that decorate the earth." The late Phraya Ariyanuwat Khemachaarii, a native of Isan and a Buddhist monk and scholar, called this the ritual of "celebrating rice crops that are growing up densely and plentifully all over the land" (1983, 46). The ritual is performed approximately two months after the rice is transplanted, well into the rainy season, when the shoots have survived the crucial stage of transplanting and grown into green crops. By this time the rice will have already gone through a period of seasonal drought and been exposed to insects, diseases, and other threats.

Isan farmers celebrate the survival of the rice crops by preparing two important offerings. The first of these is called the "rice package" (*hoo khaaw*) and contains steamed glutinous rice, grilled dry fish, grilled meat, and seasonal fruits. This package was traditionally wrapped in a piece of banana leaf; however, plastic bags are now widely used (fig. 3.10). A second package is called the "cigar package" (*hoo yaa*) and contains a homemade cigar and areca nut. Representatives from each household (usually a grandmother, mother, and daughter) bring their packages to the village temple, where Buddhist monks perform a communal ritual to "pass" these offerings to ancestral spirits, "vagabond" spirits, and rice field guardians. A number of the packages will also be brought to the rice field and offered to the guardians dwelling there (fig. 3.11). In the temple hall, the Buddhist monks consume the meals contained in the packages, sharing the leftovers with the lay participants (Smutkupt et al. 1991b).

This ritual can be seen as a celebration of the availability and abundance of food and natural vegetation that occurs in the midst of the rainy season. After suffering

3.10   Food, homemade cigars, and areca nuts and betel leaves are offered during the ritual of "Celebrating the Survival of Young Rice Shoots" (Bun Khaaw Pradabdin). Photograph by Suriya Smutkupt, Khon Kaen, 1991.

3.11   A Isan man offers a libation to the spirits as part of the ritual of "Celebrating the Survival of Young Rice Shoots" (Bun Khaaw Pradabdin). Photograph by Suriya Smutkupt, Khon Kaen, 1991.

through a long dry season, Isan farmers look forward to their growing rice crops and to the plentiful fish, frogs, bamboo shoots, wild game, and vegetation that may be gathered in their rice fields and the surrounding forests.

Embracing the Rice Mother's Pregnancy

Farmers refer to the Bun Khaaw Pradabdin as the "small rice package" (*hoo khaaw nooj*), while the "Rice Mother's Pregnancy" (Bun Khaaw Saag), held in the tenth lunar month, is known as the "big rice package" (*hoo khaaw jaj*). Mae Phosop, the Rice Mother, is a figure central to Tai cosmology (figs. 3.12, 3.13). Anthropologist Lucien M. Hanks has aptly described the ways in which she personifies the growth cycle of the rice crop and the conflation of the maturation of rice and human development.

> In many parts of Thailand today rice is deemed animate and so grows like animate creatures. The rice mother, Mae Phosop, becomes pregnant when the rice flowers bloom, and as the grain grows, she, like any pregnant woman, delights in scented powder and bitter-tasting fruit....
>
> According to some Thai villagers, man's body itself is rice, and eating rice renews the body directly. Babies grow within their mothers' stomachs where they sit eating the food that the mother eats. Tissue is made of rice because it derives from rice. Mother's milk is blood purified to a whiteness. Just as mothers give their food and bodies to nourish children, so Mae Phosop, the Rice Mother, gives her body and soul to make the body of mankind. Thus the rice growers' image of man becomes rice itself; perhaps, according to this vision, man differs slightly from other living creatures, largely because of the diet that sustains him. [Hanks 1972, 22][13]

In performing the ritual of the "big rice package," Isan farmers not only offer their ritually prepared food and cigar packages to Buddhist monks in the village temple, they also perform the rite of "Calling the Soul of the Rice Mother" in their own rice fields, a ritual that is the exclusive prerogative of women. They offer sweet and sour fruits to the pregnant Rice Mother and also bring a set of clothes, baby powder, homemade perfume, a mirror, and fragrant flowers. They then "dress" selected rice plants with these items. In some areas they bind the Rice Mother's soul by wrapping cotton threads around certain rice plants.[14] The leaves of rice plants, thought of as the Rice Mother's hair, are neatly combed and arranged. The fact that this ritual is performed solely by women is a reflection of the belief that that Rice Mother hated men, who had found her irresistible and attempted to violate her. It is said that she so despised men that she once flew away, leaving humans to starve for years, and that she returned only when they had learned how to respect and worship her.[15]

In addition to celebrating the "Rice Mother's Pregnancy," Isan farmers in some areas offer blooming rice flowers to Buddhist monks as part of the ritual ending Buddhist Lent (Bun Oog Phansaa). It is believed that fresh scent and beauty of blooming rice flowers represent the farmers' faith in Buddhism. Blessings from Buddhist monks also serve to ensure an abundant harvest.

When milky young rice grains form, Isan farmers harvest some of them and prepare a kind of sweet called *khaaw maw*, made from immature grains husked with a hand mortar. When the *khaaw maw* is ready to serve, the juice and meat of a young coconut or some sugar is added. Isan farmers offer this special rice sweet to Buddhist monks to receive merit in the season before the harvest.

3.12   A large crowd of participants observes the ritual commonly known as *hoo khaaw jaj*, or the "big rice package." Photograph by Suriya Smutkupt, Khon Kaen, 1992.

3.13   Isan women divide food after participating in the ritual commonly known as *hoo khaaw jaj*, or the "big rice package." Photograph by Suriya Smutkupt, Khon Kaen, 1992.

Celebrating the Abundant Harvest: Calling the Rice Mother's Soul, Tying the Buffalo's Soul, and the Kite-Flying Festival

The Tai believe that everyone

> has inherently in his or her body an attendant spirit or genius called "*khwan.*" If a person has a sudden fright or unexpected illness, it means that the *khwan* has taken flight and left its abode in the body. Sometimes it wanders and loses its way in the forest and cannot come back. Then the person whose *khwan* is not with him or her will continue to suffer illness, and if the *khwan* does not come back in time, that person will die…. a ceremony for calling back the *khwan* is [therefore] performed with an invocation and food offering to the truant *khwan.* [Anuman Rajadhon 1987, 13]

The Tai also hold that certain animals and inanimate objects possess their own *khwan.* Isan farmers have subscribed to this popular belief for generations, performing rituals to call the *khwan* of their two most precious commodites, rice and water buffalo.

Lucien Hanks has similarly described how Thai farmers in Bang Chan (an area located in the central plain near Bangkok) believe that the Rice Mother's "offspring, the rice, has its *khwan* (soul), like other animate beings. Along with the harvest, the soul too must be ritually gathered and taken to the storehouse. Then when the crop is sold, the buyer from the rice mill takes the raw grain, but in a handful of grain carefully returns the soul to the farmer to impregnate next year's crop"(1972, 21).

The ritual of "Calling the Rice Mother's Soul" (Bun Khuun Laan/Suu Khwan Khaaw) is held around the first lunar month (December) or right after dry rice stalks are piled up high on the prepared open ground (*laan*). After the rice threshing is finished—an operation increasingly performed by machine—the farmers tie a few reserved bundles of rice stalks to a pair of bamboo poles and strike them on the hard, dry ground to separate the rice grains from the straw. They use these grains to perform a blessing ritual before the bulk of the threshed rice is transported to the storehouse in the village.

The ritual is held in the morning, and Buddhist monks are invited to the site. The host prepares the *bai sri* tray, which contains flowers, a boiled chicken, a pair of boiled chicken eggs, candles, joss sticks, seasonal fruits, and rice sweets (figs. 3.14, 3.15). A special meal, usually of beef or pork, is also prepared for the monks and other participants. Offerings are made and Buddhist prayers are recited. Monks usually prepare the holy water that the host will splash on relatives, rice grains, the rice field, and water buffalo. The holy water brings luck, good health, happiness, and prosperity (Sarathasananan 1995, 1–5).

In the past, Isan farmers gave special thanks to the water buffalo whose labor was essential to the rice crop. This ritual is vanishing as the tractor gains ground, and only those farmers who still use buffalo participate. Known as "Tying the Buffalo's Soul" (Suu Khwan Khwaai), the ritual gives farmers the opportunity to apologize to their animals after a year of hard work. Isan farmers regard their buffalo as co-workers in the rice field and regret the times that they may have treated them badly. It is for this reason that a "soul tying" ceremony is held at the end of the rice-harvesting season (see Phinthong 1991 and Srisoong 1992).[16]

The annual harvest season lasts from around November to January, during which time Isan farmers, especially those in Burirum, Khon Kaen, Nakhon Ratchasima, and Surin, fly their kites in honor of Phayaa Thaan and Phra Inn, the supreme gods in the local mythology. This is a means of thanking the gods, who are believed to generate

3.14   A poster of the Rice Mother,
Mae Phosop, appears on top
of a pile of new rice and
Thai banknotes in the ritual
of "Calling the Rice
Mother's Soul" (Bun Khuun
Laan/Suu Khwan Khaaw).
Photograph by Pattana
Kitiarsa, Khorat/Nakhon
Ratchasima, 2001.

3.15   Rice farmers offer new rice, fruit, food, and money in the ritual of "Calling the Rice Mother's Soul" (Bun Khuun Laan/Suu Khwan Khaaw). Photograph by Suriya Smutkupt, Khorat/Nakhon Ratchasima, 2001.

rainfall and safeguard rice fertility throughout each cultivating cycle. Local kite makers invent noise-making kites from bamboo, paper, and other locally available materials, and the northeast monsoon wind usually reaches this region during the harvest. It is, therefore, a perfect time for farmers to enjoy kite flying. The Kite-Flying Festival, held in November, has been promoted as a major show in Burirum Province in order to attract local and international tourism (Smutkupt and Kitiarsa 2000a, 127–35).

Completing the Cycle

After the rice is transported to the family granary, an individual member of the household, usually a grandmother or the mother, performs a private ritual to inform the granary guardian. In the area surrounding Nakhon Ratchasima inhabited by the Thai Khorat people, the granary guardian is called Taa Pook and is made from dry rice straw tied into the shape of a sitting doll (figs. 3.16, 3.17). The term *Taa Pook* suggests the placement of the doll, who sits atop the rice grain in the granary. In other parts of the region, farmers call the granary guardian Phii Lao (lit., "the guardian spirit of the household granary"). Taa Pook and Phii Lao require similar ritual care. Farmers generally place a pair of candles and flowers at the door of their rice granaries as an expression of respect. These offerings are replaced on every Buddhist holy day.

3.16   Offerings are set out before Taa Pook and Yai Pui, the male and female guardian spirits of the granary. Photograph by Suriya Smutkupt, Khorat/Nakhon Ratchasima, 2000.

In the third lunar month (around February), Isan farmers earn merit by offering baked glutinous rice to Buddhist monks. This ritual is called Bun Khaaw Cii. *Khaaw cii* literally means "grilled rice lump." A handful of rice with a small amount of salt added is grilled over hot coals, and raw egg is applied to the lump a little at a time until it is well cooked. Farmers are always conscious that the first and best portion of the harvest be offered to Buddhist monks as a means of gaining merit and giving thanks to the rice field guardian, the ancestral spirits, the Rice Mother, and other spirits involved in the rice-cultivating process.

3.17   A female rice farmer displays
        the figure of Taa Pook, the
        male guardian spirit of the
        rice granary. Photograph by
        Suriya Smutkupt, Khorat/
        Nakhon Ratchasima, 2001.

3.18   A banner displayed at the "Bun Phawes/Jataka Festival" held in Roi-et depicts the entire Vessantara Jataka, highlighting key events. Photograph by Pattana Kitiarsa, 2000.

The end of the annual cycle of rice festivals is marked by Bun Phawes (fig. 3.18), which honors Prince Vessantara. This ritual offers the greatest opportunity to gain merit. According to the Buddhist Jataka stories, Prince Vessantara, who was famous for his acts of charity, was the Buddha's last incarnation prior to his rebirth as Prince Siddhārtha.

In most Isan villages this festival lasts two days and nights. On the first day, women prepare food, drinks, and other offerings, while men work at the village temple. The whole temple ground is decorated and portions of it made to resemble the royal palace, city, forest, and other locales encountered in the stories relating to the life of Prince Vessantara. In the afternoon, villagers join a parade to invite the spirits of Prince Vessantara from the forest outside the village to the temple hall, where the ritual is centered.

At about two o'clock in the morning of the next day, the parade of a "thousand rice lumps" is organized following a prayer session. Villagers join the parade to collect cooked glutinous rice from every household and make it into a thousand lumps (fig. 3.19), which are offered to ancestral spirits, vagabond spirits, and Buddhist monks (see Smutkupt et al. 1991a). The remainder of the second day is devoted to giving alms to Buddhist monks and listening to a series of prayers that relate to the life story of Prince Vessantara. Buddhist laypersons share the belief that they gain maximum merit if they listen to a thousand Vessantara prayers in a single day.

3.19   Isan women offer cooked glutinous rice to the spirits as part of the festival of Bun Phawes held in Roi-et. Photograph by Pattana Kitiarsa, 2000.

Buddhist monks recite prayers from morning to the late afternoon. At the end of each prayer session, the villager sponsoring it offers alms—robes, money, and other necessities—to the monk who has recited. Bun Phawes, the last great ritual, completes the cycle of rice-cultivating rituals in the northeast. It is also observed in other regions of Thailand and neighboring Buddhist countries.

3.20   A giant, grilled, sticky rice lump is transported by truck at the Bun Khaaw Jii Yak festival. Photograph by Suriya Smutkupt, Borabue District, Maha Sarakham Province, 2000.

Isan Rice Rituals in the Modern World

Rice festivals in northeast Thailand are dynamic. Over the past three decades they have been, altered, redefined, and sometimes invented to conform to rapidly changing conditions. As Thailand becomes increasingly modernized, Isan rice rituals often come to assume the role of repositories of traditional culture and wisdom and lose their former function as guides to the actual practices involved in rice cultivation.[17] Two recently held festivals are noteworthy in terms of their relation to and departure from the celebrations that inspired them.

In March of 2000, we attended the two-day Bun Phawes held in Roi-et, the market town located in the heart of the Isan region's great plain (see Smutkupt et al. 2000b). Unlike the traditional version described above, this event was organized by groups of local officials led by the governor, the mayor, and a prominent national politician, Anurak Jurimas, who was then a member of parliament from the coalition Chart Thai Party and the deputy minister of cooperatives and agriculture. Recalling the great acts of merit performed by Prince Vessantara in the ancient Jataka stories, Jurimas provided a huge feast for those attending, which included tons of rice noodles with curry sauce, Isan-style green papaya salad (*somtum*), sweets, and beverages.

Politicians from opposing political camps who were running in various local elections also gave attendees free food and made contributions to the organizing committee. They set up large billboards and printed posters featuring their names and pictures and inviting the people of the town to join in this great event. Their aim was clearly to derive political gain by credentialing themselves as good-hearted and generous leaders, using as a model a prince drawn from Buddhist mythology.

Despite the obvious political overtones, the Bun Phawes in Roi-et strictly followed the ritual process. A group of laymen and monks was appointed by the governor and the provincial education office to supervise the religious aspects of the celebration as well as the more worldly components. On the second day of the event (March 13), after the many chants and recitations, there was a distribution of alms to ninety-nine invited monks.

Around noon, there were parades and dancing processions organized by officials from many government offices, politicians, and numerous neighborhood and kin groups. People participating in these processions danced and sang with joy, donating money to the organizing committee, which was kept for public use. During the two-day ritual the province had made nearly one million *baht* in cash and goods. In short, this concluding ritual act transformed the sacred celebration of Prince Vessantara's life and the celebration of the rice harvest for village farmers into a modern form of resource mobilization for the provincial administration.

On April 6, 2001, we observed Bun Khaaw Jii Yak, a new version of a traditional festival featuring a giant, grilled sticky rice lump and held in the Borabue District of Maha Sarakham Province. In this festival the organizers decided to make one giant grilled rice lump and one giant round, flat rice cracker. Both items were made using large quantities of rice, flour, sugar, and egg. The rice lump and the cracker were so enormous that they had to be supported on metal frames, and each was carried into the ritual parade by a six-wheel truck (fig. 3.20).

Managerial skills and financial and other resources for this occasion were provided by Chanchai Chairungreung, a member of parliament representing the Borabue constituency. Assisted by his wife, a former movie star from Bangkok, Chanchai and his supporters convinced the local district chief to popularize this traditional rice festival and transform it into an extravaganza. Groups of housewives from every subdistrict were invited to join and dance in the parade procession, which was led by traditional drum bands. They were given cloth for their attire and money for lunch and transportation. These women played key roles in running the local cuisine contest, singing and dancing competitions, and other activities throughout the day. Most importantly, Suwat Lippatapunlop, the minister of industry from the Chart Pattana Party and several top national-level politicians from the same party were invited to preside over the opening ceremony.

It was apparent that this festival had nothing to do with the rice-cultivating season. Buddhist monks were not invited to join in, and we witnessed no ceremonies related to the earning of merit. Village housewives were clearly targeted in this case as potential voters, and the festival was structured with the aim of gaining their support. Isan village rice farmers seemed to be left out of this rather blatant instance of political maneuvering in contemporary Thailand.

Judging from these two recent "rice" festivals held in northeast Thailand, it is apparent that some traditional ceremonies have been transformed by political and economic factions within the Thai bureaucracy and by the impact of the regional tourist industry. Isan farmers, however, also modify and simplify their own rice rituals due to time constraints, the shortage of labor, drought, and the increased availability of modern agricultural technology. Rice is now a commodity in a market economy, and market forces alter people's perceptions as well as agricultural techniques. Isan farmers nowadays rely less on water buffalo and their manure, and belief in the Rice Mother is rare among the younger generation. Young people increasingly turn away from rice cultivation, finding jobs and earning wages in cities such as Bangkok and in other provinces in the central plain and the southern region. They send money back to their parents, who hire labor and invest their cash in improved agricultural technology.

Despite such trends, there is a new movement afoot, spurred by nongovernmental organizations and local leaders, to revive and preserve the rice rituals. This movement attempts to invent appropriate rice-cultivating techniques, using more manure and an integrated farming system in which rice, poultry, fish, and vegetables are produced in the same areas. As part of this new environmentally conscious movement and the promotion of self-sustained agriculture, traditional rituals are being incorporated at least to some degree.

•

4

Body Art and Cyclic Time:
Rice Dancing among the Tharu of Nepal

Gisèle Krauskopff

As the essays in this volume will make clear, many artistic productions emanating from Asian societies are ephemeral, some made simply to please the eye of god and mortal. Combining a visual aesthetic, a musical tradition, and a corpus of oral mythology, the rice dances performed by the Tharu peoples of Nepal are an entirely transient component of the arts relating to rice. They exemplify the deep association between the ripening of rice and human fertility, and between the entire agricultural cycle of rice and human sexuality.

A Marshland Civilization

The Tharu peoples inhabit the Terai—the lowlands bordering the Himalayan Mountains and the northern edge of the Ganges River Valley in Nepal. The region has been somewhat romantically described as a forest wilderness, a virgin jungle, and a no-man's-land devoid of civilization. It is, however, more apt to characterize it as a marshland that has been inhabited—albeit sparsely—since ancient times. According to legend, the Buddha was born in a well-developed region of the Terai, which later reverted to forest. The Terai was sparsely populated throughout much of the last millennium, due largely to political factors.[1] In some ways, it is a surviving remnant of the forest that once covered the Middle Ganges Valley before it was cleared and urbanized a few centuries before the common era. Depending on their economic and land ownership status and on the political situation, Tharu farmers were able to move to new agricultural sites within the region.[2]

Tharu is an enigmatic ethnic label, one that is applied to groups in the Terai region who until recently were reluctant to intermarry, had different traditions, and spoke different languages—albeit all within the Indo-Aryan family. From the lowland of Kumaon in western India to the low valleys of Assam in eastern India, affiliated peoples—the Buxa, the Rana, the Danuwar, the Meche, the Dhimal, and the Rajbamsi—share the same ecological niche. The Tharu live between the Mahakhali River in the Naini Tal District of India and the Kosi River in eastern Nepal on both sides of the Indian border, but primarily in Nepal. They have built up a specific resistance to malaria over the centuries,[3] and this has enabled them to succeed in tilling this inhospitable marshland and forest.

Although those labeled "Tharu" may be culturally very different, they share a way of life based on the extensive cultivation of rice and fishing. Until recently in some areas of the Terai, rice was broadcast in the fields, as opposed to being transplanted. The period when rice is cultivated is also the time for fishing "in the fields," the ponds, and

4.1 The cycle of Sakhya Nac, or "Dances of Friends," opens at the time of the Green Ritual (Harya Gurai), which marks the sowing of the rice crop. The dances are performed nightly until the harvest. The dance leader, known as the "mouth" (*mohorinya*), appears at the right in this photograph. She is most versed in the "Garden of Friends" (Sakhi Phulwar), the sacred mythical women's song that accompanies the dance cycle. Photograph by Gisèle Krauskopff, Dang Valley, 1986.

4.2   Tharu girls fish in the rice fields and irrigation canals. In their hair they wear the barley sprouts consecrated during the auspicious ceremonial exchanges of the festival of Dasain, which coincides with the performance of rice dances. Photograph by Gisèle Krauskopff, Dang Valley, 1982.

4.3   Harvested rice is carried to the threshing ground. Photograph by Gisèle Krauskopff, Dang Valley, 1982.

irrigation canals. This brings to light the strong association between fish and rice, as staple foods and as auspicious symbols.[4] fishing expeditions are an integral part of all rainy season and rice rituals (fig. 4.2).

Whether in Champaran, as illustrated by the photographs of William Archer,[5] or in the Dang Valley of far western Nepal, as I will illustrate here, Tharu culture is widely known for its song and dance (McDonaugh 1989). There is a pan-Tharu pattern whereby the timing of the dance cycle is adjusted to the rice cycle and so becomes a physical incarnation of the calendar itself (Krauskopff 1996). Dance is a visual performance of the rice cycle, and the dances embody prosperity. An essential dimension of this cultural construction is the sacred aspect of the dances, which in the Dang Valley (home of the Dangaura Tharu) are linked not only to the agricultural calendar but also to a cycle of rituals framing this calendar.

The Season of Rice in the Dang Valley of Nepal

In the Dang Valley, the *dhurya,* or dry season, and the *barka,* or rainy season, are set apart by *saha,* the beautiful season of abundance—the time when the rice is ripe and ready to be harvested, the weather is clear, the gods residing in the Himalayas come to the valley,

4.4 The village shrine is prepared for the Harya Gurai, or "Green Ritual," which opens the rice dances. The carved wooden planks represent the village goddess and her mate. In front of the planks are five large wooden pegs placed in descending order according to size. These represent the five Pandava brothers who guard the territory and are also honored during the harvest period of Dasain. Photograph by Gisèle Krauskopff, Dang Valley, 1983.

and the rice granaries will soon be full (fig. 4.3). The *barka* and the *dhurya* resemble opposite faces of a coin, each marked by a village ritual: the "Green Ritual" (Harya Gurai) and the "Dry Ritual" (Dhurya Gurai).

The Harya Gurai takes place when the rice is sown, usually at the beginning of August, and is intended to protect the young rice sprouts from the diseases that are brought to the fields via irrigation canals along with the much-welcomed water.[6] This ritual also opens the cycle of Sakhya dances and "unbinds" the drums, which were silent during the dry season. These dances are performed nightly until *saha,* the season of abundance, and the harvest—which fall during the rejoicing that accompanies the great pan-Hindu festivals of Dasain (Durga Puja or Dashara in India) and Tihar (Devali). During this short but intense period—the apotheosis of which is the dance performance—the gods are awakened, the granaries filled, and social rules and hierarchy are reinforced.

Most village rituals occur during the rainy season, stressing the importance of the rice cultivation. Even the ritual of "Binding the Cattle" (Gauri Bahana)— performed to protect the livestock on an altar composed of plows fixed in the ground—indicates the prime importance of rice cultivation as cattle are above all to be valued for tilling the rice

4.5   The "unbinding" of the drums occurs during the Harya Gurai, or "Green Ritual." After the unbinding, a possessed man dances on burning coals in front of the young male drummers. Photograph by Gisèle Krauskopff, Dang Valley, 1983.

4.6   The Sakhya Nac, or "Dances of Friends," reach their highpoint and are performed in the daytime during the festival of Dasain. Girls dance in a circle moving counterclockwise and accompany themselves with small cymbals. Photograph by Gisèle Krauskopff, Dang Valley, 1986.

fields. By contrast, the dry season is devoid of purely agricultural rituals, although certainly not free from agricultural labor. As in all of South Asia, it is the time to cultivate dry crops (oilseeds, lentils, and so on), many of which bring the Tharu significant income. The Dhurya Gurai is an exorcism designed to rid the community of evils of all kinds. It serves to protect against fire and malign spirits, particularly those of people who experienced an untimely or violent death, or of "the foreigners" roaming the fallow land and forest. While the Harya Gurai deals with life, the Dhurya Gurai deals with death.

Rice cultivation is the only sacralized agricultural activity inscribed in the ritual cycle. Some aspects of the Dhurya Gurai, such as the emphasis put on fire, fallow land, and forest, may be vestiges of an older calendar when the dry season was the time to burn and clear new lands for the monsoon rice crop.

4.7   In the Paiya dance the male drummers run in a frenzied rhythm that leads them toward and away from the row of female dancers. Photograph by Gisèle Krauskopff, Dang Valley, 1992.

Fertility, Love, and Rice Dance Performance

Prior to the commencement of the Harya Gurai, which will in turn start the rice dance cycle, the village shrine, composed of carved wooden planks representing the village goddess and her husband, is decorated with leaves (fig. 4.4). Very early in the morning, young people go fishing. Older women stay at home to make rice flour cookies in various shapes. The first sacred act to occur takes place at the shrine. It is the recitation by a knowledgeable man of the mythical origin of the Harya Gurai, which incorporates accounts of the creation of the Dang Valley; of the first Tharu, who was born as a pumpkin floating in the primeval lake; of the transformation of his body; of the lake's settling into firm ground; of the marriage of the first Tharu with Devi, the goddess personifying female forces; and of the conflicts that arose between the couple and led to the foundation of sacred rituals. This mythical song also emphasizes the creation of agricultural tools—the yoke and the plow—the digging of irrigation canals, and the diseases that can be spread through them.

In the dead of the night, after offerings are made to the numerous gods residing in the village shrine and to the ambiguous forces linked to the water and the forest, all the men line up at the shrine for the "unbinding "of the drums. After the priest opens the drums, a possessed man dances on top of burning coals, facing the line of young male drummers and making overtly sexual and provocative gestures (fig. 4.5). At this point, the ceremony shifts to the women who are not present at the shrine. An offering of rice beer is sent to the lead dancing girl who, as the "mouth" (*mohorinya*) of the gods, is the one person who best knows the "Garden of Friends" (Sakhi Phulwar),[7] the sacred mythical song of the women that accompanies the "Dances of Friends"(Sakhya Nac), which have now been ritually opened.

The Sakhi Phulwar recounts the day-to-day life of Jasu and her husband, Isaru, within the mythical framework of the creation of the world. As it is a women's song, it is sung from a female perspective. The rhythmic pattern is strophic, and the two lines devoted to each episode are repeated several times until the refrain introduces a new strophe. Following the creation of the world, Jasu commands Isaru "to go and make the plow." The song describes meal preparation before Isaru leaves the house to join the other men in their search for the right red wood with which to make the first plow. Isaru's father then sends his son to bring the bull and tie the yoke before going to till the land. Inside the house, Jasu, who has prepared the meal, puts it in a basket, places the basket on her head, and goes to the rice fields. She walks by the riverside, where "doves are playing in water" (*dahara daphul*).[8] The motif of doves at play is used to refer metaphorically to sexual intercourse in all mythical songs of origin, a genre known as "gardens" (*phulwar*). Ten months elapse, and then the birth attendant is summoned. Kanha, alias Krishna, the child born to Jasu and Isaru, is then bathed

As with the initial ritual recitation of Harya Gurai—performed by a man at the village shrine—agricultural work is a central motif in the Sakhi Phulwar. In this case through the making of the plow, the placing of the yoke on the bull, and the tilling of the land. In the Sakhi Phulwar, however, the agricultural work is linked to the motif of doves at play and leads to the birth of a child. Rice cultivation and the conflicting sexual roles of men and women infuse all Tharu *phulwar*, or origin myths.

During the Harya Gurai, fertility of the rice crop and human fertility are stressed. Young men and women who dance together embody this idea. The *bathinya*, the girls who are allowed to dance, may be engaged or even pregnant, but they must not yet have given birth. Traditionally they wear colorful blouses that tie in the back and are decorated with numerous pendants. Once women become nursing mothers, they dress in white, their blouses tie in front, and they may no longer dance.

The Rana Tharu of the westernmost Nepalese district of Kanchanpur perform the Holi dance. Photograph by Gisèle Krauskopff, 1982.

From the beginning of the ritual, young boys and unmarried girls dance together every night. The dancing usually takes place in the courtyard of the chief's house. The pleasant sounds of the drumming and singing are typically heard at night during the rainy season until the great festival of Dasain in October or November. The dances then reach their apotheosis and are performed in full daylight in the courtyard of the chief's house with the girls dressed in the family jewels and their best attire.

There are two main dances. The Sakhya Nac proper is performed strictly by women, and its main choreographic feature is the circle (figs. 4.6). The *mohorinya*, the oldest, tallest, and most experienced of the dancers, leads the performances (fig. 4.1). She is followed by the rest of the girls, some very young, all of whom line up according to age and size. Some of the older girls may be pregnant, making this the last time they will dance. They may have returned to their maternal village specifically for this event. The dance is slow and moves in a counterclockwise direction with the girls bending down and clapping the cymbals held in their hands. Either there is a very slow drumbeat with the drummer standing in the center of the circle or no drum accompaniment at all.

The Paiya is a more frenzied dance performed in two parallel lines: one of girls (dancers) and another of boys (drummers; fig. 4.7). While the girls' line moves forward and back, the boys playing the drums run toward and away from the girls who begin to echo their movements. The drumbeats give the Paiya its rhythm. While I haven't collected an exhaustive repertory of these rhythms, I have heard a very slow two-beat rhythm, a four-beat, and a faster five-beat.

The Paiya expresses the sexual polarity of women and men, as well as that existing between singing and drumming. The dance illustrates how the union of man and woman accompanies the maturation of rice until the climactic moments of Dasain and the harvest. The public dances of Dasain are performed during the season of abundance, *saha*, a word that also suggests desire and love. The cycle of rice cultivation is thus equated with the cycle of human love—both leading to fertility and prosperity.[9]

A Cycle of Prosperity

Rice dances are common in many communities in South Asia and Nepal.[10] Among the Tharu groups, however, they are central aesthetic forms and are framed in a cycle adjusted to the rice cultivation calendar. In his unpublished notes of 1940, William Archer described the dances of the Nawalpurya Tharu of the Champaran District of Bihar in India. As in the Dang Valley, these dances started when the rice was being transplanted in August and ended sometime in November. Archer described four named dances: the Bengtharia, which was danced by women only in the form of a circle; the Jhamta, a dance with women forming a circle around men and singing in a peculiar, shrill style; the Charthaparya, which was performed by women forming a circle, facing inward, and moving toward the center; and the Dhekui wherein men and women danced in parallel lines.[11]

The Holi dance cycle of the Rana Tharu of far western Nepal and the Naini Tal District in India provides an excellent contrast to the dances of the Nawalpurya and the Danguara. The Holi dance is a sacred and very elaborate cycle in which men and women dance together and women also dance separately. But it is performed in February and March, it continues for two months and coincides with Holi, the biggest festival of the Rana Tharu (fig 4.8). Holi is a time of sexual freedom, and the dances provide young people with abundant opportunities for courting. The choreographic patterns (the circle for women, the parallel lines for the mixed performances), the elaboration of the aesthetic tradition, the performance at the main annual festival, and the sacred dimension of the dances in this ritual cycle are comparable to the rice dances of the Nawalpurya. The Rana Holi dances, however, are framed within a totally different calendar.[12]

From Dance to Painted Rice Granaries

Song and dance are dominant aspects of Tharu artistry; however, other forms of art related to rice also exist. In the Dang Valley during the period of Harya Gurai and on a fixed date commemorating the birth of Krishna, a related festival takes place. A young man paints the wall of a rice granary, usually in the house of the chief (see fig 9.6). The painting is called *astimki*, as is the ritual of which it forms part. Within the painting the following are usually depicted: a line of women and a line of soldiers dancing in parallel rows; a marriage palanquin; animals such as fish, peacocks, and elephants; motifs representing the sun and the moon; and the tree, a symbol of Kanha or Krishna.[13] These elements are always drawn inside a decorated, molded frame. Stylistic differences may be found from one painting to another, but the general structure and internal organization remains consistent. It is an evocation of love, marriage, and birth and is influenced greatly by the mythology relating to Krishna. Krishna may, in fact, be depicted in some paintings, depending on the painter's knowledge and ability. Throughout the night, women of all ages sing the Sakhi Phulwar in honor of Kanha conceived of as Krishna, despite the fact that the lyrics of the song have little, if anything, to do with Krishna.

The Astimki festival is very different from other Dangaura Tharu festivals. Centered around a full day of fasting, it ends with an uncooked vegetarian meal. Deeply influenced by the Vaishnav faith, Astimki is the only vegetarian ritual in the calendar. The connection between the rice granary paintings, the Sakhi Phulwar song, and the cult of fertility associated with the rice dances is noteworthy, however, as it exemplifies an interesting shift in the means of representation, from the human body to wall painting via the use of graphic techniques linked to a pan-Hindu tradition.

The Great War

Any account of the cycle of rice dances and songs of the rainy season would be incomplete without the mention of a song and dance known as the "Great War" (Barki Mar). This is more than a strictly fertility-based song as it has a martial aspect that is equated with prosperity. It is a male song of the rainy season. An epic, the Barki Mar is in fact a Tharu version of the great Hindu *Mahabharata*. Men sing it at any time, but they

4.9   The "Great Dance" is
performed by men to honor
the five Pandava brothers
during the festival of
Dasain. Photograph by
Gisèle Krauskopff, Deokhuri
Valley, 1992.

4.10   The marriage of Draupadi,   *and Danced by the Tharu of*
        a central scene of the "Great   *Jalaura, Dang Valley, Nepal.*
        Dance," is performed in the   Photograph by Gisèle
        village of Jalhaura during   Krauskopff, Dang
        the filming of *Mahabharata:* Valley, 1998.
        The Barka Naach as Sung

4.11   Men dress as women for
the Hundungwa dance
performed by the Kusumya
Tharu. Photograph by
Gisèle Krauskopff,
Taphwa Village,
Banke District, 1993.

4.12   The joker character peforms
during the "Great Dance."
Photograph by Gisèle
Krauskopff, Dang
Valley, 1998.

particularly enjoy singing it to sustain their energy while they perform collective free labor in the field for their landlord. The landlord, who wants the work to be completed quickly, typically serves them rice beer while they sing. By the end of the song, the singers are often drunk.

Like the Sakhya Nac, the Barki Mar is formally performed at the time of the festival of Dasain during the harvest period. It is performed in a sacred enclosure, and it is referred to at this time as the "Great Dance." In some villages of the neighboring Deokhuri Valley, the men performing wear beautiful long white garb with peacock feathers attached on the back (fig 4.9). For a time in Dang, the Great Dance flourished under the patronage of the Kanphata Yogi Shivaite monastery of Jalhaura, which owns much of the valley. It was performed on an immense scale with all of the valley's Tharu priests coming to protect the dancers. With the loss of this patronage, however, it has ceased to be performed, although it was recently staged and filmed with the assistance of private funds.[14] It is a sacred homage to the five Pandava brothers, deities who protect the land and insure the fertility of the rice fields. Although it is a male dance and linked to war, through the propitiation of the weapons of the Pandava brothers, it is an evocation of fertility and marriage.

4.13   Men from the Sihara District of Nepal dress as women to perform in a semiprofessional dance troupe. This photograph was taken during a Tharu pan-ethnic conference held in the Saptari District. Photograph by Tordis Korvald, 1994.

One of the favorite—and central—scenes of the dance is the performance of the marriage of Draupadi (fig. 4.10), wherein marriage is associated with war instead of with the maturation of rice as in the Sakhya Nac. This performance, like the cult of the five Pandava brothers, is uncommon in Nepal and in the Terai. The Great Dance of the Tharu is much more reminiscent of the traditions of "dancing" the *Mahabharata*, observed in the western Indian Himalayan districts of Kumaon and Garwhal.[15] A striking characteristic of the dance is that male dancers perform the female roles. In the Great Dance weapons are the primary metaphors as opposed to agricultural implements and love making in the Sakhi Phulwar.

There are sexual elements in the Great Dance, but they are expressed in a manner very different from those observed in the Sakhya. Outside the sacred enclosure, a joker character bearing an enormous penis encircles the sacred dancers and charges the audience (fig. 4.12). In his unpublished notes, Archer devoted considerable attention to the pantomime traditions of the Tharu. These remain alive to this day. They are put on solely by men and are even seen in modern staged "folkloric" performances. In addition to the banter about relationships between men and women, these pantomimes always have an overtly sexual aspect. They accompany another kind of Tharu dance often called Jhumra but known throughout the Terai districts under a variety of names (figs. 4.11, 4.13). Contrary to the sacred cycle of women's dances inscribed in the ritual calendar, these dances are considered completely profane and may be performed at any time just for pleasure. Jhumra are characterized by a male dancer dressed as a woman who performs to the accompaniment of cymbals and drums. This dance always ends in a sexual pantomime. As with the Great Dance, it is taboo for women to participate in a performance involving such overt sexuality. As purely male dances, the Great Dance and the Jhumra pertain to a social and symbolic order very different from the Sakhya rice dances.

Archer noted that unlike the Nawalpurya, who migrated from Nepal during the nineteenth century, the local Rautar Tharu women were forbidden to dance. This taboo is enforced to a much greater degree in Muslim-dominated northern India. Among the Tharu of Dang and other remote parts of the Terai, however, rice, agriculture, and sexual intercourse have become interchangeable metaphors of prosperity, made manifest through mixed dances and songs. This conception permeates all spheres of Danguara Tharu mythology and is the core motif of song and dance performances. The dancing bodies of young men and women are the main artistic representations of the fertility of the rice crop. Ideas are mediated by living bodies in an ephemeral art linked to the cycle of time and not to material objects.

•

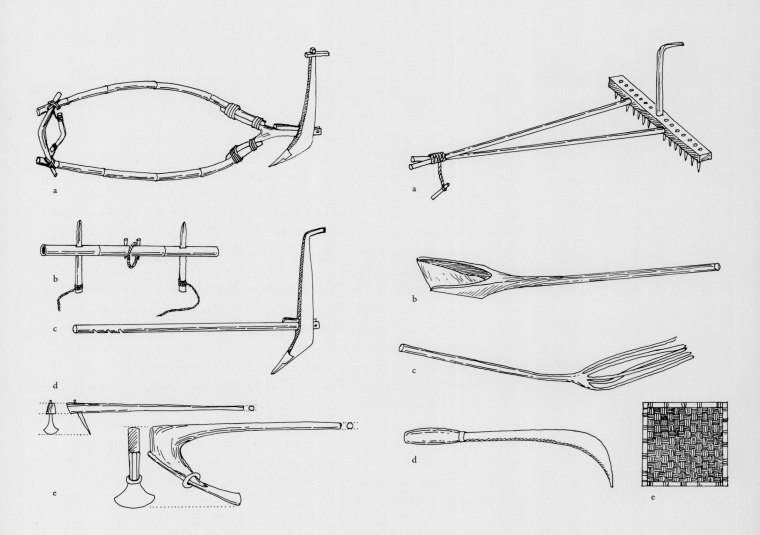

5.1a–e   These agricultural implements are used in plowing: (a) plow prepared to be set on bullocks, (b) yoke, (c) plow, (d) hoe used for planting, and (e) hoe used in a terraced field. Drawing by the author.

5.2a–e   These agricultural implements are used in soil preparation (a) and harvesting (b–e): (a) wooden harrow (*samjet maya*), (b) wooden shovel (*phouinthok)*, (c) threshing implement (*cheirong*), (d) sickle (*thangol atingbi*), and (e) winnowing fan (*humai*). Drawing by the author.

5

Rice Culture in Manipur, India

Mutua Bahadur

The state of Manipur is located in the northeastern part of India and covers an area of 22,327 square kilometers. Its eastern neighbor is Burma. Encircled by nine hill ranges, Manipur is distinguished by a picturesque, oblong central valley of 2,238 square kilometers. The entire valley is heavily populated. The Meitei peoples, who are predominantly Hindu but also practice traditional religion, constitute the majority of the population. There are also a sizeable number of Meitei Pangals (Manipuri Muslims), who are of mixed Meitei and Bangli descent, as well as ethnic communities from other parts of the country. The hilly regions are inhabited by a mosaic of ethnic groups, twenty-nine of which are recognized by the Indian government as scheduled tribes. The provisional census report of 2001 showed the total population of Manipur to be 2,388,634.

Rice is the staple food of the region and plays an important part in the religious and cultural life of the people. The systematic harvesting of rice in the valley is believed to have been practiced since the early part of the first century of the common era. At present, about eighty varieties of rice are grown in the state, and their growing times vary. Some are harvested in two and a half to three months, some in four months, some in six months, and others in nine months. The wide range of varieties includes traditional and introduced types as well as heterogeneous hybrids of these. The rice of Manipur origin is soft and smooth with a pleasant flavor and aroma. It is either white in color or black with a reddish tinge.

The Seasonal Cycle of Agriculture
Valley dwellers practice only wet cultivation of rice, while hill dwellers practice dry, shifting cultivation. Terraced cultivation is used in some hilly areas.

Dry, Hillside Farming
After a hillside site has been selected, trees are cut, the wood is dried, and the site is set ablaze. About thirty days later, in the month of Kalel (May–June), when rain is expected, rice seed is broadcast over the ashes, and the soil is turned with a hoe. In some areas, a small hoe is used to individually plant the seeds in the soil at regular intervals (fig. 5.1d). Weeding is performed four times during the growing period. Harvesting is accomplished with a sickle and thresher (*cheirong*; fig. 5.2c) or a flail (*kada,* a toothed implement made out of bamboo cane). Harvesting is begun during the month of Mera (October–

5.3a–c   These diagrams indicate: (a) the *som thaba* style of plowing, (b)the *karal lakpa* style of plowing, and (c) a pattern for sowing seed. Drawing by Mutua Bahadur.

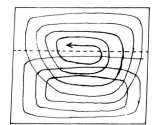
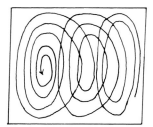
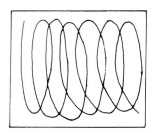

November). After two or three crop seasons, planting is shifted to another field and the first site is allowed to lie fallow for at least seven years. In some hilly areas, instead of changing fields, terraces are built into the sloping hillside for permanent use.

Wet, Valley Farming

Farmers on the valley plain begin plowing on a day determined to be auspicious in the month of Fairel (February–March). They do so with the help of a pair of bullocks, although in some areas of the valley, buffalo are used. The farmers of the plain recognize three types of rice cultivation as follows:

1. *Pumhul*—sowing seed on the plowed field;
2. *Pamphel*—sowing germinated seed in the flooded field; and
3. *Lingba*—transplantation of seedlings that have germinated elsewhere.

Plowing is generally done in a spiral pattern following a counterclockwise direction. This style of plowing is known as *som thaba* (fig. 5.3a). The field is then plowed crosswise in a pattern of concentric rings known as *karal lakpa* (fig. 5.3b). In the *pumhul* method of cultivation, the field is repeatedly plowed five to seven times—each time perpendicular to last. The seed is broadcast on the plowed field as the farmer moves back and forth across it (fig. 5.3c).

For the *pamphel* type of cultivation, plowing is carried out using the technique described above. The soil is then pulverized with a harrow-like spiked implement (*tungyan*) that is pulled by a pair of bullocks (fig. 5.4). The field is further smoothed out by the same plow team using a wooden harrow called a *samjet maya* (*samjet*, "comb"; *maya*, "teeth"; see fig. 5.2a). If the farmer does not use a *tungyan*, he levels and grinds the field using an *ukai pumnal*, which resembles a *samjet maya* but lacks spikes.

Rice seed used for sowing in the *pamphel* and *lingba* forms of cultivation is known as *phoudi*. The *phoudi* is first winnowed with a round winnowing fan (*yangkok*), to clear away unwanted foreign matter. It is then put in a sack or a vessel and soaked in water for twenty-four hours. A bed of bamboo leaves is arranged at the bottom of a basket in which the water-soaked *phoudi* is placed. The seed is then covered with another layer of bamboo leaves. The basket is wrapped in sackcloth for another twenty-four hours, and the seed is allowed to sprout in this warm, humid environment. The difference in sowing between the *pumhul* and *pamphel* methods is that with the latter the farmer moves backward to prevent his stepping on the germinated seed in the soil. In the *pumhul* and *pamphel* methods, harrowing with a *samjet maya* or an *ukai pumnal* is conducted within twenty-five to thirty days of sowing. This covers the seed and protects it from infection and also provides a weed-free environment. In the *lingba* type of

5.4   Plowing is carried out with a team of bullocks. Photograph by Th. Rabikanta, Andro, 1999.

cultivation, the transplantation of seedlings (*louhon*) takes place twenty-five to thirty-two days from the day of sowing and does not require harrowing (fig. 5.5). Weeding is done once or twice during the life span of a rice paddy. This is solely the responsibility of women, who perform the task by hand. Women who engage in this work are paid in wages or in kind (*khutlang*).

Ripe rice is reaped with a sickle (*thangol*). There are two types of sickles, which are differentiated according to the shape of their blades: (a) curved (*thangol athekpi*) or (b) straight (*thangol atingbi*). The *thangol atingbi* (see fig. 5.2d) is traditional to Manipur. Both sickles are made of an iron blade, an iron ferrule, and a wooden handle. Each, however, is maneuvered differently.

The cut rice stalks are exposed to the sun for several days. The harvesters then enter the field to begin their work (fig. 5.6). The rice plants are gathered using rope made of rice straw (*charei*) and piled on a bamboo mat. The harvesters then bind the rice again with another *charei* that is shorter than the first and beat it on a wooden frame (*keirak*). The bundles are beaten again with a *cheirong*, a curved, multi-pronged wooden flail made from a tree branch (see fig. 5.2c). Afterward, they are shaken by women known as *charu-kanbi* (*charu*, "straw"; *kanbi*, "women who shake") to release any remaining seeds. The straw bundles are then stored away.

Grain gathered in this way is then winnowed (fig. 5.7). This involves the use of other implements: a wooden shovel (*phouinthok*) used to fling the grain into the air (see fig. 5.2b); a square-shaped, woven fan (*humai*) to winnow the chaff when the seeds are thrown aloft (see fig. 5.2e); a straw brush (*hanubi*) to sweep away the chaff; and a straw indicator (*khoining*) to record the completion of a round of winnowing. The winnowed rice is ultimately packed in sacks and transported to the granary, often by bullock cart.

Cultivation Rites

Initiation of Agricultural Work (Loutaba)

Agricultural activity is initiated with the cultivation ritual of Loutaba, held on the first day of the month of Fairel. On the day before the event, a spade is carefully examined and washed so that the farmer (the head of the household) will not encounter any difficulties on his first day working in the field. The morning of Loutaba begins with a simple offering made by the head of the household to the household deities Sanamahi and Leimarel. Items used in the observation of Loutaba include rice, flowers, eight buds of the *langthrei* (*Eupatorium burmaniacuum*) plant, sweets, and a handful of rice placed on a piece of plantain leaf along with the above-mentioned spade. These objects are placed in the courtyard about an arm's length from the figures of the two household deities, on the right side of the plinth on which they rest.

The head of the household uses a sacred cloth and wears a loincloth (*khudei*), a special shirt (*louthang phurit*), a head sash (*kokyet tolok*), and a wrapper that is wound around the body and under one arm. After lunch, he sits on the right side of the verandah (*phamel*), assumes a contented appearance, and breathes through his right nostril. He then prays to Sanamahi and Leimarel to provide protection from all dangers and to drive evils southward. After this, he bows before the granary, and recites the following prayer: "O Lord Keirunghanba, who is addressed as Lourunghanba in the fields and Keikhong Keisangba at the granary, today our agricultural activities begin. Please protect us from all dangers." Then he goes into the field with the spade on his shoulder observing all that he sees before him. He is very focused and listens to no one, acting as an obedient messenger from Lord Keirunghanba to Lord Lourunghanba. He seeks divine intervention so that he may produce double the amount of the last year's yield and may fill his granary.

5.5   Transplantation of seedlings (*louhon*) characterizes the *lingba* type of cultivation. Photograph by Th. Rabikanta, Purul village, Senapati District, 1998.

In the middle of the field, while facing north, he makes ritual offerings. He divides the items listed above into four portions and offers them to the deities of the four corners: Thangjing in the southwest, Marjing in northeast, Wangbren in the southeast, and Koubru in the northwest. He digs into the earth with the spade and interprets his findings. Any living thing that he discovers, for example an insect or a snail, is interpreted as a good sign.

Festivals of Reaping and Harvesting

One day before the auspicious day selected for reaping, the farmer announces in the field that tomorrow the rice will be cut, and he entreats the insects to leave the field immediately. On the next day, just before reaping, he makes offerings to request a generous yield from Lord Lourunghanba.

After the rice is winnowed, the farmer places his foot on the straw indicator (*khoining*) used to record rounds of winnowing and kicks it behind him. He then walks toward the north, climbs on the top of the pile of grain, puts the woven winnowing fan (*humai*) on his head and chants for the soul of the rice, looking toward Koubru Mountain. He descends by stepping backward and leaving the *humai* on the pile. After this rite, if anything is placed on the mat upon which the harvesting has been done or if anybody trespasses upon it, the area must be purified by sprinkling it with water in which *tairel (Toona ciliata)* leaves and *pungfai (Dactyloctenium aegyptiacum)* have been dipped (Ngariyanbam 1979). Uttering "Phoureima [goddess of rice] and Phouningthou [god of rice], please may thou come home," the farmer packs a basketful of rice. Remaining impervious to the calls of others and following the mistress of the house, he returns home, bringing the rice as well as the indicator, the fan, some flowers, and a small portion of the day's refreshment on a piece of banana leaf. The couple are welcomed back into an environment perfumed with burning incense (Wangkhemcha 1996, 11).

Rites for Adverse Conditions

If rice has been stolen, if a swarm of bees has appeared, or if the rice crop should catch fire, the blessing of Phoureima is sought by means of a prayer called Phoukouba (*phou*, "rice"; *kouba*, "call"). This blessing may be sought in the months of Kalen (May–June), Langban (September–October), or Poinu (December–January). The essential items for this ritual are an earthern rice cooker, an earthern curry cooker, fifteen bunches of bananas that possess a characteristic referred to as *chang thokpa* (i.e., they contain an odd number of pieces of fruit when counted in pairs), one flag, fifteen portions of areca nut and betel leaves, two kinds of fruit, *leisang (Seligenella sp.)* leaves, two sets of *konyai pu* (a pair of coins, one silver and one gold), one cloth, one pot containing sacred water, one *sareng* (a type of fish; *Wallago attu*), one *ngapang* (another fish), one *sangbrei (Pogostemon heyneanus;* patchouli) plant, fifteen buds of the *langthrei* plant, some *kaboks* (a kind of sweet), sugar cane, two round plates of salt (made by the Meitei), the egg of a black hen, rice, *khoiju-leikham* (a type of incense), *napi* (a kind of grass), *tairel* leaves, *pungfai* leaves, and seven or nine tips of *laphut* (a kind of grass; see Ngariyanbam 1979, 1). After the prayer is said, the edibles are distributed among the participants.

If the rice has been attacked by caterpillars, a ritual offering is made to the goddess Lamphut Leima Wangdangnu Chomkhaidoisibi. This remedial ritual is held by the *pandit achouba* (head of the royal pundits) assisted by two *maibas* (male priests or sorcerers) on a convenient afternoon. The items to be offered are produced by the people of the villages who take part in this rite.[1]

5.6   Harvesters are shown gathering rice plants into bundles, beating them on a wooden frame, threshing them with wooden beaters, and shaking the seeds free from the straw. Photograph by Th. Rabikanta, Andro, 1999.

Rice in Other Religious Rites

Rice is indispensable in Meitei rites of passage. On the sixth day after a child is born, the Meitei observe a rite known as Ipanthaba. To perform this rite, rice and a wreath of dried, fermented fish (*ngari*) are needed. In the village of Andro before the performance of Ipanthaba, the infant's earlobes are pierced. The villagers present a measuring basket (*meruk*) of rice and an egg to the parents.

The Meitei marriage ceremony involves a form of prognostication known as *chengluk lubak kaiba*. On the day of the marriage, the groom's party brings a decorative basket made of cane and bamboo (*phiruk*) that contains rice to the bride's residence. It is opened on the morning of the fifth day to observe and foretell the couple's fortune (Ningthoukhongjam 1978).

For a period of fifteen days beginning from the day after the full moon in the month of Langban (September–October), a separate ceremonial offering known as *tarpon* is made to the dead. It includes rice, sesame, leaves of *tingthou* (*Cynodon dactylon*), fruits, and flowers.

In the Meitei traditional religion and Meitei Hinduism, rice is considered a magico-religious offering. Among the Meitei, the sylvan deities known as Umanglai are believed to rest under the water. At the festival of the Umanglai, the shrine is animated by calling the spirits from their underwater resting place. Those participating in the calling of the deities are received at the temple in what is known as a *phoudang* reception. They must step over a basketful of rice, a *ngari* wreath, salt, and other items. Following this, the

Umanglai festival known as Laiharaoba (*lai*, "god"; *haraoba*, "merrymaking") takes place. A part of Laiharaoba is Lai Lam Thokpa (*lai*, "god"; *lam*, "out"; *thokpa*, "going out"). The same reception rite is repeated on this occasion as well.

On the last day of the Laiharaoba festival and every Saturday in the month of Lamta (March–April), women perform a rite to expel evil spirits (*saroi khangba*) at a crossroads or local boundary. They offer rice, *langthrei* leaves, *heibi* (*Vangueria campanulata*) leaves, vegetables, and so on, that are contributed by every family. The women hold *dao* (a tool used for chopping) and dance (in the past, the dance was performed in the nude).

At Andro on the last day of the celebration of Pureiromba, a rice offering is brought and piled up in small mounds by the *maibi*, who are female divine servants, interpreters of omens, and fortune-tellers. The *maibi* are ordinary housewives and casual divine practitioners (a midwife is also known by the term *maibi*). The *maibi* construct a small landscape out of rice consisting of mountain ranges, ravines, lakes, rivers, and so on, inserting shoots of the *langthrei* plant, tree branches, eggs, and other items here and there. A man holding a stick walks around this construction singing the song of the peoples' origin and migration while others follow him, repeating the song. The rice is then taken away by the *maibi*.

As should now be clear, the offering of rice is performed at a wide variety of occasions. During the worship of a clan deity, a rice offering is made on a plantain leaf. The rice is given to the clan leader (*piba*), one of the *maibi*, and the musician who plays the *pena* (a type of string instrument), who place the rice at specific locations. Rice is sprinkled on the back of the leaf if it is offered to male deity, whereas it is put on the front of the leaf if the deity is female. This differentiated form of presentation suggests a connection to fertility.

An offering known as *khayom lakpa* is presented to supernatural beings to appease them and protect people from danger. This offering is made at many rituals and in the course of treating the sick. In a *khayom*, rice, gold and silver coins, eggs, and leaves of the *langthrei* plant are packed with seven layers of plantain leaves and tied with bamboo rope (Ngariyanbam 1998, 2).

In Cheiraoba, at the New Year's festival, rice—along with vegetables, fruit, flowers, and other items—is offered to the family deities. Following the ceremony, these are cooked and offered to evil spirits existing outside the homestead in an effort to drive them away during the coming year.

A garland of rice is used as a special offering to Lord Govindaji during Rathajatra, one of the greatest Hindu festivals, and to Lord Shri Bijay Govinda during a boat race known as Heikru Hidongba that is held on the eleventh day of Langban. In the boat race each of the two team leaders wears a garland of 108 grains of rice and a garland of reeds (*hup*) from the *heikru* (*Emblica officinalis*) plant. After the race these are offered to the deities Shri Bijay Govinda and Raseshwori. The boat race is held in the moat surrounding the temple of Shri Bijay Govinda. It was originally held during the reign of King Irengba of Manipur (984–1074).

Granaries

A freestanding granary with a single door is built either on the southern or the western side of a house. Sometimes a raised granary is built atop poles inside the house. There are also two types of portable granaries: (a) a rectangular granary (*kot tum*) and (b) a round

Winnowing is the last
stage of harvesting.
Photograph by Th.
Rabikanta, Andro, 1999.

granary (*changkot*). The walls of both types are coated inside and outside with a paste
made of cow dung to increase their durability. In the northern, hilly regions of Manipur,
use of a basket-type granary known as a *bou* and shaped like a woman's breast is thought
to ensure against famine. It has a double-woven wall.

For the Meitei, a gleaming black stone personifies Phouoibi, another name for
the Rice Goddess Phoureima. The stone is kept on a bed of rice inside an earthenware
pot. The people believe that as long as this shrine remains inside the granary, there will be
no rice shortage. The stone goddess is never placed directly on the ground, and taking rice
from the granary during the month of Wakching (January-February) is prohibited. On
the first day of the month of Fairel (February-March) the removal of rice begins after a
small offering is made. Rice may not be taken out of storage on an inauspicious day.

To ensure the abundance of rice, the granary deity, Kotlai (*kot*, "granary"; *lai*,
"to have"), is also worshiped. Paying homage to Kotlai is also a remedy for taking rice
from the granary on an inauspicious day or stealing rice. A variety of materials are used in
this offering: a pair of coins (one gold, one silver), two pieces of iron, seven rice seeds,
sareng, honey, molasses, areca nut and betel leaves, banana clusters with the characteristics
of *chang thokpa* (as explained earlier), rice, a *phou heiri* (*Citrus sp.*) plant, fruit, flowers,
and other items. The rice offering is cooked, maintaining the strict procedure of *laichak
thongba* (described below) to ensure accurate fortune-telling. The *sareng* is also cooked.
The cooked food is then offered to the deities Keirunghanba and Lairema. The seats of
nine gods and seven goddesses are also prepared and offerings made to them.

The first two of these implements are used in the steam method of cooking rice: (a) this steam cooking method is used by the Andro; (b) this steam cooking method is used by the Moyon and Monsang. The third illustration shows a basket for winnowing that is used by the Moyon and Monsang. Drawing by Mutua Bahadur.

Preparing Rice as Food and Drink
Cooking Methods

Preparatory to cooking, rice is spread out on a bamboo mat or circular basket trays in the sun. Only women remove the husks from the rice, using a mortar and pestle. The rice is then pounded with green bamboo leaves to obtain a beautiful white color with a slightly green tint. finally, the rice is pounded again, with the addition of straw to prevent the grains from being broken.

Several methods of cooking rice are employed by the Meitei. In the direct-heat method, the rice is washed, placed in an earthern rice cooker containing water, and boiled. At this stage, if there is too much fluid, the excess watery gruel is drained off. (Known as *chabon*, this gruel is used to make a tasty drink with sugar sometimes added to sweeten it.) The cooking rice is then reduced to a simmer. The heat given off from the hearth's side flame and a layer of burning embers on which the rice cooker is placed are quite sufficient. The cooker is rotated once or twice to ensure even cooking. In the meantime, other food is prepared. When the rice is done, it is soft and is eaten with at least one other dish. This method is commonly used by ordinary people to prepare their daily lunch.

Another method known as *laichak thongba* is employed for cooking the rice to be used in offerings to the supernaturals. This cooked rice is "read" to make metaphysical predictions. The following procedure is observed: (a) rice is cooked in a heretofore unused earthen pot; (b) once water is added, no more may be put in or removed; (c) the pot is covered with three or seven layers of banana leaves and is not opened until the cooking is done; (d) no stirring is allowed; and (e) once placed in position, the cooker is not adjusted. Of note, a mark is incised on the rim of the pot with a knife. The ends of a piece of bamboo twisted around the neck of the pot are brought even with this mark. The rice is cooked in front of the god's seat, while curry is cooked in front of the goddess's.

5.9   A Meitei boy enjoys the taste of rice cooked in a bamboo stem (*utongchak*). Photograph by Th. Rabikanta, Moreh, Manipur, 2002.

Each side of the mark on the rim of the pot indicates the share of the rice offering for the male or female deity. When the cooking is over, the surface of the rice is "read," and a prophecy is made. The rice cooker is buried in the ground at the close of this ritual.

The steam method of cooking rice (*nganba*) is employed by some groups, including the Moyon and Monsang, and by the Chakpa, an ethnic subgroup of the Meitei, at Andro. But the steam method of cooking is used at Andro only on festive occasions. This method keeps the rice from burning, as may happen using the direct-heat method. Two vessels are placed one atop the other (fig. 5.8). The one on top may be a conical earthen vessel (*ngangkok*) that is open at both ends or a basket vessel (*nganthu*); the one on the bottom is a pitcher. The pitcher contains water while the *ngangkok* or *nganthu* contains rice. A woven strainer placed between the two vessels keeps the rice from dropping into the water. A fire is started below the pitcher, and steam rises from within it to cook the rice in the top vessel. Rice cooked with the steam method swells less than rice cooked with the direct-heat method.

In the past, cooking in a bamboo stem, referred to as *utongchak* (*utong*, "bamboo"; *chak*, "rice"), was used to prepare lunch for an outdoor meal. The Royal Chronicles relate that the royal family enjoyed the taste of *utongchak* up until the early twentieth century. It is still sold frequently today at Moreh, the border market area of Manipur. There are at least fifty varieties of bamboo in Manipur. Of these, *poksang* (*Cephalostachyum pergracik*) is the favorite for preparing *utongchak*. A green bamboo stem about 40 centimeters in length with a single node at one end is prepared. Rice that has been soaked is placed with water inside the bamboo and the open end is stopped with a plantain leaf. The bamboo tube is placed at a slight angle, and a fire is kindled beneath it. When the rice has cooked, the outer, burnt surface of the bamboo is scraped off with a *dao*. The bamboo is then split open, and the soft rice, infused with the fragrance of the thin inner membrane of the bamboo, may be scooped out (fig. 5.9).

Other Recipes Using Rice

Although plain rice is common at special occasions, sweet rice pudding (*sangom kher*) and spiced rice with pulse (*khechuri*) are also prepared. Some other rice foods are described below.

Flattened rice (*chengpak*) is boiled and dried in the sun. It is then flattened in a *dengi*, a pit in the ground in which the rice is pounded with a simple foot-operated pestle. The flattened rice is then winnowed in a circular, woven fan (*yangkok*). Before being eaten, the flattened rice must first be resoaked and softened. Fried *chengpak* is also eaten and is included in many rituals. It is also sometimes served in a death ritual known as Sardha Karma. Popped rice (*kabok aphaba*) is made to swell and burst open by heating. It is eaten plain or mixed with molasses. For rice loaf (locally known as *tanja*), rice is first pounded into powder. This rice flour is made into a dough, kneaded, and flattened into circular cakes. These are wrapped in plantain leaves and boiled in water. These delicious loaves are initially offered to deities by hill dwellers and the villagers at Andro and are later eaten by the people themselves. Powdered rice is also rolled into balls to be offered during the birth rite of Ipanthaba, described above. It is also a constituent of *hameibon*, a gift offered to supernaturals. Rice dough is used to make colorful baked toys such as boats, parrots, horses, elephants, and so on. The toys are about 10 centimeters tall. Children enjoy playing with these toys so much that they often forget to eat them afterward.

Distillation of Liquor

Rice liquor (*yu*) is brewed in Manipur as an intoxicating beverage and as a religious libation used in traditional rites. There are two types of liquor, categorized by process: (a) *atingba* (which comprises *waiyu* and *pukyu*) and (b) *asaba*. The former is prepared by fermentation only, while the latter is prepared by fermentation and distillation.

The preparation of yeast cake takes place as follows: rice that has been soaked in water is crushed and blended in more water with the bark of a creeping plant called *yangli* or, as an alternative, the leaves of a small plant known as *manbi* or *yangli mana*. The paste yielded is made into round cakes about 5 centimeters in diameter. These are kept in a warm environment maintained by a bed of husks and two or three layers of cloth. The cakes are allowed to rest in this way for about five days in cold weather and three to four days in hot weather. The substance is now called *hamei* and can be used for up to three months.

For the preparation of *waiyu*, rice is steeped in water for about four hours and cooked using the steam technique. After it is rinsed in water, it is placed in a basket strainer to allow all the water to be drained off. *Hamei* is mixed into the rice, and the resultant mixture is put into a pitcher for fermentation. The process takes about five days in winter and four days in summer. Husks of the *waiyu* (*wai*, "husk") plant are added to the fermented mixture. The liquid is imbibed through a bamboo straw or a tube of the *yengthou* (*Arundo donax*) plant.

Pukyu liquor is prepared with sprouted rice. The grain is kept in a basket and covered with plantain leaves. Water is sprinkled over it now and then to cause it to sprout. The sprouts are then dried, water is added, and the mixture is fermented with *hamei*. The liquor is drained out.

Fermented rice is distilled to obtain *asaba* liquor. For this, a rice pitcher (*yuphu*) containing the fermented rice is placed on the hearth. Another pot (*yukok*) is set on top of it. One end of a pipe is inserted into the *yukok* while the other end of the pipe is

5.10a–c   Liquor has been distilled in different ways at different times: (a) top left: method used by Meitei in the past, (b) top right: method used by the Kharam peoples, and also by the Andro in the past, and (c) bottom: method used by the Andro. Drawing by Mutua Bahadur.

introduced into another water-filled pot (*yukhai*) in a trough (fig. 5.10a). The outer surface of the pipe is covered with a plantain rind. It is further bound with cane strips. *Yuphu* and *yukok* are sealed with cow dung. The distilled vapor is condensed into liquor in the *yukhai*.

At Kharam Pallel in the Senapati District, mud and rags are applied to a three-tiered set of pots to seal in the vapor. In this arrangement, the collecting pot is kept on a mat inside the middle pot (fig. 5.10b). The pot on top is the cooler. Steam rising from the lowest pot cools as it reaches the bottom of the cooler and droplets are collected inside the middle pot. Today metal vessels are used for this purpose. At Andro, an aluminum vessel is placed on top of a *yuphu*, or rice pitcher. A cooler, with water, is placed on top of the cone-shaped, earthenware *ngangkok*. Inside the *ngangkok* a spoon-like metal object connects to a rubber pipe that extends into the collecting vessel. Between the *ngangkok* and the *yuphu* is a perforated plate (fig. 5.10c). Bamboo leaves are placed in the *yuphu* for better distillation.

Land and Revenue System

Until 1951 the land belonged to the king of Manipur (Lairenmayum and Ningthoukhong-jam 1997, 697). At that time land for rice cultivation was given as a reward. Salary was also provided in measures of land. Thus, a *menjor* received 1 *pari* of land, a *pahila* received 6 *paris*, a *subedar* received 4 *paris*, a *jemadar* received 4 *paris*, a *kot* received 2 1/2 *paris*, a *havil-dar* received 2 *paris*, an *amaldar* received 1 *pari*, a *kotendar* received 1 1/2 *paris*, and a *sepoy* received 1 1/4 *paris*. Land was awarded for special service, and revenue was not collected on that land (Higgins 1998, 141). The revenue that was collected from villagers was in kind and not in cash. Today, the land and revenue system of Manipur is governed by the Manipur Land Revenue and Land Reforms Act of 1960.

Even today, however, no revenue is collected by the state from land assigned to the gods. Land is assigned to Sri Govindaji Temple, Bijoy Govinda, Gopinath, and many sylvan dieties. Those who serve the god receive a share of the rice harvested from these lands.

6.1   A procession of naked farmers carries the captured leader of the evil rice-eating demons (*kala*), who has been metamorphosed into an atropopaic plank (*tlawingan*). It is believed that demons are best seen at night when one is naked; they are invisible during the day. Leiden University Oriental Ms. 12.542 (11.8).

6.2   A farmer's family discusses the upcoming purification of the rice fields. Leiden University Oriental Ms. 12.542 (7.2).

6

The Purification of Rice Fields in Java with an Apotropaic Plank

Stephen C. Headley

Introduction

The simple myths from Sentolo, Java, presented here describe rituals used to banish parasites (*kala*) from fields of ripening rice (fig. 6.1). Twenty such stories, each accompanied by approximately eight drawings, form a very small part of the Moens collection at Leiden University Library. These myths are unique and are not in any sense typical of Javanese folklore. J. I. Moens (1887–1954), a Dutchman with a serious interest in Javanese history and oral traditions, collected them in the 1930s from Widi Prayitna, a *dhalang* (shadow puppet master), who involved other *dhalang* in the project.[1] These stories would never have been collected if it had not been for Moens's enlightened curiosity.[2] He commissioned their transcription and the accompanying drawings (Pigeaud 1980, 54). We do not know where the idea of illustrating the myths originated, but it is clear that drawing with colored pencils on European paper was not customary in the Javanese countryside.

The purification of rice fields is a common ritual among farmers in Java, but the use of an apotropaic plank (*tlawingan*) is unusual in this context. The word *tlawingan* refers to a board used for the display of an heirloom or sacred object. Although such boards or planks resemble those used to suspend kris blades, I am not aware of their existence in other areas of Java. Each of the twenty stories transcribed and illustrated by Widi Prayitna displays *kala* of a different form. The officiants for these rituals are either Javanese deities or local Muslim clerics (*modin*). The episodes portrayed in the drawings regularly include the following in the order given:

1. presentation of the *modin* and the family (fig. 6.2) or the gods involved in the story (fig. 6.3);
2. the leader of the *kala* (fig. 6.4);
3. agricultural work in the fields (fig. 6.5);
4. the apotropaic wooden board (*tlawingan*; fig. 6.6);
5. destruction of certain *kala* (fig. 6.7);
6. the remaining *kala* fleeing the field (fig. 6.8);
7. the *tlawingan* carried in procession through the field (fig. 6.9);
8. the ablution of the *tlawingan* (fig. 6.10); and
9. prayers and offerings being made for a ritual meal (*selamatan*) bringing together the leader of the ritual and the farmer's family (fig. 6.11).

6.3   The Javanese deity Guru
(Shiva) on the right consults
with Semar, a benevolent,
hermaphroditic deity, and
other gods concerning the
purification of the rice fields.
Leiden University Oriental
Ms. 12.542 (5.2).

6.4   The leader of the rice-eating
spirits (*kala*) is shown at the
right with his followers to
the left. Leiden University
Oriental Ms. 12.542 (10.3).

6.5   Farmers transplant rice
seedlings in the flooded
fields. Leiden University
Oriental Ms. 12.542 (7.4).

6.6 The leader of the rice-eating demons (*kala*) is metamorphosed into an apotropaic plank (*tlawingan*). The demon's head is painted at the top of the plank. Arjuna, who in this particular story captured the *kala* leader, is also represented on the *tlawingan*. Leiden University Oriental Ms. 12.542 (17.6).

6.7 Javanese deities Narada (left) and Guru (right) are shown slaying a rice-eating demon (*kala*). Leiden University Oriental Ms. 12.542 (5.5).

6.8 Rice-eating demons (*kala*) flee the rice fields after feeling the effects of the apotropaic plank (*tlawingan*). Leiden University Oriental Ms. 12.542 (11.9).

6.9   Naked farmers carry an apotropaic board (*tlawingan*), as well as incense and offerings for a ritual meal (*selamatan*) intended to ensure well-being. Leiden University Oriental Ms. 12.542 (14.8).

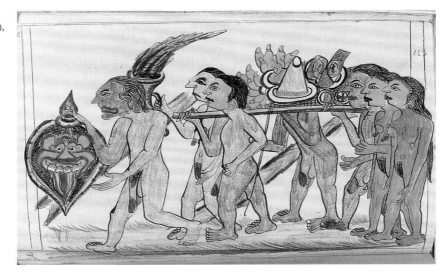

6.10   Farmers pour ablutions over the apotropaic plank (*tlawingan*). In actual rice field rituals, some of the water would have been saved to use for libations. Leiden University Oriental Ms. 12.542 (11.10).

6.11   The ritual meal (*selamatan*) is conducted by members of the farmer's family after the procession with the apotropaic board (*tlawingan*) through the rice fields. Leiden University Oriental Ms. 12.542 (6.10).

In six of the stories, the ritual action turns on the iconic metamorphosis of a rice-eating demon into the *tlawingan* (figs. 6.12, 6.13). The head of the demon—often the leader of the band and the one with the most extravagant deformities—is painted on the tip of the board, which will later be used in processions through the rice fields. Once metamorphosed into the *tlawingan*, the *kala* has the power to drive away his fellows. Thus these rituals appear to consist of an initial myth to explain the origin of the board, followed by verbal invocations recited during a procession to complete the purification of the fields and later by a meal to ensure well-being (*selamatan*). Each of these elements will be analyzed separately below.

Apotropaic rituals are those which "turn away" an approaching danger. In Javanese the word *panulak* (repeller) is used to describe them. In the lengthy eighteenth-century Javanese poem *Manikmaya*,[3] the leader of the *kala* (demon) army, Puthut Jantaka, is chased out of the rice fields not with a board but by the defender of the rice fields, Sengkan Turunan (whose name may be translated as "Rising and Falling"). To accomplish this feat, he uses a gigantic phallic whip many meters in length.[4] In the myth translated below from the Sentolo collection, the first *kala* captured becomes an apotropaic ritual board that is phallic in shape. While the form is phallic, the decorative motifs of the boards are clearly identifiable with the Kîrttimukha or Boma figures that appear over Javanese and Balinese temple gates (Rapier 1986, 207). These masks with hands look down over the temple's entryway, grasping foliage that contains the vital sap, or *rasa*, that can be identified with Guru's (Shiva's) semen, thought to be the origin of rice parasites.

According to legend, Uma, the consort of Guru, resisted the god's sexual advances. The semen from this aborted coitus fell to earth "giving birth" to Kala, the youngest son of Guru, who in general manifests the demonic aspect of the gods. For her act of resistance, Uma was exiled to earth and cursed by Guru, after which she became the demon queen Durga. In some versions of the story, she also becomes the wife of Kala, her monstrous "extrauterine son." In certain of the Sentolo stories, Durga's womb (and by association her menstrual blood) is depicted as a weapon. Her vagina also becomes a weapon that consumes those who approach her. The violence that Durga suffered, and that she in turn perpetrates, is not emphasized in the myth from the Sentolo collection translated below, but the mere mention of her name would bring her story to mind for a Javanese audience.

In Javanese tradition, Bima is often cast in the role of a hero purifying the rice fields of parasites. In Indian mythology Bima is Bhîmasena, one of the five Pandava brothers, but in Javanese oral literature he has taken on a life of his own, while continuing to be associated with his brothers in the *wayang*, or puppet theater. This role of ridding the rice fields of pests would appear "appropriate" for Bima, who is known for using his phallus as a weapon, as in his own rape of Durga.[5]

The *Aking* Kyai Barung myth translated below is the third[6] in the collection of twenty stories presently under consideration. *Aking* is a synonym for *tlawingan*, and Kyai Barung is the personification of a particular apotropaic plank.[7] The text of the story is far from literary or even fluent. It appears to be intended as an aide-mémoire and contains little of the expression that is possible in an oral style of narration. An attempt has been made to translate the entire text without additions or paraphrase. The scene breaks have been deduced from the text but are not explicit in the original. It should also be noted that the references to illustrations do not indicate where they appear in the original text but where they fall in terms of the logic of the narration. Question marks appear in the translation to indicate that my informants, Bambang Byantoro and Soetrisno Santoso, the grandson of Widi Prayitna—not to mention myself—have not understood what was

6.12   The leader of the rice-eating demons (*kala*) is subdued. Interestingly, the *kala* is represented here as an insect. *Kala* are rarely portrayed in the form of one of the actual threats to the rice crop. Leiden University Oriental Ms. 12.542 (13.3).

6.13   The rice-eating demon (*kala*) shown in figure 6.12 has metamorphosed into an apotropaic plank (*tlawingan*). Deities stand on either side. Leiden University Oriental Ms. 12.542 (13.4).

written. It is essential to remember that this story describes a genuine ritual once practiced in the fields around Sentolo. The story may be seen as a prescription for the ritual, complete with the proper incantations. Soetrisno Santoso has confirmed that he himself witnessed the last peformances of such rituals before they died out in the 1970s.

Translation of the Myth for Purifying the Rice field

*This is the explanation of Kyai Barung who serves to repel the pests [*hama*], when they enter Bima's rice field until Bima repels the pests, by reciting* mèl[8] *prayers twice and the [other] actions explained below [this opening text appears in fig. 6.14].*

        Once upon a time in the Minangsraya forest, who has the intention of entering the forest of Minangsraya? None other than hero, Adhipati Radèn Harya Werkudara [i.e, Bima; fig. 6.15]. Radèn Harya Werkudara entered the forest of Minangsraya to conjure up magical invulnerability. Once inside the forest of Minangsraya, a panunggi *[?] spell is necessary in order not to be tormented, [in order that] these attacking [demons] do not strike one. If one knows that in the forest of Minangsraya there are many armies of jinn, the action of this* panunggi *spell is [known] to be indispensable.*

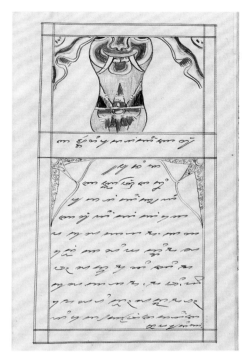

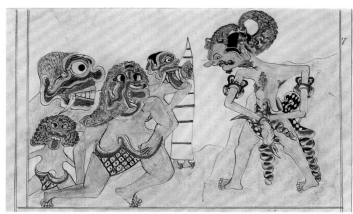

6.15   The Javanese deity Bima
(right) confronts a group of
rice-eating demons (*kala*) in
the field. Leiden University
Oriental Ms. 12.542 (3.2).

6.14   The title page of the third
story—as with most stories
in the Sentolo group—dis-
plays the head of the leader
of the rice parasites (*kala*),
who is captured during the
purification. In this case the
leader is Kyai Barung.
Leiden University Oriental
Ms. 12.542 (3.1).

*Even before they are encountered, they are ready. [Werkudara] had not yet
deployed his meditative action, [but] when [Werkudara felt] uneasy, he sat down straight away.
Sitting upright [in meditative posture] and shortly after he sat down, there occurred an
upheaval [gara-gara] among the demons.*

*Having been hexed by these Bayu [Sanskrit "wayu"; here a kind of evil spirit],
Bima fought back. The Baju Barat [lit., "Great (west) wind (spirits)"]*[9] *were stricken with
various ills, then yielded [to him].*

*Shortly thereafter, Batari [Durga, the demon queen] ordered that the forest of
Minangsraya be inspected, [to see] whether it was true that a hero was now holding forth
there. In that forest Hyang Batari ordered all the Baju Barat to inspect the site of the
Minangsraya forest according to instructions. Before long more of the stricken Baju Barat who
had entered the center of the forest were startled. The Baju Barat were all astonished, because
the hero himself had entered therein. Astonished by this courageous action, all the Baju Barat
then attacked Bima with torments of all kinds in order to chase him away, so that Bima
would leave this place, but Radèn Harya Werkudara felt nothing. Due to his involvement
[?tegèn; in battle] and the strength of* panunggeng *[?panunggi] spell, Bima could not yet
meet with Hyang Batari; she had not yet taken her place. Radèn Harya Werkudara [faced]
therefore the Baju Barat's varied attacks, repelling each with magic, each undone such that he
did not feel them. Instead it was the Baju Barat who felt sick and left. If they did not leave
quickly enough, the Baju Barat became sick due to the hex of Radèn Harya Werkudara. Then
they all fled away to Hyang Batari Durga.*

6.16   Bima and the Panakawan brothers stand at the edge of the field trying to locate the rice-eating demons (*kala*) who are hidden there. Leiden University Oriental Ms. 12.542 (3.4).

6.17   The Panakawan spot the rice-eating demons (*kala*) in the field. Leiden University Oriental Ms. 12.542 (3.5).

***Next Scene—The arrival of all the Baju Barat
in front of Hyang Batari Durga, interrupting her***

Because of this great sakti *[magical power of Bima], the Baju Barat were all bent and deformed. They spoke to Hyang Batari, once they were all assembled. They said that in the middle of the forest there was a hero of exceeding great and awesome majesty and that the Baju Barat's joint attack on him was not strong enough. Instead many of the Baju Barat were felled by illnesses due to the hard command [*bantering panuwunipun*] of this hero. Hyang Batari, having received the report of the Baju Barat, permitted all the Baju Barat to retire, along with those who had been caught and later died. So all the Baju Barat, in order to avoid [Bima], retreated according to Hyang Batari Durga's orders.*

*Next Batari departed from the heavenly fields [*kayangan paśetran*] to investigate the hero Adhipati Harya Werkudara, approaching until he became visible. Once his meditation [*tapa*] was revealed [i.e., Bima returned to his original form], she sat down with him. Then Sang Hyang Durga asked, "What is requested? Why are you in the Minangsraya forest?"*

Having asked what she needed to know, Bima replied, "This adhipati's *[i.e., ruler's] [realm] had been broken into by a hidden band of thieves [*dhustha mamor sambu*]; the kingdom of the* adhipati *has been overwhelmed. The army of the* adhipati *and all the villages know about this hex [*tenung*]. How can they all become well [again]?"*

Then Hyang Batari answered, "It is possible to request help from the children of Semar [the benevolent hermaphroditic deity who is a key figure in Javanese mythology]: Pétruk, Garèng, and Bagong. He could receive help from Semar's children, who could be made to come and could repel [them]. Then Bima, along with Hyang Batari, examined the pests, which were now eating the fields of the adhipati. *He then excused himself to seek [help] from the children [of] Semar: Garèng, Pétruk, and Bagong [known as the Panakawan, traditionally the servants and advisors of the Pandava].*

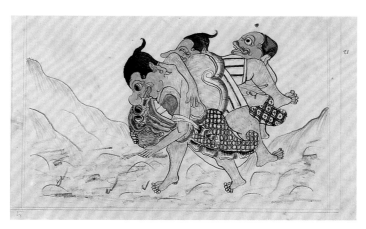

6.18   The Panakawan carry the
captured leader of the rice-
eating demons (*kala*) away
from the field. The *kala* has
been transformed into an
apotropaic plank (*tlawingan*)
and will later be used in the
procession to purify the rice
field. Leiden University
Oriental Ms. 12.542 (3.6).

After Hyang Batari had visited Bima, she also left with the Baju Barat as had
Radèn Werkudara. Once he [had returned to] the forest, following Hyang Batari's order, he
[Bima] went on to the Grengseng Mountain to the place of Garèng Pétruk, [and] Bagong so
that he could meet with Garèng.

Next Scene—[Mount Krangseng/?Grengseng] village [with Bima]
none other than the village headman Garèng [Nala Garèng]; facing him
are [his] younger siblings Pétruk with Bagong sitting
They had heard that the farmer Radèn Werkudara, afflicted with pests, was going to visit. Then
there was a great cry. What caused that? Next Pétruk spoke saying, if that's the way it is, it is
better that they be prepared [to] help later on, no matter what the circumstances. Why did these
devouring kala *do so? These pests continued to devour; everywhere the pests obstinately contin-*
ued to eat those fields. [But] before they could leave, they were interrupted by the arrival [of]
Radèn Harya Werkudara. They were all startled that he had reached Papan Padhuhan
Gunung Grengseng. Then he requested his three attendants [i.e., Panakawan] to inspect
[and] report on the dead rice plants the pests had eaten. Hyang Batari Durga then clarified
how to make the pests leave. What it would take to have them leave, repulsed. Garèng said
they would enter that rice field, and Radèn Harya agreed with Garèng's advice. Pétruk [and]
Bagong then left for that hamlet. Once they had all gone and come to Bima's rice fields, the
[story] is interrupted.

The rice fields are not described. Following the finishing of the planting, what of
the demise of all the [kala] army? Concerning the pests, three were eating. At that moment
Radèn Werkudara and the Panakawan were on the edge of the field [fig. 6.16]. From there they
could hear them, but they could not see their forms. Shortly thereafter Garèng and Pétruk both
snarled [?sami petak daya] viciously [toward] that which they understood to be without form,

6.19   Stripped naked, one of the
Panakawan siblings carries
an incense offering, while
Bima holds the apotropaic
board (*tlawingan*). Leiden
University Oriental Ms.
12.542 (3.8).

6.20   The remaining rice-eating
demons (*kala*), crippled by
the effect of the incantations
and purification ritual, flee
the rice field. Leiden
University Oriental Ms.
12.542 (3.9).

and while pushing aside [the rice], elbowing their way through, they spread out [in the rice field]. There with Garèng's own eyes along with Pétruk and Bagong, they could finally see the forms of the demons eating the field [fig. 6.17].

It was now clear that these were the ones eating that rice field. Together they [tried] to catch one in order to understand its form. What was it like for it to be bold [enough] to eat the rice? Then Garèng and Pétruk approached quietly in order to catch the pests. The pests sensed the presence of men approaching them, and the three of them ran off. Reaching the edge of the rice field, each one spread out running away from that place. Once [they grasped] the risk of their intentions, Garèng and Pétruk kept watch one [?at a time] in order to catch one of them. When one of the kala *stood up [?kajungjung], they surmised where it was. This pest ran until, ambushed, he collapsed [kèwedan], then he ran off [again], and it was not long before he was bowed [over], flatten[ed] from the rear by Pétruk. He resisted trying to run out of that field [again]; immediately the* kala *rushed quickly through the middle of the rice field. Before long there was a fight in which the pest then fought a pitched battle with the Panakawan till it was all torn apart [?rebut], it was* segelas *[?], coming to blows, shortly thereafter the pest was seized around the neck, next it was trampled on by Pétruk [and] Bagong grabbed its legs till it wailed.*

Finally the enemy was no longer capable of struggling with the allies of Bima, the Panakawan; the kala *was exhausted, nearly dead due to the determination of the Panakawan. Since they still had the same [?form] then Radèn Harya Werkudara, waiting on the edge of the rice field, and observing all the Panakawan's earlier movements [gedebug] but who had yet not understood the form the pests [had taken] was very astonished. Shortly thereafter the* kala *disappeared, at once becoming dry [aking, a wooden board] named Kyai Barung [fig. 6.18]. Because of this [change], as soon as it took the form of a board, Ki Lurah Nala Garèng wanted it to imitate the action of repelling the pests that had been eating the rice field earlier on. It was now clear that this was important.*

The pest then said to Radèn Werkudara, "With your permission Radèn Werkudara, [and] in accordance with Radèn Werkudara's desire, we have already followed the request of the Panakawan. Now how did this conform to the purpose [of] all the Panakawan? Garèng said, "Since these two pests are still in this rice field, Radèn Harya Werkudara later on will begin by saying the mèl *prayers. If then, they are unwilling [to leave], [Bima] will immediately material- ize the pests eating the plants in the rice field of the Bendara [i.e., master]. This is what to do later on if they have not yet left [*esah = késah*], driven away because of the "repeller" [*panulak*] [used] earlier, due to the repeller of the pests. Next Radèn Werkudara seized on what was ordered [?*papakenipun*]. Master Garèng, Radèn Harya Werkudara then obeyed, promising to authorize Garèng's advice, even if death occurred. Shortly thereafter he told Radèn Harya Werkudara that Radèn Harya was to strip naked before him. Next there are* mèl *prayers, the very* mèl *prayers [to be used]. The offerings were then laid out [a sort of* sadekah *or* selamatan *ritual meal as in the drawings], ready to repulse the pests eating the plants.*

*Having agreed [*ngetutaken*] to lay out [the offering] for repelling the pests eating the plants, Garèng finished preparing these offerings. When they were ready, Bima got undressed. Once he was undressed, then Gareng and Pétruk took their positions and also took off their clothes.[10] At such a moment, they [need not be] embarrassed.*

Then Radèn Werkudara carried the akik *(agate; a mistake for* aking *board?], Kyai Barung that lay behind Garèng. The various offerings were carried by Pétruk and Bagong, in prescribed order [fig. 6.19]. They went round the rice field following these numbers [see dia- gram reproduced below]. This can be explained [by following] from number one till the last number. Having arrived at the rice field, one should say the [first]* mèl *prayers starting with number one while crossing [the rice field]. Begin the* emèl *[or] prayer by "Bismilah...."*

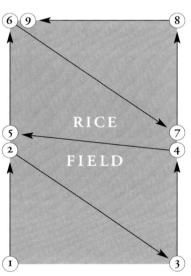

1. In Allah's name, may God bless him and grant him peace,
[I ?] am going to be a party
to an action that will repel the parasites.
Come immediately, come
[and] paralyze them.
Come [and take] pity, come mercifully, mercifully, mercifully. [Come]
by Allah's will.
There is no God but Allah,
and Muhammad is His Prophet.

Semilah Salalahu Ngalahi wasalam.[11]

Sun harsa nedhuh,
anggènipun badhé nulak ama.
Teka kèdhep, teka lumpuh,

teka welas, teka asih, asih, asih,

saka kresnané Allah.
Lah hailah lahilah
Muhamat Sun Rasululah.

2. Upright [?], a Hex   ***Salingga, Tawang Sangker***
In the name of the Lord all merciful and   *Samilah hirahkiem hirohman nirohkiem.*
compassionate.
I am going to repel parasites.   *Sun harsa nulak hama,*
Parasites, [there is] a lotus   *hama tunjung sari*
its essence creates [carves out].   *Sarining ngukir,*
tituk [?] apiece, workers [pests] heads   *tituk édhang, bahu bau*
bowed and shoulders drooping,[12]   *dhe- dhedhengkluk,*
one *narda* [?] month,   *sesasi narda*
a mantra, yes a muttered [?*umèl*],   *mi[?è]l ya umil dhek kethemil.*
for when [?*dhèk*] [?the pests] nibbled [?on
the rice].

3. *Pakantir Emèl* Prayer

In the name of Allah,
Destroyed, slack who are seen [?],
hammered by the sharp head
of an iron hammer.
Their essence is completely ground up,
Keb [?] in the monthly place [?],
the poison is spit out like the seeds
of the peeled fruit of the sugar palm.
The mount Juhar's [13]
action is outside. May Allah...

Emèl Pakantir

Semilah.
Sirna, pangluyu danarira peper,
pupuh kapulu ki panapak gada wesi.

Kulita sari rujak-rujak dheplok,
keb dudunung sasi,
wisa sembur kolang-kaling, kalingan

gunung juhar,
jahiring laku. Salalahu.

4. A Doubled-Back Spell

May God bless him and grant him peace.
Tu hirobil ngalami[n].
I look toward young brother Sang Parta
[Arjuna].
His[?] fragrant image creates
determination, prosperity, armed
strength, well-being.
Yes, well-being by Allah's will.

Emèl Nekuk

Salalahu ngalahiwasulam
tu hirobil ngalami[n].
Ningsun, mulat mara ri Sang Parta.

hungas gambaring teguh hayu,
wiyana slamet,
iya slamet saka kersaning Allah.

5. *Mèl* of the Lotus Month

I am pestering.
Come bad ones, come lame ones.
Come by the gods' will which is greater.
I am the rice here. I am *bleggudhu* [?].
At the same time as Allah desires.[14]

Mèl Tunjung Sasi

Sun lélédhang,
teka ala,teka lumpuh,
saka kersaning déwa kang linangkung.
Sun parinya, sun bleggudhu,[15]
borang-baring[16] kalamun alih, bil aobil.[17]

6. The Stretched Cord *Mèl*

[Seen] faintly in the distance,
the morning star shines like the
cotton flower,
its appearance like ivory.
Truly the parasites are going to be
totally destroyed,
crushed, becoming so many crickets.
Come mercy, come love. Love, love, love of
Bima here,
yar kamar roh kimin.

Emèl Sentheng

Kethap-kethap,
lintang krahinan, sunarira kembang
kapas,
wayahira gumadhing.
Kap temen hama bakal lebur-lebur,

ajur dadi sawalang-walang.
Teka welas, tak asih, asih-asih asiha
marang Bima iki,
yar kamar roh kimin.

7. The *Clenjer Mèl*

Emèl Clenjer

Dawn in the east. · *Bang-bang wétan.*

return to the west, · *bali mengulon,*

return to the east, · *bali ngétan*

return to the south, · *bali ngidul,*

Return to the north.[18] · *bali ngabar.*

[All] are wrong at once, · *Luput pisan, kena pisan*

[all] are hit at once.

My voice is like thunder, bolting down. · *Suwaraku gelap, ngampar.*

My position is that of the spotted elephant. · *Lungguhku gajah belang.*

My path is the gentle wind. · *Lakuku angin ngidit.*

My stride is like a bullet shot, [like the] · *Mlayuku lepasing mimit(s).*

effulgence of Allah, Allah's essence [*zat*]. · *Sir alah, dattollah,*

May my request, · *pinangka panuwunanku,*

what is arranged, be finished, · *rèngès sajati rampung.*

premonitions are made. · *jali-jalining nukir,*

Carved a sandalwood great well. · *ukir cendhana sumur Bandhung, kang*

May what is begun be sturdy, beautiful, · *awitta, teguh haju wiyana slamet,*

well-armed, well-being,

well-being by Allah's desire. · *slamet seka kersaning*

May Allah.... · *Allahu Salalahu.*

8. Completing Planting Dry Rice [*gaga*]

Mèl Putus Rancah

Gadhing creepers climbing upward. · *Tali-tali gadhing,*

[poisonous] *cubung* flowers, · *kembang cubung,*

in the future possessing the sap of the penis. · *uri-uri anyerepi[20] duduhing peli.*

[With this magic spell, *sirep*],

all is back to normal, by Allah's will. · *Sirep-sirep saka kersaning Allah*

I am the cord of life, · *Sun talining urip.*

life like night leeches, · *urip kaya ratrining lintah, lintah kaya gad-*

leeches like ivory-yellow creepers. · *hing waluh.*

Come what, come what has you, · *Teka apakaya apa*

in the form of a chant [*tembang*], · *wujudira tembang,*

a poem of a red chant [*kidung*], · *tembang ing kidung reta,*

ever the Flaming Moustache[19] · *taté Kumbala Geni,*

which destroys determination, · *kang anglebur ajuring tekad.*

which purifies the soul [*jiwa*], · *angenengna jiwangganira,*

the essence, which makes it create, · *dat ingkang murwèng dumadi,*

it seems clearly like a column, *trenjang* · *melok-melokkaya tugu trenjang lempar,*

lampur[?].

[Like] the latch of a chest, helpless, · *lampiring gendhaga lumpuh,*

it cannot succeed by acting. · *lumpuhing kelakuhan.*

The strength of the *kempul* gong reverberates, · *Daya-daya kempul,*

calling the guard. · *gemak melung.*

Salalahu, ngaliwa salam. · *Salalahu ngatewa salam.*

Analysis of the Myth

In order to understand these unusual Sentolo myths and images, it is useful to bear in mind the notion that one cannot contain violence without sacrifice (Scubla 1999, 135–70). Ostensibly, the violence in this case would be the devouring of the rice plants by the parasites. The generic name of these pests, however, is *kala*, or evil, and, as noted before they are the spilt seed of Guru's attempted rape of Uma/Durga.[21] This would suggest that containing rice parasites requires sacrifice commensurate with Guru's violent act.

This predicament seems to refer inchoately to a primordial taboo that is frequently translated into myths and their accompanying rituals: the prohibition against associating menstrual and sacrificial blood (Héritier 1979, 209–44). To clarify, at the highest level of generality this means that a woman capable of bearing children cannot be a sacrificer because as the creator of offspring, her blood should not be confused with that resulting from sacrificial murder. In the Javanese myths dealt with here, the menstrual blood is that of Durga, the cursed victim of Guru's advances.[22] While the sacrificial blood is that of the Rice Goddess Dewi Sri, another victim of Guru's ardor.

The sacrifice of Sri, however, is a more "fruitful" violence, a rape that occurs within the myth concerning the origin of rice plants. In the illustrations for the Sentolo myths, the pests attacking the rice fields are not represented as insects or rats but as an army of monsters (*kala*), deformed hungry human beings. The hunger of these numerous "spilt seeds" is emphasized by their intriguing physical deformities visible in older collections of *wayang* puppets used in exorcisms. To put these myths in their original context, we must recall the larger narrative from which they are extracted. In recalling one episode in the widely known Javanese rice myth of Dewi Sri, one is always, in a sense, retelling the whole. The *Manikmaya*, a compendium of myths, features several consecutive versions of the Rice Goddess myth.[23] In it we read how Sri, a maiden in heaven (*junggring salaka*), is desired in marriage by many but refuses all suitors. She is eventually raped and killed by Guru (identified variously as her husband, consort, or father). His violent use of his phallus has the effect of "shooting" her off to earth. The rape that kills Sri separates her from heaven. Because of its beneficial effects, however, her murder may also be perceived as a sacrifice. From her tomb—hence her body—issue forth all the edible plants, a gift to humanity that can never be repaid. In some versions of the myth, each part of her body is associated with a specific cultigen. The earth in which her body is buried has received her life-giving blood. Society's sustenance has its origin in the lifeless body of this raped victim.[24]

At this stage, humanity has yet to make sacrifices to the gods. In another version of the same myth, Sri seeks to marry her younger brother, Sedana, who has fled heaven. By leaving heaven in pursuit of Sedana, Sri will ultimately assure that human reproduction involves the two sexes, implicating mankind in networks of marriage. Although Sedana and Sri return to heaven, they leave doubles on earth. Furthermore, another couple descends from heaven to marry, and their offspring are seen as being descended from the union of Sedana and Sri (see Headley 1983). (This is in contrast to the situation of Kala, who born outside his mother's womb is destined to marry her.) During her time on earth, Sri, pursued by a demon army of *kala*, offers protection to the peasant couples she encounters in exchange for hospitality in the household granary and weekly offerings there. Escaping from a monster sent from her father's (Guru's) kingdom, Sri eventually hides by transforming herself into a rice-field snake. She is brought in from the field and granted refuge in the granary of a village household. There she teaches the householders, a childless peasant couple, how to make offerings to protect their rice fields, and she grants them a newborn child, a double of herself.

Only after this unilateral gift of the deities, the murder/sacrifice of Sri for the nourishment of humanity, will sacrifices begin to be offered to the gods by humans. The first peasant couple learned from Sri how to prepare these offerings, which protect rice and their child, who is a gift of Sri herself.[25] This is where the myth of the rice parasites enters in, for these *kala* would not only devour the rice that grows from Sri's tomb, but also give newborn children convulsions. Mankind needs Sri's protection, just as it needs to eat rice. Given the complexity of the different episodes of these origin myths, what seems initially like a apotropaic ritual—the use of the phallic *tlawingan* to combat demons originating from Guru's spilt seed—may actually be a case of identicals (penis and sperm) repelling each other. Whatever the case may be, the myths summarized here form a small part of a full-blown origin myth of Javanese society itself.

The gods who have descended to earth offer a model for kinship relations. Sri, the elder sister of Sedana, is a nubile young woman "full of heat" (*tisnawati*), elsewhere a concubine (or consort or daughter) of Guru. Sri and her younger brother, Sedana, who have fled to earth to marry, are eventually forced to return to heaven. Héritier (cf. Barraud 2001, 32–33) has shown that this aymmetrical brother-sister relationship—a cipher for several dimensions of alterity—characterizes marriage, giving birth, filiation, and collaterality (Héritier 1981, 47). The valence of the sexes is different in our myth, and this inequality for Héritier is a social construction. Here the kinship network of the mythic figures institutes a paradigm for the substitution of elements of the myth of origin of rice and that of the origin of rice field parasites. A look at some of the episodes drawn from variant versions of the two origin myths is revealing.

Three Variants of the Javanese Myth of the Origin of Rice

1a. Guru rapes and kills Sri, losing his seed. Having fallen to the earth, these seeds become *kala* and attack the rice harvests.	1b. From the cadaver of Sri, buried on earth by order of Guru, grow all the cultigens consumed by mankind.
2a. Sri, full of heat (*tisnawati*), flees heaven for earth, seeking to "marry" her younger brother, Sedana. Although they finally return to heaven, their "doubles" descend to earth to marry.	2b. In the first human village, Mendhang Kamulan, offerings are made to the rice-field snake—the form Sri assumed while in flight—to protect the rice and new-born children from attacks by the gods in demonic form.
3a. The consort of Guru, Uma, resists his attempted rape, and Guru's seed falls to earth becoming Kala. Kala later marries his "extrauterine mother," Uma, who, cursed by Guru, has become the demon queen Durga.	3b. Durga reigns as queen of the demon armies in a forest to the north of the palace of the Javanese king, and she is propitiated by Javanese kings who solicit her support.

The permutation of the roles of the goddesses (Sri, *tisnawati*, Uma, Durga, etc.) in these origin myths helps to establish the relatedness of the episodes. There was a hierarchy of codes in Javanese society in which the strongest marriage taboo was that of the elder sister with the younger brother. Henceforth, this would be reserved for kings, who in Surakarta (Central Java) were ritually wed to Lara Kidul the Queen of the Southern Ocean, a same "generation" heirloom (*pusaka*) bride (see chapter 26). Elsewhere in Southeast Asia, as well as southern China and Taiwan, "broken grain" myths[26] permit the distinction and separation of animal and human activities, with fire and thunder serving as the disjunctive motifs. Here, however, the initial code is giving birth and planting (rice) seeds. These Javanese myths construct their initial codes around the taboos concerning planting and human reproduction. While marriage of elder sister and younger brother is reserved for kings, a mother's marriage with a son, like that of Durga and Kala, is reserved for monsters.

Purification with a *Tlawingan*: The Narrative Segmentation

As suggested above, the purification with a wooden board might be viewed as a reply to the transformation of Guru's spilt sperm (*kama saleh* or *kala*) into rice parasites that devour the harvest. As one of two identical poles, the phallic *tlawingan* would repel the parasites emanating from the spilt semen. The motif decorating the top of the board is a variation of the apotropaic Banaspati (Kîrttimuka or Boma).[27] This god, whose name means "lord of the forest," is called Barung in the Sentolo myth and may be identified with *barong*, the mythical feline dragon endemic to Southeast Asia. In Bali, despite its demonic appearance, this figure is considered benevolent and combats Rangda, who is associated with Durga, the goddess of death (Beatty 1999, ch. 3). In the myth translated above, however, Barung engages in battle not with Durga, but with the *kala* themselves.

The codes to Durga's behavior seem to fall apart here. She guides Bima to Papan Padhuhan on Mount Grengseng. When the sons of Semar search the *sawah* for the *kala*, allowing Bima to remain on the sidelines, they are helping as members of the same pantheon as Bima and Semar. Locating the parasites and capturing the first rice pest is done with their assistance. Capturing the others seems to pose different problems, however, since this requires that the ritual continue with a procession of naked farmers accompanied by Bima, carrying the first *kala* transformed into a *tlawingan*.

This apotropaic ritual puts an unrelated succession of characters on stage—the rice parasites arisen from the fallen seed of Guru; Bima, the most valorous of the five Pandava; Durga, the queen of the forest demons; and the three sons of Semar—none of whom seem to require much character delineation in order to play an essential role. The main character is the board itself, and the famous figures from Javanese mythology who surround it have more or less walk-on parts. The formulaic speech of the invocations appeals to a register of peasant prayer that is no longer entirely Javanese but not yet really Muslim. Like the myths, it is a genre in mutation, not to say degeneration.

The most straightforward part of the ritual is the procession. Nocturnal nakedness is considered conducive to meeting spirits, and the paths indicated for crossing the fields recall various plowing patterns used to maximize the fertility of the soil. In several of the illustrations of the final episode of this sequence, a meal is held—not on the banks of the rice field, but inside a well-disposed village house. The world of nature is well behind the participants. The fact that the meal to ensure well-being (*selamatan*) is led by the village *modin* (as we suppose from his dress) does not mean that there is really an

Islamic counterpoint to wind up the ritual. As Beatty has shown (1999) participants are free to understand the Arabic prayers mixed with Javanese in the way that best suits them. This ceremonial meal with its own proper offerings and formal speech is able to subsume any and all lesser communal rites. It allows everyone invited to participate giving a semblance of holism that disguises the real divergences that have certainly appeared in Sentolo since the 1930s, as they have elsewhere. In the 1930s in Sentolo, however, things had not yet reached the point that they had in Banyuwangi in the 1990s, where according to Beatty: "The slametan is a communal affair, but it defines no distinct community;...while purporting to embody a shared perspective on mankind, God, and the world, it represents nobody's views in particular" (1999, 25).

In this rice field purification ritual from the 1930s, Widi Prayitna's imagination is in no way curtailed by a Muslim point of view. It is the strongest feature underlying the four facets of this ritual (myth, *pusaka* [heirloom objects], procession, and invocation). What his imagination does suffer from, however, is the lack of capacity to integrate the four aspects into a coherent whole. Perhaps Widi Prayitna was a poor observer of peasant practices. The cosmological categories evoked in the *mèl* prayers have no better fit, even though down to the present-day exorcisms (*ruwatan*) continue to use these categories cogently (Headley 2000, ch. 6). What captures our imagination most clearly are Widi Prayitna's drawings. Awkward as they are, they represent an "inartistic," and therefore uninhibited, statement not of what peasants fantasized, but of what they saw as happening in their rice fields. The ritual life of the village in the coming seventy years of the twentieth century would lose its centrality. Muslim calendars would displace local village ritual dates. Peasants would be asked increasingly to adopt and understand urban norms and values. Nothing in the way Widi Prayitna presents his version of these peasants allows us to imagine that they would be prepared to defend their ancestral vision. In fact, as just shown, its stitching seems already to have been coming apart from the inside.

•

6.21   In one of the Sentolo drawings, farmers plant rice by laying whole stalks of seed rice into the flooded field. Leiden University Oriental Ms. 12.542 (8.2).

7.1   Black Tai villagers construct
        a symbolic rice granary
        (*xelong*). Photograph by Eric
        Crystal, Bản Đốc, 2000.

7

Rice Harvest Rituals in Two Highland Tai Communities in Vietnam

Vi Văn An and Eric Crystal

During the past sixty years the northern part of Vietnam was subject to war, revolution, and considerable social upheaval. Yet at the same time that Vietnam was often at the vortex of world events, isolated northern highland villages remained largely insulated from the profound dislocations experienced in the lowlands. The consequences of national integration, cultural assimilation, and economic globalization were not as keenly felt in these areas as they were elsewhere in Southeast Asia. Anticolonial war, social revolution, and the effort to sustain the Democratic Republic of Vietnam during the conflict with the United States tended to enhance the self-sufficiency and cultural integrity of the highland peoples of northern Vietnam, such as the Tai.[1]

During what is known in Hanoi as "the American War" (1964–1975), the majority of upland villages in the southern part of Vietnam were obliterated, either through direct assault or as a consequence of forced resettlement programs. In the northern part of Vietnam, most upland villages escaped this type of devastation with the exception of those situated near bridges or strategic crossroads astride the Ho Chi Minh trail.

With the implementation in 1985 of the Vietnamese government's policy of economic renovation *(đổi mới)*, the cultural and economic isolation of the northern highlands has begun to erode in the face of free-market initiatives, enhanced communication links, and rising expectations. If ever there were an important time to investigate the traditional material cultural and indigenous belief systems of highland minority peoples in Southeast Asia, then the time is now and the place is the northern mountains of Vietnam.

In this project, we elected to look at rice rituals in two separate Tai communities in Nghệ An Province, Vietnam, in order to gauge the way in which generalized beliefs and practices relating to rice elsewhere in Asia might be expressed in these two select upland locales. The mountain Tai of Vietnam were selected because of the expertise of coauthor Vi Văn An on the one hand and because of the centrality of rice in Tai culture on the other. As Richard O'Connor has emphazied:

> Tai are wet-rice specialists par excellence. In their niche none are better. Local life molds itself closely to rice ecology. The ritual complex I call "rice" organizes households in a cooperative community and mediates between this collectivity and the rice that they grow. To them rice is a woman, mother rice. To farm well one nurtures [and] cajoles her, avoiding anger or command. Local cooperation is also nicely served by households that value the social graces and gentle manner that rice requires. [O'Connor 2000, 47]

7.2   A field hut stands in the distance in a dry-rice field. Photograph by Eric Crystal, Bản Đốc, 2000.

Throughout Mainland Southeast Asia, lowland majority populations of the alluvial plains and highland peoples of the rugged mountains reflect distinct cultural, historical, and agro-economic adaptations. Two-thirds of Vietnamese national territory is situated in midland or upland regions. Historically, the majority ethnic Vietnamese population has been concentrated in the northern Red River and southern Mekong Deltas, and along a narrow lowland alluvial plain joining the northern and southern rice-producing heartlands. In the mountains of the north and west, an extremely diverse array of highland minority peoples has resided in upland cultural enclaves for centuries.

Today, 13 percent of the total Vietnamese population of 78 million is composed of ethnic minorities. Aside from urban Chinese and relict Khmer and Cham peasant populations of the southern coast and deltas, most of the minority groups reside in highland villages. Some 53 distinct minority groups have been recognized by the government of Vietnam. Highland peoples include those subsumed under all the major linguistic groups of Southeast Asia (Austro-Asiatic, Austronesian, Sino-Tibetan, and Tai-Kadai). This discussion of rice-related rituals is centered on a very large population of Vietnamese highlanders, the Tai peoples who are thought to number nearly 1.3 million in the northwestern and north-central parts of the country.

Tai Peoples of East and Southeast Asia
The Tai peoples are the largest single ethnolinguistic group across Mainland Southeast Asia.

> Tai speakers, including those in China, number approximately 30 million, making them, next to the Chinese, the largest and most widespread within the area presently under consideration. Considering the extent of their distribution—from 7 degrees to 28 degrees North and 94 degrees to 110 degrees East—the various Tai dialects are remarkably homogeneous." [Lebar et al. 1964, 187]

The migrations of Tai-speaking peoples into Southeast Asia are well documented because they have occurred in historic times. "The most widely accepted theory concerning the Tai-Kadai language family posits that it originated just south of the Yangtze River in present-day China. Over the past 2,000 years Tai have moved South"

7.3   The *tayleeo*, a sacred marker, signals the transfer of the rice from the field to the storehouse. Photograph by Eric Crystal, Bản Đốc, 2000.

(Gittinger and Lefferts 1992, 17). Between the years 1100 and 1400 C.E., a major migration of Tai-speaking peoples poured into the river valleys and narrow mountain plains of northern Southeast Asia. These were sophisticated communities; peoples with a writing system, close control of irrigated rice-farming techniques, well-articulated social stratification, and a strong ethnic identity and local political consciousness.

Tai migrations to Southeast Asia did not emanate from a central state in China but rather from several population concentrations. Once in the Southeast Asian region, some Tai peoples coalesced into major civilizations, while others formed federations of tribal peoples subject to the rule of long-established Southeast Asian states. By the fourteenth century, Tai Buddhist states had replaced the Hindu Angkor dynasty as the preeminent power in Mainland Southeast Asia. In what is now Laos, Tai peoples coalesced around the Mekong Plain, forming several kingdoms at Luang Prabang, Vientiane, and Savannakhet. Others made their way to the mountain borderlands separating Laos and Vietnam. These peoples were never subject to orthodox Buddhist teachings, maintaining instead the primal religious traditions of their ancestors. Nevertheless, all Tai share a close linguistic heritage, a common writing system, traditions of loom-woven textiles (most notably the Tai sarong, or *xin*), a preference for glutinous rice served in woven bamboo baskets, and a keen system of water and land management. Tai peoples characteristically maintain permanent settlements that are characterized by large houses erected on stilts. Their society is traditionally marked by an interest in accumulating and manifesting wealth and by a three-tiered system of social stratification. According to Richard O'Connor, "first, the Tai are a wet-rice people. Indeed, the spread of the Tai was at least, in part, an ecological succession: as Tai displaced or absorbed earlier Austro-Asiatics, mono-cropping wet-rice specialists who were skilled in flowing water irrigation" (2000, 37).

Tai migrants probably followed river courses in their migrations south of China. Throughout the mountains of northwest Vietnam their villages are located adjacent to rivers and small streams. Wherever possible, they farm irrigated rice and demonstrate a strong preference for glutinous varieties. Tai are also skilled terrace builders and commonly plant rice, maize, and cassava in dry swidden fields. Masters of rotational farming, they till complex highland gardens, nurture large bamboo groves, sustain fruit trees, and energetically pursue aquacultural endeavors. Southeast Asian Tai farmers may have always preferred to plant rice in highland valleys or on mountain terraces where water was

7.4   This completed harvest of unhusked rice is bundled and stacked—ready to take home. Photograph by Eric Crystal, Bản Đốc, 2000.

7.5   (Opposite) The *xang khau hach* (beginning rice sequence), a fertility structure, stands in the field of the sacred Rice Mother. A symbolic representation of a vagina appears at the top, and a symbolic phallus appears lower down on the structure. Photograph by Eric Crystal, Bản Đốc, 2000.

sufficient, but this has not always been possible. As a result dryland, swidden rice also forms an important part of the agricultural regime. Some villages are entirely dependent on dry fields, others typically exploit mixed subsistence economies, while still others focus on such commodities as sour bamboo shoots or pond-raised carp.

Black Tai and White Tai

There are two subgroups of Tai highlanders in Vietnam, the Black Tai and the White Tai.[2] Clothing traditions account for the appellations. According to some sources, the White Tai have been more closely assimilated to Vietnamese culture. Both groups are distinguished by mutually intelligible dialects of the same language, intensive irrigated rice cultivation complemented by creative dryland field exploitation, cotton and silk production and textile design, and the habitation of permanent villages with large houses raised on pilings above the ground.

Our attention was drawn to Bản Đốc, a Black Tai village of approximately 167 people, because of the annual dry-rice harvest ceremony conducted there.[3] Bản Đốc is situated about seven kilometers from the capital of the Con Cuông District. A two-kilometer dirt track connects Bản Đốc to National Route Seven, which links eastern Laos to the Nghệ An capital, Vinh. Villagers here farm both irrigated and dry-rice fields. However, as Bản Đốc villagers harvest much greater quantities of rice from their extensive dry fields than from their limited wet-rice plots, the dry-field ceremony is paramount.

Tai communities were traditionally organized into *muang* (fiefdoms controlled by powerful chiefs). Large multivillage *muang* harvest celebrations were held in this area until 1960, when Democratic Republic of Vietnam government regulations restricting the expression of "feudal" prerogatives and the wasteful expenditure of resources on elaborate rituals effectively banned such festivities. Family harvest celebrations, however, have continued unabated since that time. The people of Bản Đốc have never failed to observe the ritual we witnessed.

Black Tai Ceremony on the Swidden Slope

Throughout the Tai culture zone there is a common belief in the concept of Mae Khau, the Rice Mother, who may be portrayed in different forms and by the erection of varied symbolic structures. The belief that the Rice Mother protects the harvest, inhabits the field, nurtures the seed rice, and blesses the community as a whole is widespread wherever Tai subsistence farmers are found. At Bản Đốc village, a single hillside field has been utilized for generations to honor and respect the Rice Mother. To reach this field one must ford the Khaen Doc stream five times before ascending to the crest of a small hill. Atop the hill is situated a small bamboo field house, the type of shelter commonly utilized during the end of the growing season by concerned farmers intent on keeping wild pigs, marauding flocks of grain-eating birds, and even the occasional thief from plundering the ripening fields (fig. 7.2).

At the beginning of each agricultural year a married woman, preferably one with many children, takes down the special unhusked rice from the attic of the house where it has been stored with the rest of the grain. These sheaves of rice, thought to contain the essence of the Rice Mother, should never be pounded in a mortar. In order to extract the special seed grain to be planted in the sacred plot, she carefully rubs the unmilled rice between her hands, separating the grains from the chaff. Once germinated, the seeds are carried to the same sacred field every year by a designated couple. This rice,

7.6   The rice fields of Dong Minh village lie in a lush highland. Photograph by Eric Crystal, 2000.

which is considered sacred, is planted in a special place on the dry field directly adjacent to a sacred bamboo structure, the fabrication of which is mandated by myth and customary practice.

Long ago, perhaps six generations back, when the ancestors of the local inhabitants first came to settle these lands, it is said that a widow had great trouble farming her field. As is commonly known, widows have much experience with men, but since the departure of their husbands, they must frequently eke out a living on their own with their material and sexual needs going unmet. No one can remember the name of this widow, but her story is known by all. Although she invested much labor in her rice field, she could not produce a satisfactory crop. In Bản Đốc village, only a few notables control the limited irrigated fields of the valley; most rice is produced on the highland slopes, which press fast against the forest. Year after year the widow's dry field failed to produce grain in any quantity, and eventually she came to realize that her field was barren because it had no mate. In other words, as long as she farmed a field as a widow with no husband, it would continue to be infertile and barren. The widow, therefore, devised a way of overcoming this problem. She constructed a bamboo fertility image that would symbolize the sexual union of male and female. Only with such a union consummated in the field would her harvest prove successful.

This bamboo structure—known as *xang khau hach* (beginning rice sequence)—consisted of a sharpened stake at the top, an opening akin to a vagina, and a wooden phallus (*quay ben*) situated a bit lower down (fig. 7.5). At the base of this structure were placed four bamboo tubes to irrigate the field magically, four bent bamboo fence posts to symbolically cordon off the sacred space, and three sacred sheaves of rice. Around the base of the structure four pairs of long bamboo strands—strands that would be symbolically entwined and coupled during the harvest ceremony—were positioned.

So great was the widow's harvest after she implanted this sacred bamboo fertility structure in her field that all the villagers of Bản Đốc began to participate in the construction of this image at the beginning of each agricultural year. From this point forward, the seed rice would initially be gently massaged between the hands of a particularly fertile mother, planted at the base of the *xang khau hach*, and carefully nurtured during the long growing season. At harvesttime, villagers would return to the field to conduct final ceremonies prior to gathering the crop. Concluding ceremonies would symbolically portray the union of male and female, prior to the display of the harvest before the Rice Mother and the spirits of nature.

As we arrived at the field in November 2000, a select village couple had preceded us to the slope. They had departed at dawn, rushing to the field to be sure that they arrived in advance of the main party of village celebrants (including the ethnographic field team). The couple then proceeded to construct a new symbolic fertility structure as the original one had fallen down during the year. Once the structure was in place, they proceeded to harvest three sheaves of sacred rice belonging to the Rice Mother. Using finger knives, they cut this special rice in absolute silence, being careful not to even breathe in the presence of the Rice Mother (see fig. 2.26). They then combined the three sheaves with three more from outside the sacred space and proceeded to harvest the rest of the field. The six initial sheaves of rice were deposited near the field house, shortly to be stored in a symbolic four-pillar "granary" (*xelong*) indicating the bounty of the harvest.

As harvesting proceeded to the east of the field hut, other villagers working immediately to the west began erecting the four poles that would support the *xelong* (fig. 7.1). The sturdy wooden poles of this shadow structure are meant to symbolize the erection of an especially great storage facility that would house an unusually ample harvest. The weaving of two modest lattice-like markers known as *tayleeo* signals the beginning of the transfer of rice from field to storehouse, symbolically ending a successful agricultural year (fig. 7.3).

Two small *tayleeo* are affixed to the *xang khau hach* after the first grain has been harvested. At the base of the pole the four pairs of green bamboo strips are each intertwined, symbolizing the sacred coupling of male and female, of the widow and her consort—the successful fertilization of the field and the Rice Mother. Then a few sacred sheaves of the Rice Mother are moved from the field and affixed to the post of the *xelong*. Along the floor of the symbolic granary are piled many sheaves of unhusked rice. A sleeping mat is unrolled next to the storehouse and the *thầy mo* (priest) made ready to invoke prayers to the spirits of nature. Before the prayers begin, small symbolic sheaves of rice, some tied as if to be carried singly by individuals and some prepared as if to be carried by two people, are arrayed in front of the offering place. This rice is arranged in special harvest bunches called *phat nung* that consist of eight sheaves of rice. finally, before the offering place at least two Tai sarongs, several lengths of homespun cotton, old coins, and silver jewelry were arrayed. Prayers were invoked to *phi* (spirits)—the Phi Phu (Mountain Spirit), the Phi Paa (Forest Spirit), the Phi Fah (Sky Spirit), the Phi Dinh (Land Spirit) and the Phi Houei (River Spirit).

After the offerings are made, participants partake of a ritual meal. Most of the rice will be left to dry at the symbolic granary, but the Rice Mother will be brought back home and will be given a special place of honor high in the eaves of the home. She will rest there until the following year when it will again be time to extract the seed between

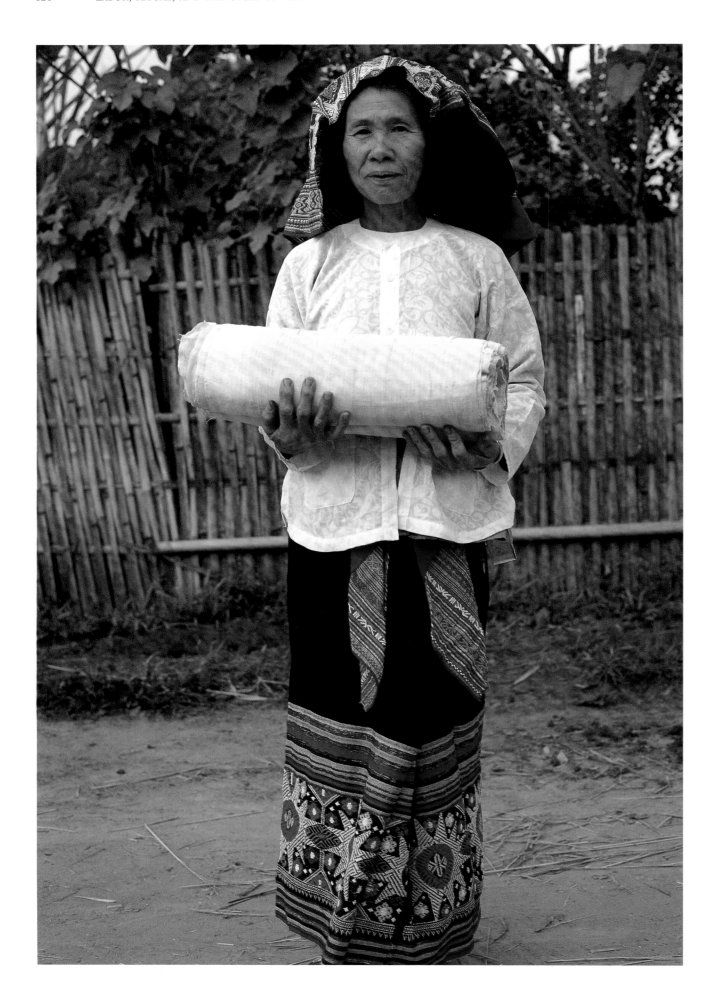

the hands of a fertile woman, implant the bamboo fertility symbol first used by the mythological widow, ensure the fertility of the fields through the union of widow and phallus, and again repair to the field for the annual building of the symbolic granary and the celebration of the harvest (fig.7.4).

The concept of the Rice Mother is common throughout Tai communities, even though the means of honoring her differ from mountain village to mountain village. It is a fertile woman who extracts the seed at the beginning of planting time. It is her husband who implants the bamboo fetish in the ground as the seed is placed in the soil. As the harvest is symbolically taken to the storehouse, strands of bamboo in the vicinity of the sacred rice are intertwined symbolizing union, fertility, and procreation. The *tayleeo* latticework taboo sign is important here and throughout Tai rice rituals. This sign warns of ownership, wards off malevolent spirits and influences, and clearly marks the field as sacred and as one where special rice-related taboos apply.

White Tai Rice Mother in Quỳ Châu fields

In order to further pursue Tai concepts of the Rice Mother, we also visited Quỳ Châu District, home of a large population of White Tai. Here the ceremonial focus was on wet-rice fields. The Quỳ Châu District is located adjacent to newly designated Pumat National Park. This large forest tract continues to serve as home for elephant, rhinoceros, and several species of large wild deer or elk. Quỳ Châu is well known for its textile traditions, which include ikat silk as well as graphic representations of the natural world embroidered on silk sarongs (*xin*). The lower band of *xin* often depicts animals of the forest or such mythological representations as water dragons or rising-sun disks. The diverse natural world of the Quỳ Châu area is also graphically represented on such textiles as headcloths, shawls, blankets, and shrouds.

Dong Minh village is directly accessible by an all-weather road. As is often the case in Vietnam, rice fields are located several kilometers away in well-watered highland valleys. Dong Minh villagers spent considerable time gathering to prepare for the ceremony that would honor the Rice Mother. Women with specified ritual tasks secured the requisite rice wine, fetched the straws, and filled baskets with rice and condiments. They were also careful to dress appropriately for the occasion. Their sarongs depicted the *tahng in*, or rising-sun motif (fig. 7.7). Participants agreed that indeed this motif was mandated by the ceremony honoring the Rice Mother because it is from the sun that the rice receives its energy and vigor.

In procession, a small group of adults made their way from Dong Minh village to the fields. Within minutes the dusty barren roadside village had been left far behind as the procession made its way through bamboo groves, across narrow bridges, and down into a lush valley of rice fields heavy with ripening grain (fig. 7.6). Several chickens, some bolts of handwoven white cotton cloth, silver jewelry, several folded sarongs, areca nut, and glutinous rice wine in a ceramic jug were deposited adjacent to a small bamboo field hut. Here again the field hut served as the logistical ceremonial center. An offering of nine areca nut slices and nine betel leaves was initially prepared to recall the original nine mountains and nine rivers that existed on earth before human beings came to populate the planet.

As the chickens were dispatched and the rice was cooked, several large mats were opened up on the ground facing the field to be blessed. Offerings were prepared to the five spirits of nature (mountain, forest, land, river, and sky). Silver jewelry was arrayed on homespun cotton cloth as several participants began to work with rice straw. Soon they had twisted and bent the straw into a simple but graphic image of Mae Khau,

7.7   A White Tai woman carries a bolt of handwoven cotton cloth for the Rice Mother ritual. She wears a sarong (*xin*) with the rising-sun motif. Photograph by Eric Crystal, Dong Minh, 2000.

7.8   The symbolic straw Rice Mother appears atop a latticework *tayleeo* sign. Photograph by Eric Crystal, Dong Minh, 2000.

the Rice Mother (fig. 7.9). This rice figure was attached to a vertical bamboo, below which was tied the latticed *tayleeo* ownership sign (fig. 7.8). Local artisans were careful to weave this latticework sign so that nine holes appeared, again symbolic of the first mountains and rivers.

Once the Rice Mother figure was complete, it was taken to a corner of the rice field and implanted in the field (fig. 7.10). After this, the *thày mo*, or priest, arranged offering trays of meat and betel nut before him as others arrayed white homespun cotton cloth, silver jewelry, and sarongs also to be symbolically offered to the Rice Mother. As prayers were invoked, all participating villagers gathered on the mat behind the priest facing the image of the Rice Mother implanted in the irrigated rice field. After the meal was complete, rice wine was then imbibed by all through reed straws. In Tai culture both adult men and women must partake. The rule calls for everyone to take three long drafts while the wine manager carefully measures the intake by refilling the ceramic jar with pure stream water from a perforated buffalo horn. This water should be pure spring water, as opposed to water drawn from a well—the same type of natural water, running down from the mountain peaks, that floods the high valley fields.

Participants returned home without harvesting the field this day. Tomorrow they would begin to cut the rice. When the field was harvested, the image of the Rice Mother would be given a special place of honor under a house gable on the upper level where the rice is stored. The figure would not be returned to the field at the beginning of the next agricultural year but would instead join a collection of these figures, which remained in the house. Each year a new Rice Mother figure is made.

7.9   (Opposite) A White Tai woman from Dong Minh village makes the symbolic Rice Mother from rice straw. Photograph by Eric Crystal, 2000.

Conclusion

In this essay we have briefly discussed rituals performed in two separate Tai communities in the highland Nghệ An Province in Vietnam. Although government policy has had some impact on traditional culture here, it would appear that the core of the Tai belief system remains intact and viable. Subsistence agriculture continues to be the major preoccupation of most villagers, and great energy and time are invested in maintaining irrigated fields as well as in cultivating highland swidden plots. In both the dry-rice field in Bản Đốc village and the wet-rice field in Dong Minh village, we encountered graphic representations of the Rice Mother. In both villages, we have observed that concepts of a rice spirit or a rice goddess are vibrantly alive, renewed in annual rituals at which family members gather in the fields to celebrate the harvest and to thank the spirit of rice—the Rice Mother—who has bestowed fertility and the bounty of the harvest. Whether the image is one of the widow and her consort in the dry field or the graphic straw Rice Mother in the irrigated field, it is clear that fertility, wealth, and procreation are closely associated with the production of rice in the highland fields of Nghệ An. These findings suggest, once again, that across the rice-producing areas of Asia, common beliefs and practices relating to grain production survive today both in isolated highland minority villages and in flourishing lowland peasant communities.

In Black and White Tai communities in central western and northwestern Vietnam, the concept of the Rice Mother, as well as the linkage between rice and community, rice and fertility, and rice and well-being, remains strong. Whether the ceremonies celebrating these concepts are carried out on high swidden fields or in flooded rainfed plots, whether the image is a bamboo representation of sexual union or a straw depiction of the Rice Mother, the underlying beliefs remain common throughout Tai culture.

The fertility of humans and the fertility of the fields are closely linked. Fecund mothers are selected to handle the sacred seed rice. The field of a sexually inactive widow begins to produce only when the male and female sexual members depicted on the sacred structure are firmly implanted in the sacred soil of the highland field. When the time comes to harvest the sacred Rice Mother, a couple blessed with many children is dispatched to the field to reap the first grains quietly and efficiently. At the base of the sacred fertility structure are four bamboo plants through which runs the life-giving rainwater sent from on high to irrigate the upland fields. When the first grains are harvested four sets of paired bamboo strands are twisted together. This indicates the symbolic coupling of the male and female, suggesting that the climax of the harvest ceremony is equivalent to the climax of sexual union. Now the harvest can metaphorically be arrayed before the field house for the gods of nature to behold. In Dong Minh village, the image of the Rice Mother is graphic, dynamic, and compelling as it is implanted in the corner of the rice field now filled with heavy grain-bearing plants. The Rice Mother seemingly glides across the field, caressing and guarding the sacred crop now about to be harvested by the community.

The Tai of the Vietnamese Trường Sơn Range maintain the pre-Buddhist religious traditions of their early ancestors, people who very well may have been the first to farm rice in Asia. The spirit of rice remains vital in Tai culture and communities in upland Vietnam, reinforced with annual rituals and observances that date back to the earliest cultivation of this most sacred grain of Southeast Asia.

•

7.10   The Rice Mother is situated in the rice field. Photograph by Eric Crystal, Dong Minh, 2000.

Part Two

The Granary:
A Home
for the Rice Spirits

8.1   A Sa'dan Toraja woman
       retrieves sheaves of rice from
       a granary and places them
       into a large basket.
       Photograph by Eric Crystal.
       Piongan, Rindingallo
       District, Sulawesi, Indonesia,
       1971.

8

The Granary: A Home for the Rice Spirits

Roy W. Hamilton

8.2   Rice granaries in a Minangkabau community. Batipu, Sumatra, Indonesia. Circa 1900. Koninklijk Instituut voor de Tropen, Amsterdam, album 285/26b.

In villages of the Tai Yong people in northern Thailand, the houses are built above the ground on pilings, with different sections of the house raised to different levels. The higher the floor, and the closer to the auspicious east side of the house (east being the direction of the Buddha and the rising sun), the higher the status of the space and the activities conducted in it. The most sacred space in the entire house is a shelf on the east wall, raised well above the heads of the inhabitants, where the altars for the Buddha and the matrilineal ancestor spirits are located (Trankell 1995, 114). In front of each house is a rice granary, which is raised to the same height as the human house. The granary too has an altar shelf on the east wall, reserved for the special lineage of rice that has been passed down from the ancestors (Trankell 1995, 115). That the human home and the granary are fashioned as parallel structures reflects the belief that among all organisms, it is primarily humans and rice that possess a special type of living spirit (*khwan*). Just as the house is the home of the humans, the granary is the home of the rice spirits.

Specialized granary forms that mirror the human house, like those of the Tai Yong, are (or once were) widespread especially in Southeast Asia (fig. 8.2). Villages of the Sa'dan Toraja in Sulawesi and the Toba Batak in Sumatra both consisted of a line of houses facing a matching line of granaries (fig. 8.4). The most elaborate Toraja granaries feature the same beautifully carved and painted facades as the famous Toraja lineage houses (fig. 8.3). Batak granaries (*sopo*) were built high off the ground, with open-sided lower levels that served as work areas, gathering places, and young men's sleeping quarters (Sibeth 1991, 52). The grain was stored overhead and accessed via a small door (fig. 8.5) in the gable, under a sweeping roof like those of Batak houses. In Bali, the granary is one of many structures located within the walled family compound, each with its own specific function. In the Ifugao region in northern Luzon, the house and the granary are combined into a single structure. The Ifugao *bale* is essentially a granary, raised on pilings, to which a hearth and living space have been added. After the harvest, the bundles of grain are stored both in an attic space and in the main room: "In a good year the sleeping space inside a house may be completely covered with such stored bundles, leading household members temporarily—but with never a complaint—to sleep on the house ground space below" (Conklin 1980, 34). In all of these examples, granaries comprise an integral part of the village plan. They add to the community's architectural character, and in some cases even serve to define the household as an independent unit.

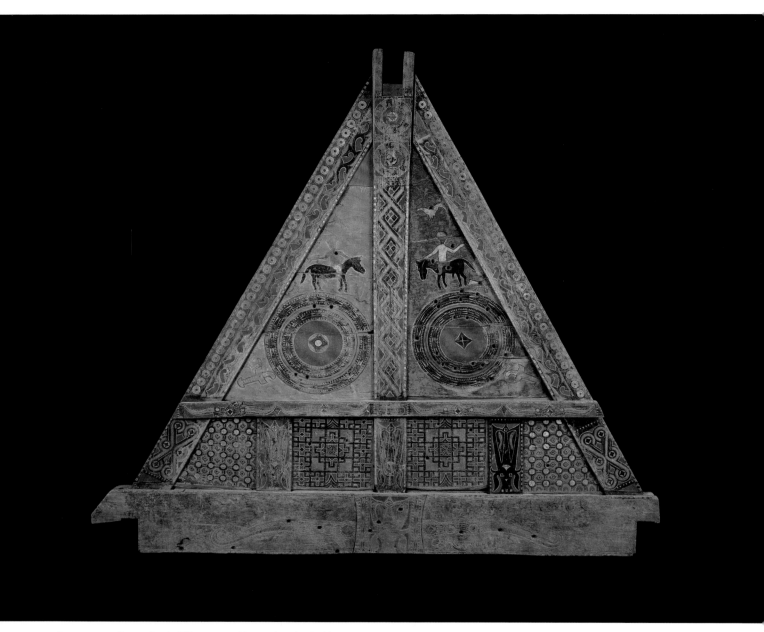

8.3

Granary facade. Sa'dan
Toraja peoples. Tondon,
Sulawesi, Indonesia.
Carved and painted wood.
H: 211 cm. FMCH X85.855;
Gift of Dr. and Mrs.
Robert Kuhn.

Toraja rice granaries are
thought of as extensions of
the "face" or northeastern
part of the house.
Appropriate rituals are carried out prior to storing the
harvest to magically expand
the quantity of the grain so
that it will last as long as
possible. The granaries are
often decorated with incised
and painted images of the
rising sun (associated with
life, rice fields, and the
female gender) and with
fighting cocks (associated
with courage and the male
gender). The two large circles that appear on this
granary represent the sun,
while three water buffalo
appear below. Both are intimately connected with rice
agriculture and their appearance on the granary suggests
fertility and abundance.

8.4   This small Sa'dan Toraja village consists of a row of houses facing a row of rice granaries. Photograph by Ursula Schultz Dornburg, Tikala District, Sulawesi, Indonesia.

Throughout Southeast Asia, special rituals may be conducted when the rice is brought in from the fields and ceremonially installed in the granary. In some communities this ritual marks the biggest festival of the year, such as in the Naik Dango rites of the Kenayatn people of West Borneo. This ceremony indicates that the spirits of the rice have taken up their abode, where they will remain until the following planting cycle. Another indication of spiritual presence in the granary can be found in the special figures sometimes placed inside. The best-known examples are the carved wooden *bulul* figures of the Ifugao, consisting of a male and female pair (fig. 8.6). These figures, which are propitiated with libations, sacrificial blood, and the reciting of epics when the rice is installed in the granary, are credited with power to protect the grain, or even to make it magically increase in quantity after it is placed in the granary. In Bali, the doll-like Nini Pantun figures (see fig. 2.24), which are made from the first panicles cut during the harvest, are afterward placed high in the granary to preside over the rice. Similar figures made of rice straw include the Phii Lao or Taa Pook (see figs. 3.16, 3.17) placed as guardians in granaries in

8.5   Granary door. Batak peoples. Northern Sumatra,
Indonesia. Carved wood. H:
115 cm. FMCH X79.477.

This door opens into the
upper level of the granary,
where the rice is stored high
above ground level. Lizard
motifs are common in
Indonesian art and are associated with prosperity. A
Batak proverb says, "A house
without a lizard is a house
without happiness"
(Cameron 1985, 83; Collet
1925, 355).

8.6   Pair of granary figures
(*bulul*). By Taguiling.
Kababuyan, northern
Luzon, Philippines.
1900–1925. Carved wood. H:
51 cm. FMCH X86.5522;
FMCH X86.5523.

Ifugao carvers of *bulul* are
rarely known by name, but
the work of the carver
Taguiling, who made *bulul*
during the first quarter of
the twentieth century, is easily recognizable by his characteristic style (now also
widely copied).

northeastern Thailand. An interesting variation is found among the Gaddang of northern
Luzon, who cook a bit of the harvested rice and place it in a clay pot inside the granary
for the spirits of the rice (Lambrecht 1970, 98).

   In many societies there are prohibitions about who can enter the granary to
take the rice out, or when and under what conditions this may be done. Often only the
female head of household is considered appropriate for this important responsibility. The
movement of the rice from the granary to the human domain may mark a symbolic transition. In Tai Yong villages, for example, the rice must not be brought into the house before
it is milled because the rice spirits are still alive and their proper home is the granary. After
milling, the rice spirits no longer inhabit the grain and it is placed in a celadon jar by the
hearth, where it has entered the human world and awaits cooking (Trankell 1995, 115).

   The large, freestanding Southeast Asian granaries are related to the practice of
harvesting rice by cutting entire stalks and arranging them into sheaves. The bulky sheaves
require a sizable granary (fig. 8.1), but the rice will keep in good condition longer in this
form. Threshing takes place on a daily basis, with just enough rice removed from the
granary at a time to meet the day's needs. Today, as more and more rice production in
Southeast Asia moves away from subsistence farming and into commodity production,
the grain is more likely to be threshed and bagged in the field and then taken by truck to
commercial warehouses. The Southeast Asian granary, once a proud example of vernacular
architecture, is rapidly disappearing.

   In other regions, where rice has long been cut with a cycle and threshed in the
field, only the grain itself needs to be transported for storage. This is a somewhat less
arduous task, and the granary can be smaller. A large bin suffices in many cases rather
than a freestanding structure. After it is threshed, the grain will store better if left in the

8.7   Untitled print showing rice granaries (*kurayashiki*). By Katsukawa Shunzan. Japan. 1788–1792. Woodblock print. W: 27 cm. Fine Arts Museums of San Francisco; Museum Collection, 1964.141.881.

In this wry print, Daikoku, one of Japan's Seven Gods of Good Fortune (Shichi-fukujin) tallies grain with a rat, while another rat loads a bale of grain into a granary. More bales (*tawara*) are stacked to the right, and a row of granaries (*kurayashiki*) lines the water-way in the background.

husk because it is more resistant to insects and microorganisms that way. In traditional Chinese villages, however, most rice was milled and polished prior to storage so that it required less space in the granary (Bray 1984, 382).

Granaries in South Asia are often large vats made of clay, which can be sealed up for prolonged storage. Frequently leaves of the neem tree (*Azadirachta indica*) are closed up with the rice to prevent insect damage. When it is time to retrieve the grain, a hole is bored near the bottom of the container. The clay granaries are often built inside the house, an idea that reaches its fullest expression among the Tharu, a longhouse-dwelling people of lowland Nepal. Tharu granaries are not only built indoors but they serve as the room dividers that separate the individual family compartments within the longhouse. The Tharu granary then becomes the focal point of domestic ritual and artistic expression. Tharu granaries are described in more detail in chapter 9.

Many types of granaries have been used in China, including large cylindrical containers made of basketry or clay. A remarkable record of ancient Chinese granaries is preserved in the form of ceramic models buried in tombs (fig. 8.8). Some granary models from southern China in Han times (206 B.C.E.–220 C.E.) show complex structures elevated

8.8

Granary model. China.
Han dynasty, probably
Eastern Han (25–220).
Ceramic with green glaze.
H: 25 cm. Collection of
Christina C. Yu.

Models of objects common
in daily life were placed in
Han tombs to provide familiar surroundings for the
deceased in the afterlife.

8.9

Rice tester (*kome sashi*).
Japan. Edo period, late
eighteenth century. Bamboo
with bone netsuke. L: 22
cm. ©The Seattle Art
Museum 35.571; Eugene
Fuller Memorial Collection.

This device was used to test
the quality of grain stored
in bales (*tawara*). When the
pointed end is thrust
through the straw sides of a
tawara and then pulled out,
a sample of the grain
comes with it. The attached
netsuke served to personal-
ize the object, which
presumably belonged to an
overseer in the rice trade.

off the ground and entered with a ladder, in the manner of the raised granaries of
Southeast Asia. Even the overhanging stones at the top of the supporting poles that prevent
the entry of rats can be seen in some models, a feature still used in Southeast Asia today.

Japan also once had granaries called *takakura* that were raised on stilts in the
manner of Southeast Asian granaries. These used to be widespread but are now found
only on remote Hachijo Island and in the far southwestern islands of the Ryukyu chain.
The better-known form of granary in Japan is the *kurayashiki*, a rectangular warehouse
with plastered walls. In former times these were owned by feudal lords who collected taxes
in the form of rice. During the Edo period (1603–1868) they lined waterways in the Osaka
region, in those days the most important center of Japan's rice trade (fig. 8.7). With their
rectangular lines, *kurayashiki* look surprisingly modern. They in fact represent an early
form of commoditization, complete with special testing equipment for quality control
(fig. 8.9). Inside the *kurayashiki*, the rice was stored in large bales called *tawara*, made
from rice straw. Until very recently, *tawara* were such a common symbol of rural life that
they frequently appear in Japanese woodblock prints. Daikoku, one of Japan's seven gods
of good fortune, characteristically is depicted standing on two *tawara* representing bountiful abundance. Today *tawara* are no longer routinely used, but they are still made in small
quantities by community groups trying to keep rural craft traditions alive.

9.1   The head of a family
performs ritual worship
(*puja*) in front of a granary
(*deheri*). Women apply the
handprints on a special day
of family worship.
Photograph by Kurt W.
Meyer, Parau, 1994.

9

The Granary of the Tharu of Nepal

Kurt W. Meyer and Pamela Deuel

Over half of the people of Nepal live in the Terai, a narrow ribbon of land along the southern border. The Terai, most of which lies at less than 200 meters above sea level, forms the northern extension of India's great Gangetic Plain. Numbering about a million people, the Tharu make up the largest single group living in the Terai (see chapter 4). The designation "Tharu" is a collective name that includes several ethnically distinct subgroups whose origins are unknown and whose belief systems and languages vary from place to place. Trappers, fishermen, and farmers, the Tharu have lived for so many centuries in the mosquito-infested Terai jungle that they have developed a resistance to the region's endemic malaria. In the past they were seminomads who cleared forest land to farm, moving on to create new settlements when the land became depleted.

During the nineteenth and twentieth centuries, the Tharu became the principal cultivators of the Terai, converting jungle into farmland so successfully that the region is known today as the breadbasket of Nepal. When the government's DDT mosquito abatement project in the 1950s and 1960s practically eliminated malaria, other ethnic groups streamed down from the hills to the now-habitable Terai, usurping most of the Tharu's land, which had been held and managed by the community as a whole. The illiterate Tharu were easy victims of better-educated immigrants who knew how to use the new land registration laws to their own benefit. Proud, free peasants became tenant farmers overnight.

While various groups of Tharu are found along the Nepal-India border, this essay deals only with the Dangaura Tharu, who have retained their traditions to a greater extent than any other Tharu group. While other Tharu groups have adopted many practices of Hinduism, the Dangaura have kept their traditional spirit beliefs and maintain only a casual connection to Hinduism. The most notable difference is the absence among the Dangaura Tharu of any large permanent temples or any sculpted or painted deity images (except for the devotional painting called *astimki*, which honors the Hindu god Krishna). The Tharu deities are believed to exist everywhere: in the forest, in the rice, in rocks, but there are no images of these Tharu gods, unlike the profusion of statues and paintings of Hindu gods found throughout South Asia.

The Dangaura village character is defined by long, narrow houses facing east and lining both sides of the central village lane (fig. 9.3). All houses are built on a north-south axis; the family worship *(puja)* room is always located in the northeast corner of

9.2   A house under construction
with *deheri* in place.
Photograph by Kurt W.
Meyer, Gobardiha, 1997.

9.3

A traditional longhouse.
Photograph by Kurt W.
Meyer, Gobardiha, 1994.

the house and contains the family shrine. The extended Tharu family could easily exceed fifty family members of four generations. To accommodate the growing family branches, additions are made to the house, thus turning it into the Tharu longhouse *(laamoghar)*. The house is elongated, shortened, or divided into two parts according to the family's changing needs. When brothers separate and divide the family holdings, they also divide the house either by constructing a division wall or by physically slicing the building in two.

Farmers in the Terai depend on three elements for survival: plentiful monsoon rains to nurture the crops, the blessings of the gods for a plentiful harvest, and adequate facilities for long-term storage of the rice. Without any one of these, they may starve.

The granary serves some of these needs. It provides rice storage, a place for worship, and interior wall dividers. Safe preservation of the main food crop, rice, is critical because this crop feeds the family. A good monsoon produces two crops a year, but a weak monsoon will yield only one crop, barely enough for the survival of most farmers and their large families. A storage system is needed that will protect the grain for at least a year from one annual crop to the next. For the family's survival during times of extreme drought, it must hold and preserve enough rice to feed the family for several years. Is it any wonder that the Tharu have given so much attention to the preservation of rice?

The Granary for Storage

As subsistence farmers, the Tharu rely on the natural world about them for materials: clay, dung, grasses, and scarce timber. Using these simple materials, they have developed sophisticated structures and tools. The rice granary is a good example of this sophistication; its design has evolved over hundreds of years. Called in Nepali a *deheri,* it provides for the storage of rice for up to three or even four years. The shape and form of the *deheri* emerged from its use and environmental conditions. Constructed of a mixture of clay and dung, frequently combined with chopped straw, it stands about two meters high. The narrow throat at the top in which the husked rice is poured is easily sealed

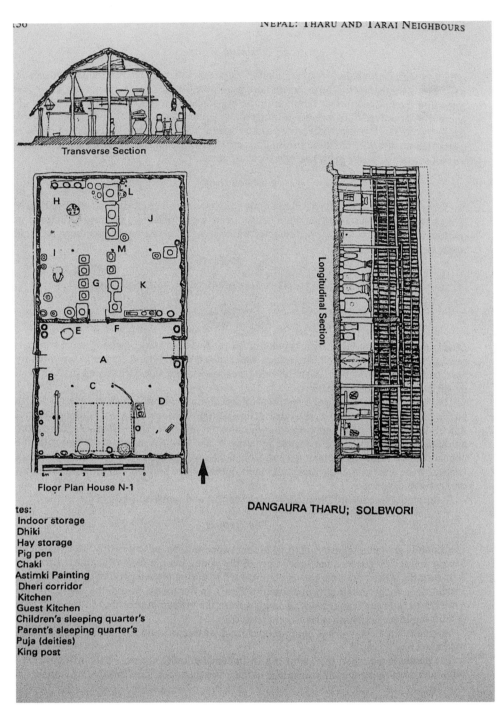

Transverse Section

Floor Plan House N-1

Longitudinal Section

DANGAURA THARU; SOLBWORI

tes:
Indoor storage
Dhiki
Hay storage
Pig pen
Chaki
Astimki Painting
Dheri corridor
Kitchen
Guest Kitchen
Children's sleeping quarter's
Parent's sleeping quarter's
Puja (deities)
King post

9.4   Floor plan of a Tharu house
in Solgwori. Drawing
by Kurt W. Meyer, 1994.

9.5   Central hall formed by
deheri. Photograph by Kurt
W. Meyer, Parau, 1994.

with mud. To protect the contents against rodents and moisture that might seep through the mud floor of the house, the granary is raised about twenty centimeters off the ground. When rice is needed, the farmer knocks open a small hole near the bottom of the *deheri*, and the rice spills into a basket sitting below. When enough rice is collected, the hole is resealed with fresh mud.[1]

The Granary as Room Divider

The granaries are so large and heavy that, once built, they cannot be moved. Therefore house construction begins only after the granaries are built in situ (fig. 9.2). The *deheri* serve multiple functions. They protect the stored grain from the rain and weather. They also make economic use of the interior space; these bulky *deheri* are used as wall partitions between the various rooms and activity areas.

The floor plan in figure 9.4 shows the layout of a typical interior. The large rice *deheri* form a central corridor and serve as room dividers. The northeast room is dedicated to the family shrine. While customs vary from village to village, the male family elder who is responsible for family worship often also sleeps in the *puja* room. The *deheri* that divides it from the kitchen, located in the northwest corner of the house, traditionally becomes the backdrop for the family place of prayer. The south end of the longhouse is set aside for family gatherings and work. Small animals (chickens, goats) are kept here, and the *deheri* located closest to the main entry at the south end of the house is an informal "canvas" for the *astimki* dedicated to Krishna.

The large *deheri* store the rice while smaller granaries contain seed and other grains. The wealth of a family can be gauged by the amount of stored rice, indicated by the number of *deheri* inside the longhouse. A poor family may have only one or two, whereas a wealthy family will have a dozen or more large granaries.

The interior of the longhouse features a central walkway, lined on both sides with granaries that form room dividers (fig. 9.5). The largest granaries, up to about two meters tall, are freestanding. The entire extended family lives in this longhouse, including great grandparents, grandparents, parents, and children. Often the *deheri* are connected with clay walls.

The Granary in Worship

To assure plentiful crops and to drive away evil spirits, the Tharu appeal to their supernatural powers to bring the rains, prevent spoilage, and protect the crop and its storage. A large *deheri* in the sanctuary, located in the northeast room, serves as the background for the family shrine (fig. 9.1). Here the family elder worships his family's ancestors.

The Tharu create no images of their gods in painting or sculpture; their deities are represented by twigs and stones stored inside a pouch that hangs on the face of the sanctuary granary. On the floor beneath these sacred objects, the elder prepares an altar for prayer. He lights oil lamps and arranges clay horses that serve symbolically as the vehicles of various Tharu gods. The gods are assumed to be in the room when their mounts are present to hear the elder's prayers. A small basket is used to receive rice offerings, and other objects represent various gods and ancestors, or are ritual tools used for worship. Small clay pots in the rear hold rice beer, another offering to the gods. Prints of the right hand on the granary *(pithagurin)* are associated vaguely with the female spirits. The women apply the handprints on the day the family worships its ancestors during the autumn Dasain festival.

9.6   A devotional painting (*astimki*) decorates a rice container for Krishna's birthday. Photograph by Kurt W. Meyer, Baibang, 1994.

9.7   *Astimki* painted on a clay
partition. Photograph by
Kurt W. Meyer, Musepani,
1994.

There is a close relationship between rice storage and the deities: the prayer room is located in the rice storage area, and the backdrop to the shrine is a large rice container. The Tharu address the essence of rice as *sahi bharkat,* the life force that can be stolen. In some areas this is referred to as *saha,* which might be equated with the soul of the rice. If the rituals are not performed properly or are omitted, the *saha* departs. Many of these beliefs and rituals are slowly disappearing.

The Granary as Canvas

In the common space at the south end of the longhouse is typically found the Krishna *astimki* painting, sometimes painted on a *deheri* and sometimes on the interior wall of the house. It is painted or refreshed annually by the family elder or village priest for Krishna's birthday during the monsoon season. On Krishna's birthday, after fasting all night, the village women line up to give *tika* to the figures in the painting, daubing them with a mash made of rice and red powder. Each elder who is responsible for the creation of the *astimki* creates his own design but always stays within the traditional outlines.

Some *astimki* designs feature a line of women and a line of soldiers dancing in parallel rows (fig. 9.6). Others show the five Pandava brothers from the *Mahabharata* on the top row and the Kaurava cousins on the lower row. In some cases the entire design forms an obvious human figure (fig. 9.7). The womb-like lower portion of the design indicates that the figure is female, reiterating the theme of Krishna's birth. The themes of the *astimki* paintings and the role of the granary in Tharu social life make it abundantly clear that no household is complete without the *deheri.* Sheltered within the dim longhouse, the sturdy but elegant granaries are the essence of Tharu life. In both practical and symbolic ways, they nourish the family—an indispensable sign of the inward and outward well-being of the Tharu.

•

Part Three

Rice Festivals:
Community
and Celebration

10.1   The summer Kanto festival
in Akita, Japan, features a
procession of men balancing
tall bamboo poles set with
lighted lanterns. The poles
symbolize giant stalks of
rice, and the lanterns the
rice grains. Akita Prefecture,
in northern Honshu, is one
of Japan's most productive
rice-growing regions, and
the festival expresses the
wish for a good harvest.
Photograph by Bruce R.
Carson, 1990

10

Rice Festivals: Community and Celebration

Roy W. Hamilton

10.2   Balinese women make harvest figures called Nini Pantun (Rice Mother) from selected stalks at the beginning of each harvest season (see fig. 2.24). More elaborate versions of the Nini Pantun figures appear at rare festivals called Ngusaba Nini, which may happen in a given community approximately once in twenty-five years. Each family in the community prepares a Nini Pantun figure and brings it to the main irrigation temple, where all the figures are assembled for a grand procession to the village temple (*pura desa*), residence of the Rice Goddess Dewi Sri. The purpose of the festival is to vouchsafe the fertility of the rice crops. Photograph by Garrett Kam, Sayan, Bali, Indonesia, 1989.

Participants in rice-transplanting festivals in Kagoshima Prefecture, on the southern Japanese island of Kyushu, carry poles so that the deities have a means of descending to earth (Plutschow 1996, 118), where their presence will bode well for a bountiful harvest. In the version performed at Nitta Shrine, Sendai City, the carrying of the poles has evolved into the *yakko* dance, said to have originated when a farmer attached rice straw to the top of a halberd and waved it in the air to attract the attention of the deities. As a result, a long string of crop failures came to an end. The form of the pole, with the rice straw at the top, in fact seems to mimic a huge rice plant. This auspicious imagery reaches its ultimate expression in the famous Kanto festival in Akita, in northern Honshu, where every August over a million people jam the streets to watch a procession of hundreds of gargantuan "rice plants," each consisting of a long bamboo pole hung with row upon row of lanterns. The spectacular show of the giant glowing grains both pleases the deities and expresses the human wish for a bumper crop.

If rituals are the small devotional acts that accompany each step of the rice-growing process, events such as the Kanto festival are the grand celebrations that punctuate the annual agricultural cycle. While most festivals have a specific ritual at their core, they also typically include a range of additional elements such as visits to relatives, processions, competitions, public entertainment, music, dancing, feasting, and drinking. Rituals have a limited audience; festivals are open to the public. Rituals are usually quickly concluded; festivals may go on for days. Rituals are sharply focused; festivals are broader cultural expressions. Festivals, above all, are a time when a community comes together, sets daily routines aside, and celebrates.

The great majority of festivals in Asia are in one way or another related to the agricultural cycle, and therefore to rice. Sometimes they are linked to a particular event in the cycle, such as a planting festival or a harvest festival. More often they are couched in a broader framework provided by the prevailing religious tradition. A harvest celebration in the Philippines may take the form of a Roman Catholic saint's feast day, for example, while in Java thanksgiving offerings may be carried in a procession on Muhammad's birthday, or in Thailand the first of the new rice may be offered to the Buddha at the local monastery. Typically festivals serve multiple objectives. The Philippine saint's day

10.3   New pots, decorated for Pongal, enclose the fire and support the *pongal* pot as it boils over, symbolizing abundance. The upturned pots have been filled with rice straw, which burns to create an intense and even heat. While many families buy new pots for Pongal, this celebration took place at the home of a potter, Mr. M. Palani of Menalur, so a supply of pots was readily at hand.

10.4   The rice and other ingredients have been added and the *pongal* cooks in the pot, covered with a tin plate. One of Mr. Palani's daughters creates new *kolam* after purifying the space in front of the hearth with a fresh application of cow dung.

10.5   Mr. Palani ladles the cooked *pongal* onto fresh squash leaves in front of an impromptu altar constructed in front of the family home. The altar includes a lamp representing Ashtalakshmi (see fig 18.7) and a mound of cow dung representing the elephant-headed god Ganesha. The tray lined with leaves (bottom) contains *pongal* and fried bananas, which will shortly be offered to Surya, the Sun God.

10.6   A coconut and a burning lump of camphor have been added to the tray, as the Palani family prays and offers the *pongal* to Surya. The offering expresses thanks for the past harvest and the desire for a bountiful harvest in the year to come. Afterward the now-sanctified contents of the tray are divided. Some are taken to the backyard and crows, who represent ancestors, are called to eat. Another portion is given to the family cow. Finally, the remainder is consumed by the family.

A Tamil Family Celebrates Pongal

The three-day celebration known as Pongal is the most important festival of the year for families in Tamil Nadu, South India. On the first day, Bhogi, the house is given a thorough cleaning, unwanted or worn-out items are burned, and new earthenware pots are purchased for domestic use for the coming year. The core rituals take place the following day, which is formally known as Thai Pongal. (*Thai* is the name of the month according to the Tamil lunar calendar, and *pongal* is a special dish of rice cooked with milk, jaggery, and sometimes additional flavorings such as ghee, cardamom, cashews, or dal). After decorating the yard with rice flour drawings (*kolam*), the women of the house set a pot of milk over an outdoor fire. A key moment comes when the milk boils over, spilling down the sides of the pot (fig. 10.3). This symbol of overflowing abundance is greeted with jubilant cries of "*pongal-o-pongal, pongal-o-pongal.*" The rice and other ingredients are then added to the pot. When the *pongal* is cooked, an impromptu altar is assembled and the *pongal* is dished out before it, together with other food offerings (figs. 10.4, 10.5). The male head of household leads the family in prayers, offering the *pongal* to Surya, the Sun God (fig. 10.6). Thai Pongal is reckoned as the time when the sun begins its journey back toward the north, initiating the new agricultural cycle. While every family conducts its own offering ritual, a strong sense of community is generated when the rituals are conducted simultaneously in front of every home and the shouts of "*pongal-o-pongal*" echo through the community. The following day is Mattu Pongal, a time for offering thanks to animals, and especially cattle. Families who own cows decorate them with flower garlands, paint their horns in bright colors, and sometimes add spots or other patterns with dye to their coats.

Photographs by Roy W. Hamilton, Cheyyar Taluk, Kanchipuram District, Tamil Nadu, India, 2000.

10.7   For the Tamil festival of
Pongal, when the first of the
new crop of rice is offered
to the Sun God, Surya,
elaborate drawings (*kolam*)
made of rice flour are
created in front of village
houses. Making the *kolam* is
a form of devotional prayer
(*puja*). While many Tamil
women make fresh *kolam*
every day of the year, special
care is taken on Thai Pongal
to make elaborate designs.
Photograph by Roy W.
Hamilton, Damal,
Kanchipuram District,
Tamil Nadu, India 2000.

provides an opportunity to show off marriageable daughters in a beauty pageant. The procession in Java, while celebrating the Prophet's birthday, also showcases the generosity and grandeur of the sultanate. The offerings to the Buddha in Thailand also feed the monks of the monastery for the coming year and earn religious merit for the donors. While each individual experiences the festival in his or her own way, the mass of celebrants together reaffirms the community and its values. Nor do the humans act alone, as in most festivals the deities also are called into attendance.

No matter what form a festival takes, or what time of year it is held, the stated goals very often include giving thanks for harvests past and expressing the desire for successful crops in the future. Because the fertility of the rice crop is closely associated with the fertility of humans, the festival also becomes an expression of the wish for continuity of the human community. Chapters 11 through 17 describe a wide variety of festivals in several different countries, yet these same goals underlie every one of them. Often there is a conscious attempt to perform the festival in exactly the same manner year after year as an expression of this continuity, to do it, in short, in the same way the ancestors did it. This fiction may be maintained even in the face of plainly evident changes as the performance of the festival inevitably evolves over time.

While these wishes for a good harvest and the perpetuation of the human community may be expressed in festivals at any time of the year, they seem tailor-made for New Year celebrations in particular: as the cycle turns from one year to the next, out with the old, with thanks for its success, and in with the new, with great hopes for the future. New Year is tremendously important in many Asian societies and often is the single most

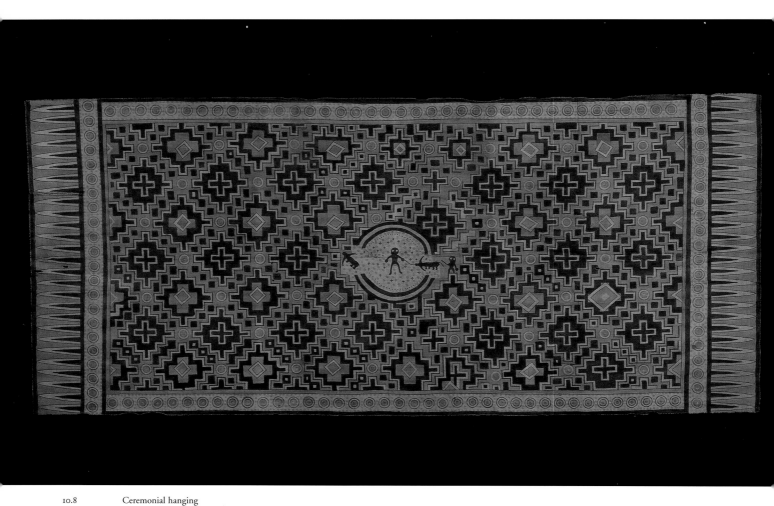

10.8   Ceremonial hanging
(*mawa*). Sa'dan Toraja peoples. Sulawesi, Indonesia.
Cotton cloth with hand-
painted pattern. L: 178 cm.
FMCH TR2002.20.25; E.
M. Bakwin Collection.

Mawa textiles are hung at
Toraja festivals as a key element in the creation of a
sacred atmosphere. In many
mawa designs the rectangular field of the cloth represents a rice field, identified
by the circular buffalo wal-
low at the center of the field.
In this cloth the rice field is
filled with geometric motifs
borrowed from highly val-
ued Indian trade textiles, a
fittingly honorific treatment.
Another rice field *mawa*
appears in figure 15.18.

10.9   The central ceremonies of the Erau festival, in Borneo, take place in the former palace of the sultan of Kutei, now a museum in East Kalimantan, Indonesia. The festival is based on an annual practice established in the time of the sultanate that involved representatives from the various groups of people living in the interior of the island coming to the palace to give witness to the rites of the state. In 1959 the Indonesian government abolished the last of the Borneo sultanates remaining in Indonesian territory, and the last sultan of Kutei, Parikesit, relinquished authority. Since the 1970s the Erau festival has been supported by the provincial government, and the rites in the former palace have been conducted by a nephew of Parikesit. They closely follow the rites as conducted during the late colonial period, which were in turn heavily influenced by Javanese rituals of state. The most important rituals are conducted on a mat decorated with dragons or serpents (*naga*, a royal symbol) and other motifs made of brightly colored uncooked rice, surrounded by various offerings. Photograph courtesy of Mulawarman State Museum, Tenggarong, circa 1994.

important celebration of the year. Asian New Year celebrations are usually held at times that differ from the date specified in the Western calendar (Japan is an exception, having adopted the Western calendar after 1868). In the more temperate regions they tend to fall during the cooler season from November to February, when one year's rice crop has been harvested and preparations for the next crop are about to begin. Some examples include the Hmong harvest and New Year celebration in November, the Tamil celebration of Pongal in mid-January (see box, p. 156), and Vietnam's Tết, which usually falls in February according to the Vietnamese lunar calendar. Chapters 13, 16, and 17 document other rice-related festivals that occur during this period.

There are many exceptions to this timing, however. Due to the different monsoon cycle on the west coast of India, the people of Kerala celebrate their New Year and harvest festival, Onam, in August or September. The closest equivalent to New Year in Thailand is Songkran, held in April at the hottest time of year when the changing monsoon winds are about to unleash their life-giving rain. Bali follows its own 210-day annual cycle, which means that Nyepi, the "yearly" day of silence and renewal, comes at a different time in the Western calendar every cycle. Moreover, the double and triple cropping that is now possible with fast-maturing modern varieties of rice means that there is often no single annual harvest or renewal period. Many farming communities have accommodated this change by pegging their harvest festivals to the largest harvest or the traditional harvesttime for their indigenous varieties of rice.

It is hard to imagine any festival in the rice-growing regions of Asia that does not in some way involve rice. At the most basic level, it is nearly certain to be a part of the feasting that accompanies the festival, no matter what the purpose. Very often it also fills a more central role as a sacred offering, either in the form of some type of prepared rice dish or as rice wine. If the festival is related to the agricultural cycle, rice is in fact likely to be a major focus of the activities. In the Japanese ceremonies known as "rice field play" (*ta-asobi*), held in some shrines following the New Year, the complete rice-growing cycle is ritually reenacted. In countless ways, then, rice becomes a focus of celebration and a means of enacting community.

•

11

The Pahiyas Festival
of Lucban, Philippines

Roy W. Hamilton

The Pahiyas festival is held annually in the town of Lucban to celebrate the harvest of rice and other crops and to honor San Isidro Labrador, the patron saint of agriculture. San Isidro (also known as Ysidro, Isedro, or Isidore) is a Spanish saint who is thought to have died around 1130 and was canonized by the Roman Catholic church in the sixteenth century. An illiterate but pious plowman, he became the patron saint of Madrid as well as of agricultural workers wherever they toil. Versions of Isidro's story differ in character, depending more on the political views of the teller than on any known facts of the saint's life. By some accounts, Isidro was a simple tenant farmer so pious that he spent long hours in the fields praying on his knees. His piety was rewarded by an angel, who plowed the saint's field for him. In more radical versions of the story, the overworked plowman collapses from exhaustion after being hard pressed by his cruel and greedy landlord. When he awakens, he finds an angel of mercy in the field finishing the work he has been unable to complete. In either version, the landlord visits the field and witnesses the miracle. He either falls on his knees in adoration of the pious plowman or to beg forgiveness for the error of his ways.

San Isidro's story has been depicted repeatedly in European artistic traditions, typically as a rather standardized set of the three figures: Isidro, the landlord on his knees, and the angel with the plow. This imagery was carried abroad by Spain's colonial expansion, especially to Mexico and the Philippines, and depictions of San Isidro are common in Philippine art in the forms of paintings, *retablos*, and carved *santos* figures (fig. 11.4). Prior to the mid-nineteenth century, the Philippine renditions tend to be quite idiosyncratic, with a strong local or regional flavor. Later in the nineteenth century, Philippine portrayals of the saint become more Europeanized as a result of increasingly close communication between the provinces and the metropolitan culture of Spain. The painting of San Isidro from the island of Cebu (fig. 11.2) can be dated on this stylistic basis to about 1880.[1] It shows a fairly typical European depiction of the Isidro story, with an important departure: the fields are not the dry grain fields of the Iberian Peninsula. They are bunded rice fields, and the angel drives not an ox but a water buffalo. This seemingly straightforward painting is in fact a remarkable cultural hybrid attesting to the transplantation of a European Roman Catholic saint into the fertile rice lands of Southeast Asia.

11.1   A Lucban house decorated with rice garlands and colorful hanging "lanterns" (*aranyá*) made of rice flour. Photograph by Roy W. Hamilton, 2000.

11.2   *San Ysidro.* Obtained from
a house in Carcar, Cebu.
Circa 1880. Oil on canvas.
H: 48 cm. FMCH
X2001.4.1.

11.3   Two girls in dresses decorated with beans echo the harvest theme of the festival as they ride on a float of vegetables. The floats are part of a competition, each one organized by a different neighborhood (*barangay*). Photograph by Roy W. Hamilton, Lucban, 2000.

11.4   *San Isedro Labrador.* Philippines. Carved and painted wood, plant and animal fibers. H: 72 cm. Collection of Don R. Bierlich.

Although it is missing the plow, this *santo* figure is beautifully made, with expressive features and even fine eyelashes. The artist and exact point of origin are unknown.

11.5   Santa Cruzan processions consist of a number of "queens" (*reyna*) in formal dress accompanied by their escorts. This nighttime procession was organized by a *barangay*. Portable arches bearing neon lights—strung together with electric cable and powered by a mobile generator—were pulled behind the procession on a cart. Photograph by Roy W. Hamilton, Lucban, 2000.

The saint's day for San Isidro is May 15, when European fields might well be under plow. Agriculture in most parts of the Philippines is year-round, but May marks the hottest time of year, locally called "summer," before the rains set in. If there is any time of year associated with the bounty of harvest, it would be this season when the first crop of rice is harvested and the production of vegetable crops reaches its peak. In the rich volcanic uplands of Quezon Province, San Isidro's day has evolved into the most famous harvest festival in the Philippines. Versions of the Pahiyas festival are found in other towns as well, but the largest celebration takes place in Lucban. This old colonial town was founded at the foot of Mount Banihaw in 1578, and the massive cathedral that dominates the center of the community was built in 1639. Many people from Manila enjoy traveling the 130 kilometers by road to Lucban for the festival, as the town is viewed somewhat nostalgically as a center of southern Tagalog culture with strong roots in the colonial and agricultural society of the past. Lucban today is densely populated and urban in its feel, but the rice fields still begin at the edge of town and many among the town's residents work daily in agriculture. In addition to rice, the town specializes in the production of market vegetables.

Lucban's festival has been growing in popularity in recent years, attracting ever larger crowds from Manila. The sequence of events varies somewhat from year to year depending on the circumstances. The festival of 2000 featured a series of ancillary events

11.6   San Isidro presides in front
of an artificial church
facade made entirely of rice
garlands. Photograph
by Roy W. Hamilton,
Lucban, 2000.

11.7   *Kiping*, with their leaf-like form evident, show their colors best in the bright sunlight. Photograph by Roy W. Hamilton, Lucban, 2001.

over a three-day period and drew a huge crowd despite unusually rainy weather. The full roster of activities included competitions for agriculturally themed floats (fig. 11.3) and three separate processions known as Santa Cruzan. Part fashion show and part beauty pageant, Santa Cruzan are stylized reenactments of the legendary fourth-century search for the remnants of the true cross in Jerusalem by Queen Helena (Reyna Elena), the mother of Constantine I. Traditionally neighborhoods (*barangay*) organized the processions in part as a means of publicly announcing girls who had reached marriageable age (fig. 11.5). The Pahiyas festival of 2000 included one such Santa Cruzan and two others that were more unusual: one for senior citizens and another consisting of young men and women from California who were the descendants of Lucban emigrants. In 2001 no Santa Cruzan was performed, and most people agreed that the entire festival was much quieter because it fell right at the time of a critical national election.

11.9   Urban tourists pose with an imagined rural landscape. These miniature rice terraces, complete with running water, are presided over by San Isidro. Photograph by Roy W. Hamilton, Lucban, 2001.

11.8   San Isidro, the angel, and the kneeling landlord, surrounded by garlands of rice. Photograph by Roy W. Hamilton, Lucban, 2000.

11.10   Water buffalo are honored in the parade for their vital role in agriculture. Photograph by Roy W. Hamilton, Lucban, 2000.

What draws the biggest crowds to Lucban, however, is the way in which the houses that line the streets of the town are decorated to celebrate the saint's day. The name *Pahiyas* is in fact derived from the root word *hiyas*, which means "gem" in standard Tagalog. Thus *pahiyas* is "to make gem-like" and is also related to the Malay root *hias*, or "decoration." The decorations are the most dense along the parade route designated for the festival each year by the organizing committee. The route is changed from year to year so that every few years the parade will pass down any given street, giving all residents of Lucban equal opportunity to participate in the decorating, as well as a chance to win the 30,000 *peso* (U.S.$600) grand prize for the best-decorated house. The prize generates a great deal of excitement and competitive spirit, as homeowners vie to decorate their houses in the most spectacular manner (fig. 11.1).

The basic ingredients for the decorations are garlands of rice and colorful leaf-like wafers of rice dough called *kiping* (fig. 11.7). *Kiping* are an edible snack, but they appear in the festival primarily as decoration. They are made by dipping large leaves into a thin gruel of rice flour that has been brightly colored with food dye. The leaves are then steamed, and after they have been dried in the sun, the wafer can be readily separated from the leaf that has served as its mold. The veins of the leaf can still be seen in the *kiping*. Most *kiping* are made on the oval leaves of the *kabal* tree, but sometimes square shapes are made using cut pieces of banana leaf. The *kiping* are suspended from the eves of the houses in tiered or spiraled clusters known as *aranyá*, or lanterns.

The masses of colorful *aranyá* lend a festive atmosphere to the street, but much of the decorative genius is devoted to smaller installations celebrating the themes of the festival, especially the story of San Isidro (figs. 11.6, 11.8). Rice straw and harvested vegetables are used to fashion other decorations in imaginative forms. For most of May 15th the parade route is thronged with local residents and tourists, who often pose for photographs with the decorations (fig. 11.9). In late afternoon, the crowd parts to let the official parade through the streets. The parade features sleds drawn by buffalo (fig. 11.10), more beauty queens, and *gigantes* (performers on stilts). Many observers fall into line and join the procession as it moves through the packed and sweltering streets.

For the duration of the festival a boisterous atmosphere prevails, with vendors stalls and food stands erected in the town plaza and the surrounding streets. A life sized grouping of *santos* figures representing San Isidro, the landlord, and the angel are on display inside the portal of the cathedral. These figures are privately owned, however, and were not carried through the streets as part of the procession in 2000 or 2001.

•

12

The Gods Walk on Rice in Selat, Bali

David J. Stuart-Fox

There are no Balinese rituals, not even grand cremations, more spectacular than the great periodic temple festivals held at intervals of five or ten years in some parts of Bali. Certain villages in the easternmost Karangasem region, on the flanks of Bali's highest mountain, the volcano Mount Agung, enact these rituals in their own unique ways. Fittingly for such a productive rice-growing area, rice plays a prominent role in these rituals. In Selat, gods walk along a path of rice laid on the ground; in Sibetan, rice is fashioned into fantastic wishing trees of heaven hung with hundreds of brightly colored rice cakes. Rice, the most important foodstuff, is the medium through which devotees offer gratitude to their gods and request prosperity and physical and spiritual well-being and fulfillment.

Selat, whose decennial festival is illustrated and discussed here, is one among several very old villages along the southern flank of Mount Agung. Ancient bronze inscriptions are found in several villages; that concerning the village Kanyuruhan, the old name for Selat (where the inscription is still kept), dates from the year 1181. The territory of these villages extends from the bare peak of the 3014-meter mountain, through high-altitude dry farming gardens, down to rich irrigated rice-growing lands at lower levels. In earlier times, rainfed rice was also grown at the dry farming levels. The core settlement of these villages lay within or close to the lower irrigated areas, with small subsidiary hamlets at higher altitudes that retained close links with the original settlement.

The old core villages of this region each possess, besides many others, two key temples: *pura puseh*, the "navel" temple, often glossed as "temple of origin," and *pura dalem*, or "death temple" from its special connection with funerary rites. The main shrine of the *pura puseh* is usually a multi-roofed wooden pagoda (*meru*). In Selat and some other nearby villages, however, the main shrine is a three-roofed structure called *kehen* or *gedong kehen*, like the *meru* but rectangular rather than square in shape. There, it is situated on the east side of the temple, with its shrines facing west. Another important building is the long open-sided pavilion (*bale agung*), where full village members sit on special ritual occasions. These village temples are the focus of yearly and multiyear ritual cycles. A major festival is generally held every year, although sometimes a smaller one alternates with a larger one. The multiyear cycle culminates in great festivals held every five or ten years, such as those in Sibetan and Selat.

In Bali, and indeed elsewhere, oftentimes particular ritual features, over and above the basic ritual, give a ceremony its meaning or purpose. A common feature is the empowering of a substance or object associated with a certain ritual. As we shall see, the

12.1 For the Pacayan procession when the deities are moved from the Pura Bale Agung to the Pura Puseh, the pathway is prepared with uncooked rice that has been dyed yellow with turmeric. Led by a girl bearing offerings, the procession walks along this pathway of yellow rice. Photograph by David J. Stuart-Fox, Selat, 1978.

12.2   Before the Pamijilan procession, a special offering called *titih mamah*, consisting of the hide of a sacrificed buffalo, is laid out at the Pura Bale Agung in front of the shrine where the deities will take up their temporary residence. Here it is seen the following morning before the procession begins. When the procession reaches this point, the deities will be carried over the hide and up into the shrine. Photograph by Ann Thompson, Selat, 1998.

12.3   Groups of participants organized by neighborhood committees prepare the pathway over which the deities will travel in the Pamijilan procession, covering a white cloth with cooked white rice and offerings. The prepared pathway stretches for 350 meters and is entirely covered with cooked rice and offerings. Photograph by Roy W. Hamilton, Selat, 1998.

12.4   Just before the Pamijilan
            procession leaves the Pura
            Sakti, the observers sit
            reverentially along the sides
            of the rice pathway with
            their faces turned in expecta-
            tion of seeing the gods make
            their appearance. Photograph
            by David J. Stuart-Fox,
            Selat, 1978.

12.5   The Pamijilan procession
of 1998 was led by a group
of children from the local
community, dressed in formal
Balinese attire. Photograph
by Ann Thompson, Selat.

12.6 During the Pamijilan procession, a special offering that precedes the inscriptions is carried on its own palanquin. Photograph by David J. Stuart-Fox, Selat, 1978.

"rice pathway" of the Selat ritual is a case in point. In Bali, ordinary substances that are ritually empowered include water (there are various sorts of holy water), certain food offerings (or other components of offerings), and, in the case of Selat, rice. One constant theme of Balinese ritual is the request or supplication for prosperity and well-being, both those of the worshiper and those of the supporting natural and material world, in particular the production of foodstuffs.

The rice-growing cycle is accompanied by an intricate series of rituals. Besides these, empowered substances from a variety of village and temple rituals are brought into contact with field shrines or the fields themselves. Often these empowered substances have a rather general purpose, but sometimes they have a more specific function. For example, there is a ritual (Usaba Nyungsung) at a particular shrine in one of the Besakih temples that is expressly intended to protect fields against insect pests. After the ritual, specially prepared rice porridge, holy water, and other ritual objects are offered at field shrines or simply cast over the fields.

The essence of such empowering is the contact made, through ritual means, between deity and substance. Holy water, for example, is sometimes called *wangsuh pada*, or "water from washing the feet of the deity"—a metaphor for bodily contact. A similar metaphor comes into play in the rice pathway of Selat where a deity, through the bearers, comes into contact with the rice, which is then considered to have divine power.

The decennial festival of Selat is held when the Saka year (seventy-eight years behind the Gregorian calendar) ends in a zero. No one knows when it was first held. The festival held in 1938 (Saka 1860) was described in detail by Roelof Goris (Goris 1939; Goris 1969).[1] The author witnessed it in 1978 (Saka 1900); see also Setia (1978). It was most recently held in 1998 (Saka 1920); see Supartha (1998).

The festival consists of a series of major rituals, themselves each composed of subrituals. The ritual of the rice pathway, the most significant and spectacular among them, should be seen in the broader context of the whole festival, which lasts for some two months. The festival involves, above all, two deities whose material symbols are bronze inscriptions (*prasasti*). These inscriptions were sanctified and over the centuries became the symbols of deities. The most important deity is Bhatara[2] Angerta Bumi,

whose name is derived from *kerta*, meaning prosperous or flourishing, and *bumi* or *gumi*, meaning world or earth, i.e., to make the world prosper or flourish. Bhatara Angerta Bumi is the deity of the older of the two inscriptions, consisting of seven bronze plates, issued by King Jayapangus in 1181 (Saka 1103). In this inscription the village is not called Selat (or Baledan, another old name of the village), but Kanyuruhan. This name reminds one of the Selat official called Ki Panuruhan. In the *Raja Purana Pura Besakih*, the important lontar palm-leaf text dealing with many aspects of Pura Besakih, Bali's paramount temple, he is responsible for the administration of Besakih's rice lands (*pelaba*), which were mostly located within Selat territory. Bhatara Angerta Bhumi is the main deity of the temple Pura Sakti (or Pura Sakti Bingin, after its location in the village ward called Banjar Bingin). The inscription is brought out from the Pura Sakti at festival times, though normally housed elsewhere for safekeeping. In charge of this deity and temple, and indeed, through control of special village lands, of the whole festival itself, is the *kabayan*. His home is located right next to the temple. The *kabayan* is one of the most important members of the traditional village leadership.

The other deity, Bhatara Ratu Putra (*putra* means "son"; he is thought of as the "son" of Bhatara Angerta Bhumi) is the deity of the more recent inscription, which consists of five plates. It is apparently undated, though no earlier than the fifteenth century. The wording of this inscription is for the most part similar to a section of the same lontar text dealing with Pura Besakih. At least since the time of the Gelgel dynasty (fifteenth to seventeenth century), if not earlier, Selat has had a special and close relationship with Pura Besakih.

The festival consists of four major rituals. The opening one is the Tabuh Gentuh, a sacrificial and purificatory or cleansing ritual conducted by Brahmin high priests. A buffalo is sacrificed, among other animals. In the hierarchy of such sacrificial rituals (Bhuta Yadnya), a buffalo is used only in the larger and more important rituals. The aim of the ceremony is to cleanse the village territory of impurities.

The second major ritual is the Pamijilan (from *mijil*, to come out, to come forth), or in full, Pamijilan Bhatara Angerta Bumi, referring to the coming forth of this deity. The rice pathway is part of this ceremony. The name of the deity Angerta Bumi (and by extension the ceremony that bears this name) reflects one of the most frequently expressed aims of Balinese ritual, the bringing about of prosperity. For the Balinese, prosperity depends on the success of the harvest, particularly of rice, the staple food crop. The laying down of rice along the gods' pathway is a direct way to ensure this prosperity, for the rice, blessed by the gods, is returned to the rice fields.

On the day before the main day of the Pamijilan, the inscriptions of Bhatara Angerta Bumi and Bhatara Ratu Putra are brought to Pura Sakti from their respective places of storage. There, the two inscriptions are ritually purified using a variety of purificatory substances: different kinds of holy water, a specially prepared rice flour mixed with other ingredients (*tepung tawar*), and ritually empowered coconut oil.

The highlight of the Pamijilan is the procession carrying the two deities, embodied in the inscriptions, from the Pura Sakti, along the main village street that runs downhill, to the Pura Bale Agung. The Pura Bale Agung is situated at the downridge (*kelod*) end of Selat, in the market area. It is hardly a temple in the normal sense of the word; it has no surrounding wall and no elaborate gateways, and it consists only of the *bale agung* with its own shrine at the end toward the mountain.

For this ceremony the road from the Pura Sakti to the Pura Bale Agung—a distance of about 350 meters—is lined with decorative bamboo poles (*penjor*). Elaborate offerings, fashioned in the ritual kitchen attached to the Pura Puseh, are laid out on the

12.7   The *kabayan*, ringing a bell, heads the dense phalanx of white-garbed priests who bear the inscriptions down the pathway in the Pamijilan procession. Photograph by Roy W. Hamilton, Selat, 1998.

12.8   In the Pacayan procession, a villager carries one of the holy inscriptions that symbolize the gods, shaded by umbrellas of honor. Photograph by David J. Stuart-Fox, Selat, 1978.

12.9   The gods are installed in the *kehen*, the main shrine of the Pura Puseh. Photograph by David J. Stuart-Fox, Selat, 1978.

12.10   As in the earlier procession, after the gods have passed, the villagers scramble to gather up the sanctified grains of rice. Photograph by David J. Stuart-Fox, Selat, 1978.

ground beside the Pura Bale Agung; these include the buffalo offering called *titih mamah* (fig. 12.2) and meter-high male and female figures, called *pering*, fashioned from sugar-palm leaves. In the early afternoon, villagers lay down white cloth and palm leaf mats along the whole length of the road (fig. 12.3). On top of the cloth and mats they place boiled white rice (*nasi*), meat dishes (*olahan*), and small offerings (including the everyday offering *canang*, consisting of betel leaf, areca nut, and lime, along with flowers, and *panca pala*, whose ingredients vary by region). There, squatting by the roadside in deference to gods, they wait patiently for the procession to pass (fig. 12.4). Led by women and men bearing offerings, and the *kabayan* ringing his bell, the palanquins of the two deities—the inscriptions within wrapped in white cloth—are borne along the rice pathway, into the Pura Bale Agung, over the buffalo *titih mamah* spread out at the steps to the temple, between the *pering*, and into the shrine at the end of the *bale agung* (figs 12.5–12.7). As soon as the procession passes, villagers rush and jostle each other to gather up the rice and meat dishes, for these are now blessed by the gods who have walked upon them. Much of the rice is eaten on the spot, or it can be carried home as a blessing or scattered over the fields to promote their fertility. Within a few moments, the pathway is completely clear of the elaborate offerings that have been laid out. At the Pura Bale Agung, a Brahmin high priest, sitting in the *bale agung*, conducts a ritual for the deities who have taken up residence.

The gods remain in residence in the Pura Bale Agung for a period of thirty-five days until the third ceremony takes place. This is the Pacayan, derived from the word *caya*, meaning splendor or radiance. For this ritual a new pathway of mats and white cloth is laid down from the Pura Bale Agung right to the foot of the *kehen*, the central shrine of the Pura Puseh, a distance of more than a hundred meters (fig. 12.1). Instead of the cooked white rice used for the Pamijilan pathway, the villagers spread out uncooked yellow rice (*beras kuning*), the yellow color being derived from turmeric (*kunyit; Curcuma domestica*), together with Chinese coins and small offerings. The yellow and white rice in these two rituals reflects the widespread use of these two colors in objects and offerings associated with or intended for deities. God-figures are commonly wrapped in white and yellow cloth, and certain important offerings contain white and yellow rice cakes or cookies (*jaja*). The use of both cooked and uncooked rice is likewise commonly found in many offerings and suggests completeness.

As in the Pamijilan procession, the cloth-wrapped inscriptions are once more carried over the path of rice (fig. 12.8). Villagers again scramble madly to obtain some of the rice over which the two gods have walked (fig. 12.10). After the two gods are installed in the *kehen*, temple priests and Brahmin high priests conduct their respective rituals, and then the assembled villagers offer the gods their worship (fig. 12.9). The day's events end with a ritual meal in the Pura Puseh's *bale agung* for village leaders and those men who have full village membership (based on special landholdings).

The final ceremony, relatively simple compared with those already held, is the Panyimpenan, the closing ritual. The two gods, this time without the rice pathway, are carried back to the Pura Sakti, and the final farewell is made.

•

13.1   Monks and novices build
          the bonfire in front of the
          wihăn at Wat Paá Daád. The
          wood is bound in place with
          lianas, and then fresh
          bunches of green leaves and
          lengths of bamboo are
          inserted. As the fire burns,
          the bamboo will explode like
          firecrackers and the leaves
          will crackle merrily. Photo-
          graph by Roy W. Hamilton,
          Mae Chaem, 2000.

13

Offering the New Rice to the Buddha in Mae Chaem, Northern Thailand

Roy W. Hamilton and Thitipol Kanteewong

13.2   A maker of *khanom niap* places a lump of grated coconut sweetened with brown sugar into a shell of glutinous rice flour dough. Photograph by Roy W. Hamilton, Mae Chaem, 2000.

Mae Chaem is a rural district in Chiang Mai Province occupying a fertile valley surrounded by forested hills not far from the Burmese border. The villages of the valley floor are inhabited by the Khon Muang, the majority population of northern Thailand's plains and valleys.[1] Mae Chaem is reached by a three-hour drive from the city of Chiang Mai, along a road that winds over the shoulder of Doi Inthanon, Thailand's highest mountain. The descent from the heights provides a panoramic view of the valley far below, a scene that changes dramatically with the passage of the seasons. During the months from July to October, the intense green of the fields of growing rice covers the valley from side to side.[2] This is the season of the monsoon rains in northern Thailand, which fill the bunded, or diked, fields of the valley to overflowing.

In November, when the fields have dried and turned golden, the harvest gets under way. By the time the coldest days of the year arrive, from late December to late January, only brown stubble remains and the valley presents a barren look. This is the month of Sii, the fourth month of the lunar calendar used by the Khon Muang. The fifteenth day of the month brings the full moon, called Sii Peng, a special day throughout the valley. In the chill of the predawn hours, each community lights a huge bonfire in front of its *wihăn*, the hall in the local monastery (*wat*) that houses the huge image of the Buddha.[3] On this night, reckoned to be the coldest of the year, the radiance of the fire penetrates the *wihăn* and sets the gilded surface of the Buddha aglow, bathing it in a warm light. As dawn breaks, the people present the first offerings of their newly harvested crop to the Buddha and also to their ancestors, to the spirits, and to the monks and novices of the *wat*. It is a day awaited with anticipation, not only for the festive atmosphere but also because the new crop of rice is not supposed to be eaten until it has been ceremonially presented in this manner.

Mae Chaem has a reputation as a place where the true northern Thai traditions are still maintained. The people of the valley say that this is because there is a large population of conservative elderly people who are still interested in such traditions. In contrast to the rapidly urbanizing environs of the city of Chiang Mai, the quiet rural atmosphere indeed seems remarkable. Rites such as the ceremonial offering of the first of the new crop to the Buddha on Sii Peng would once have been common throughout northern Thailand, but they are now rare in Chiang Mai Province. Though urbanization has weakened many such traditions, closely related practices survive to a greater degree in

13.3   The *wihān* at Wat Paá
Daád, and the gilded
Buddha inside it, glow in
the reflected light of
the bonfire. Photograph by
Roy W. Hamilton, Mae
Chaem, 2000.

13.4   The bonfire burns in front of Wat Paá Daád. Photograph by Roy W. Hamilton, Mae Chaem, 2000.

surrounding regions with Tai populations that have been relatively less affected by urbanization, including Isan (northeastern Thailand), Laos, and the Shan region of Burma. Mae Chaem as yet has no substantial tourist development. People from Chiang Mai and even from Bangkok, however, occasionally come just to enjoy the traditional rural atmosphere and the surviving rituals and crafts.

The preparations for Sii Peng begin a few days in advance, when the monks and novices who reside at the monastery gather firewood in the surrounding hills. The day before the ceremonies, the wood is formed into a huge conical pyre a short distance in front of the *wihăn* (fig. 13.1). Sometimes a special packet of flowers and leaves is tucked into the pyre as a form of *sŭma*, the asking of forgiveness from the spirits before doing something—in this case the spirit of the land where the fire will soon burn. At Wat Yang Hlŭang, one of the four monasteries in the valley, a symbolic cotton string of nine strands is run from the pyre to the Buddha image within the *wihăn*. This will burn when the fire is raging, symbolically conveying its warmth to the Buddha.[4]

Meanwhile, the people of the villages are busy, catching catfish in small ponds and at home making a variety of sweet treats from glutinous rice. These foods will be major components of the offerings presented on Sii Peng. As in many parts of Southeast Asia, rice and fish are seen as complementary, although in this case the sweet rice treats are far fancier than the plain steamed glutinous rice that is the daily fare in northern Thailand. Collectively the various types of sweet rice cakes are known by the expression *khao lam khao jee*. *Khao lam* are cakes made of glutinous rice cooked in tubes of bamboo. *Khao jee* are balls of glutinous rice threaded onto a skewer and toasted over the coals. This expression, however, is a holdover that has not kept up with changing tastes. Although *khao jee* and the rites of Sii Peng are commonly associated, in fact *khao jee* are rarely made today in Mae Chaem and no longer appear on Sii Peng. Some *khao lam* appear, but most of the treats actually offered are fancier types known as *khanom niap* and *khanom jaok*. Both of these are made with glutinous rice flour, which can now be conveniently purchased in shops rather than laboriously ground in a foot-powered mill. *Khanom niap* are steamed balls of rice-flour dough filled with a lump of grated coconut and brown sugar at the center (fig. 13.2). *Khanom jaok* are made of a similar sweetened dough, wrapped into a triangular shape in a piece of banana leaf and steamed.

Four separate but interrelated rites are conducted on the morning of Sii Peng. The first, called Híng Hlŭa Phà Châo (lit., "to warm the Buddha with fire"), is the bonfire itself (figs. 13.3, 13.4). The lighting of the fire is accompanied by the burning of incense sticks and the saying of prayers, another form of *sŭma*. As the fire grows in strength, tended by the monks and novices, the community members begin to assemble at the monastery, bringing the offerings they have prepared at home and also sacks of uncooked rice and bowls of cooked rice. The uncooked rice is piled in front of the Buddha image and plates of cooked catfish and glutinous rice cakes are arranged nearby.[5]

The most elaborate form of offering consists of a triangular pedestal tray containing rice cakes, flowers, and incense sticks. The tray is surrounded by three smaller pedestal bowls that represent the three fundamental Buddhist elements consisting of the Buddha, the scriptures, and the monkhood. The entire arrangement is known as *khăn keâw*. The people who bring these special offerings position them before the Buddha on the steps of the *wihăn*, light the incense sticks, and recite prayers (fig. 13.5). This is the second formal rite, known as Sai Khăn Keâw (lit., "to place the *khăn keâw*"). It is completed just as dawn begins to light the sky.

13.5   Women arrange the *khǎn keâw* offering at Wat Paá Daád. Photograph by Thitipol Kanteewong, Mae Chaem, 2003.

13.6   Sacks of donated rice piled in front of the Buddha image at Wat Paá Daád. Photograph by Roy W. Hamilton, Mae Chaem, 2000.

13.7   A congregant holds her offering bowl of glutinous rice aloft. Photograph by Thitipol Kanteewong, Wat Yang Lúang, Mae Chaem, 2003.

13.8   Offerings placed before the monk who acts as a spirit medium. Photograph by Thitipol Kanteewong, Mae Chaem, 2003.

13.9   Plates of rice cakes steamed in banana leaves (*khanom jaok*) and other food offerings. Photograph by Roy W. Hamilton, Wat Paá Daád, Mae Chaem, 2000.

13.10   Filling the line of bowls with rice. Photograph by Thitipol Kanteewong, Wat Yang Lúang, Mae Chaem, 2003.

The third rite is quite esoteric and is more properly an act of spirit worship than an official rite of Buddhism. It is known as Taan Khan Kâo (lit., "to offer a full meal"). For this the members of each household fill a plate with a meal consisting of cooked rice and fish, plus the sweet rice cakes prepared the previous day. These plates are taken to a special monk who also acts as a medium to contact ancestors and spirits. The participants provide the medium with the names of their ancestors on slips of paper. He prays, contacting certain spirits (of rice, of the land, of the river, etc.) and also the specific ancestors whose names have been provided (fig. 13.8). At the end of the rite the medium blesses the people who have offered the food to the spirits and to their ancestors.

The fourth rite is called Yoong Khâo Bàat (lit., "to raise an alms bowl of rice"). For this, which is the central rite of Sii Peng, the entire community congregates and sits in the presence of the Buddha image. The abbot, the monks, and the novices sit on a raised platform along the side of the hall. It is still very early, before the sun has risen completely over the hills. The piles of uncooked rice have been arranged at the base of the Buddha and a large array of plates stacked with sweet rice cakes has been displayed nearby (figs. 13.6, 13.9). Each person in the congregation sits with a bowl of cooked glutinous rice in front of them. First they hold grains of popped rice and flowers between their hands, positioned with palms together as though in prayer. Then, as the monks chant from the Buddhist scriptures in the archaic Pali language, the congregants hold the bowls of cooked rice aloft as an offering (fig. 13.7). A designated layman from the congregation, who has had enough former experience as a monk to be familiar with Pali texts, recites a short benediction: "On this occasion, we present this rice and other offerings to the monks. We all wish to respect Buddhism."[6] The congregation repeats this phrase after him.

As the monks and novices continue to chant, the congregants arise with their bowls of rice and form a line reaching out onto the temple grounds, where a long trestle table has been set up. Stretching down the table is a double row of bowls. The first few bowls are genuine Thai alms bowls, but the rest are large enamel basins that will hold more rice. Each person in the line puts a small ball of glutinous rice from his or her bowl into the first bowl on the table, then moves on to put a similar ball into the next bowl, and so on, until all the participants have distributed all of their rice and all of the bowls on the table are brimming (fig. 13.10). At some monasteries the participants bow to the line of bowls after distributing their individual contribution. The filling of the many separate bowls bit by bit until they are overflowing symbolizes abundance, and also community solidarity, as all must participate together to fill the bowls.

Finally the many bowls of rice, along with plates of fish and sweet rice cakes, are all brought before the row of seated monks and novices, to whom they are donated. Some of the food will be consumed later in the day, and the remainder given to the poor. Frequently small numbers of the minority group populations from the hills come to the monasteries later in the morning to request portions of the food. The uncooked rice piled before the Buddha provides the monastery with food to be used throughout the year, and also a surplus that can be sold for cash. The congregants return home, having warmed the Buddha on the coldest day of the year and earned religious merit by providing for the monks for the coming months. Most importantly, by having offered the first of their new crop to the spirits, their ancestors, and to the Buddha, they can now eat themselves from their newly harvested store of rice.

•

14

The Ghost Festival
of Dan Sai, Loei Province, Thailand

Ruth Gerson

Rice is the staple food of Thailand, and nearly 55 percent of it is grown in the northeastern region known as Isan, a parched area of the country that suffers from frequent droughts. The farmers who live there have come to respect the power of nature in the form of raging seasonal floods, the onslaught of pests, and most of all a devastating shortage of rain. To ensure plentiful rainfall, the people of Isan try to appease the spirits of nature and elicit their help in creating favorable conditions for the forthcoming harvest. They do this through various festivals that are held in local villages, raucous affairs that dramatically play out the theme of fertility. One such celebration is the Ghost Festival held annually at the start of the rainy season in the village of Dan Sai in Loei Province.

Nestled in a remote valley surrounded by the tall mountains of Loei, close to the Lao border, Dan Sai is the stage set for the Ghost Festival (Phi Ta Khon, lit., "masked ghosts"). Characterized by elaborate, costumed figures, this event is not held anywhere else in Thailand or its neighboring countries. A blend of spirit beliefs and Buddhism, as well as a coming of age ceremony for young men, the Ghost Festival is a colorful and lively celebration.

The origin of the festival and the legend in which it is swathed can be traced back about four hundred years to the Ayutthaya period (1350–1767), according to residents of Dan Sai, as will be discussed later. As Dan Sai is situated on the border of Thailand and Laos, the festival reflects the cultures of both countries. The legend tells that a young man named Kuan and a beautiful young woman called Nang Tiem fell in love. As love stories go, Nang Tiem was promised in marriage to an older man, so her love for Kuan was doomed. But the lovers continued to meet secretly in a cave by the river, hidden under a shrine, until one day an earthquake struck and they perished. The spirits of the lovers have survived through this legend to become the guardians of the village of Dan Sai. They are accorded the much-respected name of Chao Saen Muang and are thought of as responsible for the well-being of the inhabitants of the village. A Buddhist shrine was erected over the fabled site of their meetings and is named Sri Song Rak (Sacred Place of the Two Lovers).

These two ancient spirits are regarded as harbingers of fertility and have been represented by a long line of male and female village mediums who are believed to embody them. These mediums, Chao Pho Kuan and Chao Mae Nang Tiem, are given the honored titles of father (*pho*) and mother (*mae*), and it is they who invite the fertility spirits to the village. They are thought of as descending from royal ancestry and representing

14.1   Two ghost masqueraders wearing Phi Ta Khon Lek masks and carrying phallic swords join the festivities at the Dan Sai Ghost Festival. The one on the left wears a round, earlier style of mask made from a basket for steaming sticky rice (*huad*). In the more recent style (right), the *huad* is used for the hornlike component, while the main part of the mask is made of from the base of a coconut frond. Photograph by Ruth Gerson, Dan Sai, 1999.

14.2   The head *saen* is accompanied by village elders on his way to the river to conduct a search for the magical white stone, Phra Upakut. The stone is believed to bring rain and a good rice harvest in the coming year. Photograph by Ruth Gerson, Dan Sai, 2000.

14.3   The head *saen* prays at one of four temporary shrines placed around Wat Porn Chai. Photograph by Ruth Gerson, Dan Sai, 2000.

14.4   The current Chao Pho Kuan, pictured in a portrait photograph, has been the medium of Dan Sai village since 1988. Photograph by Ruth Gerson, 2000.

14.5   A woman ties a blessed white cord on the village medium's outstretched arms during the *bai sri* ceremony. Photograph by Ruth Gerson, Dan Sai, 1993.

different groups of people. The male medium is believed to be a descendant of the Thai king Maha Chakrapat of Ayutthaya, while the female medium is regarded as a descendant of King Settathirat of Laos, both of whom reigned in the mid-sixteenth century.

The Chao Saen Muang communicate their wishes to the villagers through the male medium, the Chao Pho Kuan, who makes all spiritual decisions in the village. It is he who decides, only two or three months before the event, when the Ghost Festival will take place. The festival is usually held at the beginning of the rainy season, around late May or June, but in 2000 it was celebrated in early July.

The festival is rooted in spirit beliefs, but Buddhist elements were woven in during the last century. On the eve of the festival, pilgrims, some dressed in white, come to pay their respects to the four-hundred-year-old Buddha enshrined at Sri Song Rak temple, as well as to the spirits of fertility. Pictures of past and present mediums are propped up in both galleries that flank the compound's main shrine, or *ubosot* (fig. 14.4).

Wat Porn Chai, the central temple of the village, is festively decorated for the occasion. Banana trees laden with fruit are set up inside and outside its main shrine. Fresh flowers and colorful banners decorate the doorways and corners of the hall, while two fanciful ladders are placed on either side of the temple's Buddha image. The ladders, made of cloth remnants and topped with tiaras, are positioned to please the spirits, inviting them to run playfully up and down the rungs.

The Ghost Festival is in fact a complex, multifaceted event. It is referred to officially as Bun Luang, a royal ceremony for acquiring merit. Its components include a ritual involving a sacred white stone; a ceremony of bestowing blessings (*bai sri*); the Phi Ta Khon, the processions and festivities of masked ghosts; a parade with characters drawn from the Pawet Santhorn, the story of the last incarnation of the Buddha; and the Bun Bung Fai, the shooting of rain-inducing rockets into the sky. These collectively reflect the ancient culture of India, the ancestral spirit worship of Laos, and spirit beliefs that are bound to the land and its harvest.

Central to all fertility worship in this festival is a sacred, round, white stone kept year-round in Wat Porn Chai, tucked inside a small niche behind the main Buddha image. No larger than a man's fist, this stone is called Phra Upakut. The term *phra* is honorific, used in general for religious figures and highly venerated men. Phra Upakut was a revered Buddhist monk in ancient India, one with supernatural attributes (for a Burmese depiction of this monk, known there as Shin Upagok, see fig. 21.21). It is said that he was able to transform himself into any being or object he chose. In Dan Sai, Phra Upakut became the magical white stone, believed to bring good fortune to the people and to have the power to bestow fertility on the land. In a way, the stone is the palladium of the village.

On the eve of the two-day festival, the village medium has the sacred stone placed in the Mun River, which runs through the center of the village. Usually kept in a gauzy white cloth, the stone remains wrapped when submerged in the water. The deed is performed in the dark hours of the evening, and there are no witnesses. Only when the stone is located in the riverbed, retrieved, and returned to its rightful place can the village rest at ease and enjoy the talisman's beneficial powers for another year. After this feat is accomplished, the villagers are assured of a plentiful rain and a good harvest.

The ritual of finding the perfect white stone is enshrouded in mystery. The ceremony begins in the early hours of the morning, when most of the people of Dan Sai are asleep. The village medium assigns an old man, a member of the *saen*, the group of men who assist the medium in his tasks, to lead this unusual rite. The head *saen* walks from the medium's house to Wat Porn Chai to prepare. He purifies himself by bathing and dressing in white clothes. He then enters the temple alone to pray. The village elders

14.6   Giant male and female
fertility figures (Phi Ta Khon
Yai) prior to the Ghost
Festival parade in Dan Sai.
Photograph by Ruth
Gerson, 1999.

14.7, 14.8   Examples of the contempo-
              rary ghost masks (Phi Ta
              Khon Lek). Dan Sai,
              Thailand. Plant fiber, paint,
              cloth. cm, respectively. H:
              78 cm. FMCH X2001.14.3.
              H: 80 cm. FMCH
              X2001.14.5.

who have accompanied him from the medium's house to the temple in a jovial and informal procession wait in the temple courtyard. They play regional music on reed and bamboo instruments while beating drums and cymbals to let the people know that the
ceremony is about to begin. The music and dance continue in the darkened courtyard
until a large gong, which will be carried down to the river, is struck and the lights go on,
marking the time for the start of the ritual. Meanwhile, several of the medium's helpers
wait patiently, armed with weapons carved of pale, softwood. These are crafted into a
sword, a club, and a gun, all tied together. The weaponry may vary somewhat from one
year to the next, as all items are made by the villagers, following their moods and imagi-
nations. The presence of these armed "soldiers" ensures that no evil spirits will interfere at
the river ceremony.

Finally, the head *saen* emerges from the temple carrying a loosely woven basket that contains a lit candle. As he walks slowly toward the gate, the other people present fall in step with him, marching with determination to the riverbank, just a few
minutes' walk from the temple. It is awe inspiring to watch the silent crowd fill the
breadth of the road, turn the corner, and head down to the soggy bank (fig. 14.2). Once
he arrives there, the leader sits on a mat placed inside the roofless pavilion made of
loosely tied bamboo canes. From this site, he will pray and chant.

First he places the basket and the candle on the riverbank, facing the water.
He crouches and prays, asking the river for permission to retrieve the stone. All the spectators crouch, too, and everyone is silent, watching for the old man's next move. When
his incantations stop, he signals to another elder to enter the river and search for the
stone. It is said that the designated diver, who is nearly seventy years of age, has been performing this ritual for as long as the villagers can remember. Clad in a white loincloth, he
plunges into the river, the brown swift-moving current surrounds him. The diver surfaces, holding a stone, which he shows to the headman, who examines and rejects it with
a shake of his head. The gong is struck and the process is repeated, with a second stone
rejected as well. The diver disappears once again into the water, this time bringing up a
stone deemed to be the perfect one, and all cheer at his success. The stone is duly placed
in the basket, and the procession back to the temple begins.

14.9   Metal bells and a phallic
         sword adorn the back of a
         "ghost." Photograph by
         Ruth Gerson, Dan Sai, 1993.

14.10   "Ghosts" parade in the festival at Dan Sai. Photograph by Ruth Gerson, 2000.

At the temple, four bamboo shrines have been erected outside the ordination hall, placed at the four cardinal directions of the compound. Unlike most Buddhist temples, which face east to receive the morning sun, Wat Porn Chai faces north, toward the Mun River, emphasizing the importance of water in this agricultural village. As the procession returns from the river to Wat Porn Chai, carrying the stone to be enshrined there once again, it stops at the four temporary shrines that surround the main sanctuary. Walking counterclockwise, the worshipers first stop at the shrine on the east side (fig. 14.3). There, along with the flowers, candles, and incense that have been placed on each of the shrines, is an additional stand, also of bamboo, that holds a monk's regalia—a saffron robe, a fan, an alms bowl, and other incidental items. These are traditionally offered as gifts to monks on important occasions.

After stopping here, the stone is then carried to the other shrines with the crowd following the head *saen*. The large gong, suspended on two poles and carried by two men—the same gong that was carried down to the river for the ceremony—is struck throughout each session of blessings and prayers. The end of the ritual at each shrine is punctuated by a booming rifle shot, announcing to the heavens that the proper respect has been accorded to the forces of nature as well as to the Buddha. When prayers at the fourth shrine are completed, the gong is struck again, announcing the end of the ritual and the start of the ghosts' celebrations. This time several gunshots are dispatched skyward to confirm that the sacred stone has returned to its rightful place.

14.11   A young "ghost" brandishes
            a phallic sword. Photograph
            by Ruth Gerson, Dan Sai,
            2000.

14.12   Papier-mâché buffalo harnessed to plows parade through Dan Sai village. Photograph by Ruth Gerson, 1993.

The stone ceremony is followed by food offerings to the monks at Wat Porn Chai. Bringing rice and other food to the monks is a means of acquiring merit, which has become an integral part of Thai culture. The Ghost Festival is one of twelve special days in the year when the people of northeast Thailand can acquire merit.

As day breaks, people converge on the house of the village medium to participate in the *bai sri* ceremony, that of bestowing blessings on a person. This ritual originated in Laos but has been practiced in Thailand for many years. It accompanies any important event, particularly preparation for a journey. In this case it involves inviting the village spirits through the medium to visit Dan Sai village.

Chao Pho Kuan, the male medium, is distinguished by his white garb and his long hair, which is wrapped in a long white cloth. He is assisted by Chao Mae Nang Tiem, the female medium, who wears a white blouse, her hair knotted at the crown of her head. Members of the *saen* and their female counterparts, called *nan taeng*, assist throughout the sacred ceremonies of the festival.

The villagers crowd into the small house of the male medium, sitting with their hands clasped in reverence while the medium falls into a trance. As the helpers chant and burn incense, the villagers approach the male medium one by one and tie blessed white chords on his outstretched arms (fig. 14.5). The medium remains seated, stiff and immobile, until the last villager has performed this act of respect. The spirits of the lovers have now been invited by all the villagers to join the festival and will be hosted by the mediums for its duration. The ritual ends with the offering of boiled eggs and rice to Chao Pho Kuan, items symbolic of a fertile year and a good rice harvest.

At this time "ghosts," pretty maidens, papier-mâché buffalo, and giant papier-mâché male and female fertility figures with exaggerated genitalia, fill the streets. All these will participate in the main parade of Dan Sai, the "ghosts" brandishing phallic swords with red tips, reminding everyone that this is a fertility rite and ensuring that all will have a good time.

As most inhabitants of northeastern Thailand have their roots in Laos, it is not surprising that the tradition of dressing up as ghosts came to Thailand from Luang Prabang, the old capital of Laos. There the ancestors of the Lao peoples are depicted as an old man and a woman with large, round, masklike faces, bulging eyes, and a toothy grin. These characters appear to be the forerunners of what later became known as Phi Ta Khon, the masked ghosts that gather annually for the Dan Sai Ghost Festival.

According to the villagers of Dan Sai, the Ghost Festival can be traced back approximately four hundred years through two types of masked figures: the giant body mask known as Phi Ta Khon Yai (fig. 14.5), which measures two to three meters in height; and a later innovation, the small mask, or Phi Ta Khon Lek, designed to be worn on a person's head (fig. 14.6). The big, naked figures were used primarily to ask for rain and a good rice harvest, and they are used for the same purpose today.

It is said that in early times a fertility rite was held in an attempt to bring rain. Two naked dolls called Phi Ta Khon, one male and one female, were placed in a hole dug in the ground along with frogs and other aquatic creatures. Brahmin priests sat nearby and prayed for rain, while the dolls and animals were kept in the hole until the rain fell. Giant figures were also used for presenting rice to the temple, as well as offering candles and other items for the monks to use. Money collected for the temple was presented through these enormous figures as well; today these donations are arranged as a money tree and carried to the temple on a pedestal.

The giant male and female fertility figures of Dan Sai are constructed from woven bamboo frames that are covered with various materials. The faces are made of papier-mâché and painted with large, comical features to captivate the crowd. The hair, arms, and exaggerated genitalia are made of coconut fiber, swatches of cloth, pieces of wood, and any other material that suits the purpose. The female figures sport large, dangling breasts while the males have oversized, string-operated penises that are set in motion as each figure ambles along. Holes are cut out in the body mask to allow the young man inside to see where he is going, catch a breath of air, and eat and drink as he parades and dances along the route. Made of natural materials that disintegrate easily, these figures are discarded after being used in one festival. In former times the figures were thrown into the river, but today they are burned within a few days of the festival's close. The following year new figures will be constructed.

Although historically the small ghost masks, or Phi Ta Khon Lek (figs. 14.7, 14.8), appeared after the large figures, they date back nearly four hundred years as well. The precursor of this mask type was a small basket called *huad*, used for steaming sticky rice, a type of glutinous rice eaten throughout northeast Thailand. The basket, made of woven bamboo, was folded like a hat to fit on a young man's head.[1] Later, another feature was added: the base of a coconut frond, which remained after the leafy part was used for thatching a roof. The frond is naturally curved and lends itself well to this purpose. Holes for eyes were cut out, and a nose crafted from softwood was added. Small, pointed horns were cut from a coconut shell and attached to both ends of the basket, giving the mask a somewhat devilish appearance. The horns eventually evolved, becoming longer and more dramatic.

14.13   Villagers act out the story of the Vessantara Jataka, the last incarnation of the Buddha. They represent Prince Vessantara, his wife, Madi, and their two children returning to the city after a long exile. Photograph by Ruth Gerson, Dan Sai, 2000.

14.14   Zombie-like figures covered
in mud join the parade.
Photograph by Ruth
Gerson, Dan Sai, 1993.

The colors and patterns added to the mask over the years, as well as the shape of the nose, can date it. The first masks that we know of had round moon-like faces that became longer and narrower in later years. The earliest decorations were stripes painted across the hat in black, red, and white. The black color was obtained from soot scraped off pots, the white was derived from powdered chalk, and the red was made from the juice and spittle formed by chewing betel leaves and areca nut. The decorations around the eyes were typically Lao in style with black or red paint, or sometimes both, following the contour of the eye. This curved design is called *ee-huak* and is drawn in the shape of a tadpole with the thicker center narrowing toward the outer corner of the eye and, ending in a tail-like curve. There were two styles of "tails" the artist could choose from, one long and curved (see fig. 14.7) and the other made with a wiggly line that looks like a swimming tail. Today, many of the new masks use Thai decorations around the eyes, replacing the old tadpole designs from Laos. New designs include the *lai kranok,* the traditional flamelike Thai design (see fig. 14.8).

The nose of the early mask was simple, pointed, and short. In time it became longer and sharper, and as the years passed, the nose became so large and elaborate that today it resembles the trunk of an elephant. In addition to using softwood for the nose, new materials have been introduced, with colors and designs evolving continuously. In older masks the mouth had a wide grin, literally from ear to ear, lined with teeth that filled the entire space. The oldest teeth were scooped out of the mask in straight lines delineated in red around the white teeth. Lips were often outlined in black, with the upper lip featuring a mustache. Some modern masks have a mustache in the *lai kranok* design and other imaginative patterns. Occasionally a beard is added.

The garb of the ghosts has also changed over time. The original garments were traditionally white or saffron in color. No pants were worn, only a long robe. Wearing pants with the outfit is a recent addition adapted from Chinese clothing. The original robes were sleeveless and made of simple material, often mattress covers turned inside out that were sewn to the back of the mask. When the festival was over, the masks and clothing were discarded, never to be used again. Unlike the large figures, the masks and gowns were not burned but were put into the river, where they would disintegrate. Today some people keep their masks to be used again, and now masks are made to be sold to visitors who come to see the Ghost Festival.

The ghosts' outfits today feature belts with metal cowbells dangling from the back (fig. 14.9). These clang loudly as the young ghosts prance in the streets, letting everyone know their whereabouts. Also hanging from the belt is a gourd, a Chinese influence and a reminder of the importance of the water it carries. Traditionally, the young men create their own masks and outfits. Boys at the age of puberty may join the parade if they can make their own masks. In today's culture, however, there are shortcuts to these traditions, and even some small boys take part in the ghost parade.

Although the "ghosts" have been roaming the town since the morning, the parade is well planned and orderly (fig. 14.10). The "ghosts" jump around merrily, threatening young women with phallic swords painted red at the tips (figs. 14.11). Some of the women shriek with delight while others feign alarm. Young men enjoy the freedom of their role under the cover of the ghost outfit, but all is done in good humor. The elder women do not let the men rule the day. They join the procession carrying fishnets, with which they try to snag the wooden phalluses, and fishing baskets for stowing their catch. Several pairs of the enormous fertility dolls dance in the streets along with many farmers who guide papier-mâché buffalo pulling plows (fig. 14.12). These are interspersed with troupes of dancers and drummers from the neighboring villages.

14.15   Chao Pho Kuan, the village medium, straddles a rocket on which he is carried to Wat Porn Chai. Photograph by Ruth Gerson, Dan Sai, 1993.

14.16   A Buddhist monk is carried to Wat Porn Chai in a sedan chair made of rockets. Photograph by Ruth Gerson, Dan Sai, 1993.

14.17   A villager readies rain-inducing rockets for firing behind Wat Porn Chai. Photograph by Ruth Gerson, Dan Sai, 1993.

The parade, which starts at the village school and ends at Wat Porn Chai, is headed by the male and female mediums who are carried high and shielded by an umbrella as a sign of honor. Also in the parade are representations of characters from the Vessantara Jataka, the story of the last incarnation in the life of the Buddha, known in Thai as the Tessana Mahachat, or the "Great Life." Here, Prince Vessantara and his wife, Madi, return to their home after being exiled for several years. The handsome young couple is transported on a traditional Thai-style carriage (fig. 14.13). This tale is consistent with the tradition of summoning rain, as the charitable prince was banished for giving away the kingdom's white elephant, a figure that ensured continued rainfall and a good rice harvest. It is said that everyone was so happy to see Prince Vessantara return that even the ghosts came to meet him. Among the "ghosts" and "ghouls" in this procession is a frightful cluster of figures covered in mud, looking like zombies who have emerged from their graves. They hold short clubs in their hands and bang them on the ground as they march, singing in unison all the while, which adds to the ghostly atmosphere of the event (fig. 14.14).

The second day begins with an afternoon ceremony at the village's main crossroads where a temporary shrine is erected. After so much revelry, partying, and drinking, the villagers are subdued in the early morning, but they arrive in droves just past midday. The ceremony includes the two mediums as well as four Buddhist monks, members of the *saen*, and a statue of the Buddha. All are carried with great fanfare to Wat Porn Chai and seated on bamboo palanquins, which are actually rockets filled with explosives (fig. 14.15). The mediums and monks are jostled in their precarious seats as they hold on bravely. They proceed to the temple and walk around the main shrine three times, counterclockwise as in funerals, since the "ghosts" are no longer considered to be among the living (fig. 14.16). By late afternoon the rockets are mounted on a tall tree behind the temple and fired toward heaven one at a time. The first two rockets are those of the male and female mediums, with other rockets following (fig. 14.17). This is yet another act to ensure that the rains will come soon.

The Ghost Festival is the last of the series of seasonal rain ceremonies in northeast Thailand. It ends on a Buddhist note in the evening, with the reading of the Vessantara Jataka. The village slowly quiets down with the knowledge that another year will pass before the ghosts will roam the streets again. Assured of plentiful rain, the residents of Dan Sai look forward to a bountiful rice harvest in the year to come.

•

15.1   Participants in the Ma'Bua'
        Pare ceremony carry torches
        as they race three times
        around the sacred tree.
        Photograph by Eric Crystal,
        Sulawesi, 1971.

15

The Ma'Bua' Pare Ceremony
of the Sa'dan Toraja of Sulawesi, Indonesia

Eric Crystal

The Sa'dan Toraja, a highland people numbering some four hundred thousand individuals reside within the administrative boundaries of Tana Toraja Regency, about 310 kilometers northeast of the South Sulawesi Province capital, Makassar. Sa'dan Toraja culture embraces a distinct language, a unique architectural heritage, and a rich panoply of oral poetry, instrumental music, folklore, and dance. A highly complex religious system involving sequentially stepped ritual sacrifices pervades all of Toraja traditional culture. As elsewhere in traditional Indonesia, the arts of the Toraja have been closely linked to indigenous belief and practice. That is to say, major religious events are likely to be prescribed venues for the performance of traditional arts.

During the course of the past century, particularly since 1950, there has been a significant loss of cultural resources and diversity within Indonesia. Specifically, the small-scale religious traditions of groups such as the Dayak communities on Indonesian Borneo (Kalimantan), the Sumbanese in eastern Indonesia, and the Sa'dan Toraja are losing ground due to increasing pressure from evangelical missionaries and Jakarta administrators and the culturally homogenizing effects of contemporary mass media. The result, compounded by the pace of social change, has been a great diminution—in some cases a total extinction—of traditional religions in many areas of outer island Indonesia. Official government statistics in Tana Toraja graphically illustrate this point. In 1968, 48 percent of the population was listed as traditional believers. Yet by 2000, this figure had fallen below 5 percent. The following is a description of a Ma'Bua' Pare ceremony that was observed late in 1971 at a time when traditional believers in Piongan hamlet could still generate widespread community support for a great indigenous rice ritual.

Each Toraja village sets aside sacred ceremonial ground beyond the bounds of inhabited hamlets for its rituals. On the *pantunnuan* (lit., "[meat] roasting field") great funerals of five or seven nights' duration are carried out with thousands of villagers participating. On the *pabuaran* (fructification field) major ceremonies of agricultural renewal and celebrations of fertility are carried out. Both fields are marked by sacred megaliths. They serve as consecrated ground, sacred space reserved for rituals so large and so important that they cannot be contained within home or hamlet.

In recent years as religious conversion, economic growth, and tourism have influenced the Toraja region, funerals have become ever more grand, dominating the Toraja ceremonial landscape almost to the exclusion of so-called life-side rituals. Because Toraja Christian converts maintain the outer shell of indigenous funeral ritual (e.g.,

15.2   The *pabuaran* ceremonial ground with one of the *lantang bua*, or ceremonial huts, to the left. The small hut to the right encloses an underground megalith and a bit of the earth of the heavenly rice field. Upright stakes positioned around the sacred tree will support "sugarcane bamboos" with symbols of fertility attached to their tops. Photograph by Eric Crystal, Sulawesi, 1971.

slaughter of animals, status-linked distribution of meat, and several days of feasting at a specified funeral site), the casual observer might be led to believe that Toraja religion places almost exclusive emphasis on death. And it certainly is true that here, as in many other highland Indonesian cultures, funerals do function to confirm family rank, reinforce community cohesion, and celebrate in death the social position of the deceased. However, it is also true that the less commonly observed Toraja rituals of life are equally important in the intricate dualistic system of belief and practice.

For the Toraja, rice is the essence of life. Rice rituals are always conducted in the morning hours as the sun rises on the horizon. These events follow an annual cultivation cycle. That is, the extraction of the seed rice from the granary, the sowing in the seedbeds, the transplantation of seedlings, the first flowering of the rice plants, the filling of the panicles, the harvest, and the subsequent storage of the grain are each marked by small, family-oriented rituals. Offerings of cooked rice, chicken, and sometimes small pigs are undertaken either at the house or adjacent to the fields astride the paddy dykes.

The Toraja believe in a high god, Puang Matua, whom they regard as a powerful but remote deity. In mythological times Puang Matua could be directly approached by humans, whose acesss to his celestial abode was through an *eran di langgi'*, a ladder to the heavens. In those days when people encountered problems that only the deity could solve, they directly beseeched Puang Matua by climbing the ladder to the heavens. But one day a supplicant before Puang Matua stole the golden fire flint of the high god before descending the ladder. Thus was the trust and passageway between humankind and their deity forever sundered—enraged at the transgression, Puang Matua in a fury smashed the ladder to the heavens. The remains of the ladder are said today to constitute the Sarira ridge that separates Makale and Sangalla Districts in Tana Toraja.

From this point on people would be able to communicate with the supernatural only through the agency of spirits (*deata*). The *deata* vouchsafe rice production, assure fertility of humans and animals, and are involved with all aspects of life. On rare ritual occasions the *deata* become visible to trance dancers in the context of the Ma'bugi' ritual.[1] In daily life the *deata* are thought to protect the rice crops, but they will do so assiduously only if all requisite ritual offerings are undertaken according to form and schedule.

Deata reside in unusually shaped stones strewn across the countryside, in the vicinity of special trees and in the depths of swift-flowing rivers, and they may also be summoned from their abode in the sky. They are sometimes referred to as *deata mamase*—the all-loving spirits. Yet the *deata* may also be vengeful toward those who neglect to speak to them in refined ritual language, forget requisite chicken and pig sacrifices, or fail to adhere to the 5,555 *pemali*, or taboo actions, given in the Toraja code of proper behavior. Young people who mysteriously die are said to be *naala deata*, taken away by the spirits.

Ma'Bua' Pare—Recalling the Gift of Rice

There are three types of Bua' rites. The Great Bua' (Bua' Kasalle), carried out in northern Toraja districts, celebrates the wealth and status of an individual family. This ritual also spans an agricultural year during which time young virgins dressed in yellow are secluded in a special porch erected for them at the front of the traditional house. Two separate forms of Bua' ritual are performed in the western districts of Tana Toraja.[2] The first is the House Bua' (Bua' Tulak Padang), which celebrates the wealth of a specific household. This ritual takes place within the hamlet and is marked by a spectacular raising of sails above the family home in recollection of the mythological voyages of Toraja ancestors from other lands to their present place of residence. The final type of Bua' is the Rice Bua' (Bua' Pare or Ma'Bua' Pare), the celebration of rice. The term *bua'* has generated some controversy in the literature. *Bua'* can refer to a hamlet or traditional residential area. Another interpretation of *bua'* is "fructification" or "to fruit." In the districts where Bua' Tulak Padang and Ma'Bua' Pare are found, the Bua' Kasalle is not traditionally carried out.[3] In the west the Bua' Tulak Padang is referred to as a "male" ceremony and the Ma'Bua' Pare is referred to as a "female" ceremony. This essay explores aspects of thanksgiving, fertility, renewal, and wealth linked to rice during the Ma'Bua' Pare ceremony.

Every twelve years in several western Toraja districts, households grouped together in traditional precolonial federations (*bua'*) unite to celebrate past harvests and supplicate the heavens for well-being in the future. The ceremony that is undertaken is the Ma'Bua' Pare (lit., "to fructify the rice"). Just as funeral rituals for people of distinction may be undertaken only on special sanctified funeral grounds, so also Ma'Bua' Pare ceremonies may be carried out only on sacred *pabuaran* fields. Funeral grounds are clearly marked by *batu simbuangan*, upright megalithic stones, sometimes arrayed in circles. For every seven-night funeral undertaken on these grounds, a new stone is typically dragged from cliff-side quarries to the *pantunnuan* site. *Pabuaran* grounds are marked by three special trees—sandalwood (*sendana*), banyan (*barana'*), and *lamba'*. The sandalwood tree is especially significant to the Toraja because when it is cut, the resin that oozes out of the wood is blood red. This tree is sometimes referred to as *kayu mangrara' tau* (lit., "the tree with human blood"). Because blood is closely connected with life, sandalwood trees are linked with life-side rituals, which are associated with the east. Thus sandalwood saplings are oftentimes planted on valley paddy dikes, frequently alongside *pessungan bane* megaliths, where offerings to the rice spirits are made. Banyan are considered sacred throughout Asia. It was under just such a tree that the Buddha achieved enlightenment. Much earlier in religious history it was the tendency of these trees to send out

aerial roots that caused them to be considered sacred. The *lamba'* tree is one of the tallest in the Toraja highlands. Because the sap of this tree is milky white, it is also associated with rice, life, and ceremonies of the east. Of course, such trees may never be felled for any purpose because they serve as the anchor point for infrequently seen rituals of harvest and fertility. Ideally the three trees are clustered together at the *pabuaran* ritual site. At least one of them should be tall and imposing, serving as the focal point for the activities that will unfold as the sun sets. At the ritual described below, the banyan tree was the most prominent.

For months in advance of the Ma'Bua' Pare celebration, villagers gather, chop, and carefully store firewood. As rain can come at any time in the Toraja highlands, it is important that this wood be cured and kept dry. Normally firewood supplies are kept under house and granary floors. If, in the months prior to the execution of the ritual, someone in the participating community dies, then the rite must be postponed until at least the following year. Death is considered to pollute the ritual, compromising months of preparations.

The Ma'Bua' Pare ceremony is normally carried out in October and November, a time after the traditional harvest of rainfed rice. Around the *pabuaran*, four two-story bamboo structures are usually erected (fig. 15.2). The construction of temporary bamboo residence structures on sacred Toraja ground is common at funerals, where mourners typically reside for seven or more nights in sometimes hundreds of such temporary shelters erected just for the funeral occasion. Here, the bamboo huts will be utilized for only a single night. A gallery with benches is fabricated on the lower level at which singers will be seated. On the upper level a floor is woven of green bamboo for people to rest and observe. Atop the structure are many single bamboo poles pointing skyward.

Nighttime is normally associated with death rituals by the Toraja. Black is the color of death; funerals are always initiated in the darkness of night. The lighting of a ritual fire to initiate the funeral, the shifting of the corpse to the death axis, and the erection of a previously supine funerary statue within the house transpire at night. Funerals are also associated with odd numbers—the Toraja execute one-night, three-night, five-night, and seven-night funeral events. At funerals, food is normally prepared on cooking fires located to the rear of the temporary bamboo residential structure. Water buffalo are prominently slaughtered and pigs are offered as well.

Daytime preparations anticipate a major nighttime life-side ritual. On the morning preceding the Ma'Bua' Pare event, miniature implements of daily life are fashioned (fig. 15.3). Of special importance here is the *panganduran*, replicating in scale the bamboo vessel in which milk from lactating water buffalo cows is collected to be made into *dangke'* cheese. The implements also include representations of the basket used for storing rice seeds, the special winnow used for broadcasting grain into the seedbeds, and the feeding trough for young pigs. All of these miniatures will be bound to the tops of twenty-foot-long "sugarcane bamboos" (*tallang ta'bu*). Lengths of green, freshly cut cane are affixed to the tall bamboo poles. Sugarcane is notable because when planted it gives rise to many young green shoots and so here symbolizes a plenitude of young rice plants, an abundance of baby domestic animals, and the births of many new members of the human community. Prior to the beginning of the ceremony, each participating hamlet carries the sugarcane bamboos to the ceremonial field and binds them together to stakes set around the trunk of the central tree (15.4–15.7). At dawn the next day the implements of fertility atop the sugarcane bamboos are believed to have been blessed by the high god.

15.3   Symbolic miniatures of implements from daily life hang from the eaves of the rice granary in Piongan hamlet. These implements relate to fertility and were placed on the granary twelve years prior, following the previous Ma'Bua' Pare rite. Photograph by Eric Crystal, Sulawesi, 1971.

15.4   Prior to the beginning of
the Ma'Bua' Pare ceremony,
sugarcane bamboos with
symbolic miniature imple-
ments affixed to their tops
(see fig. 15.3) are set against
stakes surrounding the
sacred tree in the *pabuaran*.
Photograph by Eric Crystal,
Sulawesi, 1971.

15.5   The sugarcane bamboos are
positioned. Photograph by
Eric Crystal, Sulawesi, 1971.

15.6   Participants tie the sugar-
cane bamboos in place.
Photograph by Eric Crystal,
Sulawesi, 1971.

15.7   The bound sugarcane
bamboos surround the
sacred central tree in the
pabuaran ceremonial
ground. Photograph by
Eric Crystal, Sulawesi, 1971.

As preparations continue, in each household a small female pig is killed and butchered. In this case families carefully extract the *kollong*, a ring-shaped cut of meat from the collar of the pig. From the abundant sugar-palm trees (*Arenga pinnata*), tapped daily for palm wine in most Toraja villages, are harvested the supple and golden immature leaves. These leaves together with the *kollong* are arrayed on carrying poles for each household to bring ceremoniously to the *pabuaran* (fig. 15.9). Throughout Indonesia yellow palm leaves are used to embellish festivals, especially rituals related to harvest, fertility, or marriage. Here the long flowing leaves symbolize the abundant harvest. The round cut of meat symbolizes the female element and the hope that, as a result of the ceremony, not only will future rice harvests be ample but domestic animals will remain healthy and multiply. Village women put on their traditional blouses and finest sarongs. The colors red, yellow, and white prevail here. Black is not seen.

Prior to entering to the ceremonial field, female singers gather in a line to begin the verses of the *serong mundan* (lit., "nest of the wild ducks"). It is the ducks (*mundan*) that glean the last of the harvest, landing in reaped fields to search for fallen grain. Each singer (fig. 15.8) wears gold and silver jewelry and the right hand of each supports a walking stick (*tekken*). The notion here is that the golden jewelry is so bountiful and heavy that only with the support of a walking cane can the female singers make their way to the sacred field. The Toraja believe that rice buys pigs and especially water buffalo, and water buffalo indeed may be traded for gold.

The Ma'Bua' Pare ceremony calls upon the skills of the most learned and experienced Toraja ritual specialists. These are the *tominaa*, masters of life-side ceremony, skilled in the recitation of ancient myth and in the presentation of sacrificial offerings to the *deata*. At a great ritual such as this, offerings are made directly to Puang Matua. The *tominaa* are summoned to recite the origin myth of rice, to manage and control a ritual that not only recalls the gift of rice from gods to humans but also manifests clearly and dramatically the gratitude of the community to the now distant and remote high god. Not one but several such religious specialists will be summoned to partake in the ceremony.

The Ma'Bua' Pare ritual is one of the few occasions when offerings can be proffered directly to the zenith of the heavens. In the daytime, prior to the ritual itself, an offering platform is prepared on the *pabuaran* site. Here supple leaves are folded in the shape of an eagle, and raw cotton fiber is hung from the plaited bamboo platform that will hold sacrificial chicken meat. The cotton symbolizes the purity of the ceremony; the folded-leaf eagle suggests that the offering will brought on high to Puang Matua.

At the 1971 ceremony, three of the requisite four bamboo huts were prepared; a fourth was canceled because almost all members of that *bua'* region had become Christians. A small temporary bamboo enclosure was placed at the base of the great tree at the center of the ceremonial ground. The modest bamboo hut enclosed sacred ground beneath which was buried a special stone. Within the hut, two male ritual specialists called To Ndo' (lit., "the mothers"—local officials charged with the initiation of planting and the organization of rice rituals) maintained a nightlong vigil. The hut is termed the *lantang pamukuran* (lit., "measuring structure"). Under the earth of this hut resides the sacred megalith known as the *batu polo' barra* (lit., "stone of two types of rice"—male and female rice).

According to officiating *tominaa*, another megalith, the *batu pamukuran* (lit., "measuring rock"), had been implanted in the central tree when it was young. This megalith is upright, reaching skyward. The megalith under the ceremonial hut is flat and supine, anchored in the earth.

15.8   A *tominaa* priest instructs
female singers as they enter
the ceremonial ground.
Photograph by Eric Crystal,
Sulawesi, 1971.

15.9   A member of a local household ties the immature, golden leaves of the *arenga* palm to a symbolic carrying pole. These poles will ultimately be raised atop a structure in the ceremonial ground. Photograph by Eric Crystal, Sulawesi, 1971.

During the Ma'Bua' Pare ritual, *tominaa* priests recite myths of the origin of the world and of the peopling of the earth. According to Hetty Nooy-Palm, the Toraja maintain "countless traditions concerning the origins (of rice). These…fall into two categories: the first contains the myths in which rice is created in the upperworld and afterwards brought to the earth, and the second which tell how rice was created on earth" (1986, 222). One of these myths, synopsized by *tominaa* Ba'ddu Rappanan of Tampo, Makale, relates the gift of rice from Puang Matua to humankind. At one time, rice was tilled in only one rice paddy, located in the heavens. Puang Matua farmed this heavenly field (*tana rangga*), which contained the essential varieties of rice: great rice (*pare kasalle*), white rice (*pare busa*), red rice (*pare lea*), black rice (*pare lotong*), and glutinous rice (*pare pulu'*). Puang Matua determined that it would be well to instruct people in the ways of rice cultivation and invited some Toraja men to his abode to learn about rice growing. Puang Matua wanted them to master the techniques of irrigated rice cultivation so that in the future they could make prayers and offerings to him with rice. So he instructed them not only in planting and harvest but also in the ways of ceremony (*aluk*), the sacrifice of domestic animals, and the burning of aromatic incense. Prior to dispatching rice to earth, Puang Matua instructed the Toraja to circle the *tana rangga* three times with offerings of pork and steamed glutinous rice wrapped in yellow leaves (*belundak*). This celestial circumambulation of the first irrigated rice plot is said to have set the boundaries for the first *pabuaran* ceremonial ground.

Puang Matua then instructed that some soil from this heavenly rice field be placed in a basket and that, in other baskets, rice seeds, the sprigs of sacred sandalwood and other trees, and the implements of rice agriculture be taken down to earth. The journey down the "ladder to the heavens" was to take place at night—thus the need for great torches *(sulo)* to light the dark sky. Once returned to earth, the basket containing the sacred soil was buried in the ground (with valued beads and metal coins) and the sandalwood tree was immediately planted beside it. Here also was buried the sacred megalith, the *batu pamukuran*. Puang Matua had thus instructed Toraja people in the ways of rice cultivation, given the master cultivators long-lasting torches to light their way down from the sky, and allowed the sacred land, the sacred trees, and the sacred seed to be taken from heaven to earth. But mankind was instructed never to forget from whence the bounty of the rice field originated. And so, from time to time if harvests are good, domestic animals fertile, and families ever growing, the community is expected to reenact the sacred drama of the gift of rice.

The Ma'Bua' Pare ceremony lasts from sundown to sunrise. It begins with a procession from the village to the sacred ceremonial site. The procession is followed by the raising of the carrying poles bearing yellow leaf streamers and symbolic cuts of raw pork (figs. 15.10–15.12). *Tominaa* gather under the central tree to recite the origin myths of the Toraja people and recall the gift of rice from on high. Female singers are seated throughout the evening, singing once again songs in praise of rice and the harvest. Individual households maintain their separate identities during this ritual, not only by posting the symbolic carrying poles but also by the circling of the torches, when each household is represented by a torch bearer. At about 1:30 A.M. long bamboo torches that have been carefully prepared in advance are retrieved by male heads of households. On a signal from the *tominaa* managing the ceremony, the torches are lit and the participants race around the central tree (from left to right) three times (fig. 15.13). At this time the bonfires *(kaponan)* in front of the bamboo huts are stoked to their highest flame. Sparks fly across the ceremonial ground as participants race around the central tree at top speed (fig. 15.1). As they run, the bearers cry out: "Dionglele' kami kalando lappoki', kami layuk

15.10   With a carrying pole on his shoulder, a man enters the ritual field in the late afternoon. The yellow leaves on the pole are knotted in a special fashion to symbolize heavy stalks ripe with rice. Photograph by Eric Crystal, Sulawesi, 1971.

15.11, 15.12   Men raise the carrying poles with leaves and meat atop a ceremonial hut in the *pabuaran*. Photograph by Eric Crystal, Sulawesi, 1971.

pamatokki dalli tedong, dipatungga' ponanule" ("Our paddy harvest mound [*lappo'*] will be high, great will be our rice stocks to buy water buffalo, the biggest of water buffalo"). The ceremonial ground is now brightly lit with the orange glow of fires and flares; the central tree is marked by scores of racing torches. From the heavens, the high god is thought to take note, cognizant of the efforts of the village to replicate the initial gift of rice, to give thanks for recent harvests, and to solicit blessing for future yields.

After the circumambulation of the tree, an essential drama begins. The central area is now thick with churned mud, the ground soft underfoot. Bearing the type of winnows used to broadcast seed rice in planting beds, several men pretend to cast rice on the ground. Next, closely spaced reeds are stuck upright into the ground to simulate the transplantation process. Later the stalks are pulled up and spaced in the manner of moving rice from seedbed to terrace field. A mock harvest is undertaken; village actors affix yellow palm leaves to carrying poles crying, "How heavy is our load" as they exit the ceremonial stage (fig. 15.14). Illuminated by the fires in front of each hut, two male actors now enter the stage. Each is adorned with a headdress of carved water buffalo horn, similar to those worn by the officiating *tominaa* priests. But here the actors indeed are portraying water buffalo. As they enter the "planted" field, one simulated bovine makes advances on the other. To the delight of the gathered audience, soon one "buffalo" mounts the other. Spectators simulate the plaintive bleats of hungry water buffalo calves searching for their mothers as the act of procreation is repeated again and again.

As the drama expressing the linkage between rice and fertility comes to an end, the fires that illuminate the ceremonial field begin to die down. On the horizon a faint glow of orange begins to color the sky as a new day is set to break. Singers, now hoarse from ten hours of performance, make ready to return home. At dawn, household heads climb up the bamboo structures to retrieve their carrying poles of pork and yellow leaves (fig. 15.15). As a heavy gray mist hangs over the *pabuaran*, they circle the sacred tree three additional times. With the circumambulation now complete, household heads ceremoniously cut off the yellow leaves, dropping them into a pile at the base of the tree. This action symbolizes the sacred harvest, the blessing of future crops, and the success of the ritual. Now participants scramble to retrieve their sugarcane bamboos (fig. 15.16). They will return home to cook the meat of the small pig sacrificed earlier, plant the sugarcane sacralized during the previous night, and mount the symbolic implements of fertility under the eaves of their rice granaries, where they will be displayed along with those from earlier Ma'Bua' Pare rituals (see fig. 15.3).

It is significant that the blessed implements of fertility are displayed under the eaves of the rice storehouse. The granary, enclosing the fruits of the rice harvest, is wholly associated with the realm of the east. Blessed miniatures of daily life are arrayed on the granary for a generation for all to see.

Tominaa specialists, the rays of the sun now piercing through the yellow palm fronds flowing from their horn headdresses, make ready a special sacrifice. If this were a great funeral ground, then the sacred space would serve as a venue for slaughtering the largest and most valuable water buffalo. But here a small calf is bound up with yellow *arenga* palm leaves and brought to the edge of the *pabuaran*. A young calf, yet to be weaned, is offered to Puang Matua in the hope that many more such young animals (and by inference, new members of the human community) will be born into the village before the next such ceremony takes place (fig. 15.17). As the sacrificial calf is stabbed through the heart, its death cries are silenced by a muzzle of golden leaves. This sacrifice celebrates life and so the yellow palm leaves that symbolize ripened rice are used to muffle the cries of the sacrificial juvenile.

15.13   The central sacred tree is
          illuminated as the first
          torches are borne in proces-
          sion around it. Photograph
          by Eric Crystal, Sulawesi,
          1971.

From the time of transplanting to the time the rice panicles fill with grain, no one may eat eggs in the village. A taboo sign consisting of numerous broken eggshells is posted at village entryways to warn villagers of this restriction. To break and cook eggs during the growing season would compromise the integrity of the living rice field and jeopardize the success of the forthcoming harvest. At harvesttime, finger knives rather than sickles are used so as not to disturb the *deata* spirits guarding the rice. The finger knives are especially suited to traditional long-stemmed varieties of rice selected by the Toraja for offerings. Harvesters may speak only in hushed tones and never use coarse language lest the *deata* spirits in the vicinity of the rice field be offended. Even after the rice has been stored in the granary, certain taboos apply. Women are uniquely empowered to enter the granary to withdraw rice for daily needs. Ceremonies are undertaken after the harvest not only to assure that the granary will safely store the grain but that the grain will be magically increased while in the storehouse. Specially shaped stones gathered from river beds (*balo'*) in the form of chickens, pigs, and water buffalo are often placed inside the granary to ensure that the rice will remain fresh and indeed multiply.

Everyday restrictions clearly indicate that a supernatural power guards the rice, that people must show respect and deference to the guardian spirits, and that the rice field itself is a sacred space. Each step along the path from germination to harvesting is marked by small family rituals. The association of rice with women and femaleness in general is apparent also in everyday life. Only women extract rice from the granary— and when they do they should not wear a blouse (these days they simply tie their sarongs under their armpits). Eggs and milk again link the female directly to rice. The taboo concerning the eating of eggs during a portion of the rice growing cycle has been discussed above, as has the use of a miniature container for water buffalo milk as part of the preparations for Ma'Bua' Pare.

15.14   A "mock harvest" is performed with village members pretending to carry home the heavy rice crop. Photograph by Eric Crystal, Sulawesi, 1971.

At the Ma'Bua' Pare ritual, past good harvests are celebrated not so much by individual families as by traditional village communities that come together no sooner than every twelve years to gather at the sacred *pabuaran* ground. The great tree that anchors this space is truly an axis mundi, a cosmic linchpin linking the Toraja celebrants directly with the high god responsible for the gift of multicolored rice. The evening drama reenacts the gift of rice, manifests the gratitude of the celebrants, and transpires precisely at the spot where the first heavenly seeds and soil for rice land are thought to have descended from the heavens.

Analysis: The Spirit of Rice in Toraja Culture

The Ma'Bua' Pare ritual dramatically demonstrates the meaning of rice in the culture of the Sa'dan Toraja. In this highland Southeast Asian region there is no graphic representation of the mother of rice, no fabrication of anthropomorphic rice goddesses from paddy straw or glutinous rice. Yet the cognitive link between the cultivation of the staple grain and concepts of fertility, procreation, harvest, wealth, and continuity is vital and strong. The ceremony described here reflects quintessential Toraja beliefs relating the cultivation of rice to the realm of the sacred and to notions of fertility and wealth.

All offerings are directed to the sky in the belief that they will be received by Puang Matua without the normal mediation of the *deata* spirits. The two To Ndo' officials coordinate prayers and offerings in a small hut that encloses what is essentially a female megalith as celebrants circle the great tree inside of which is implanted a male megalith, the upright *batu pamukuran*. Male and female, sky and earth, human and high god, night and day are here conjoined, all for the sake of ensuring the fertility of the fields, domestic animals, and mankind.

The noted historian of world religions Mircea Eliade has written of the concept of *illo tempore*, referring to the recapturing of mythological time in religious rituals:

15.15   At dawn the carrying poles
with *arenga* palm leaves
and meat are taken down.
Photograph by Eric Crystal,
Sulawesi, 1971.

15.16   Men break down the sugar-cane bamboo poles at dawn to bring the implements of fertility that were attached to them home to the grana-ries (see fig. 15.3).

"by its very nature sacred time is reversible in the sense that, properly speaking, it's a primordial mythical time made present. Every religion festival, any liturgical time, repre-sents the re-actualization of a sacred event that took place in the mythical past" (1957, 68–69). The *pabuaran* is activated as ritual high ground after more than a decade has passed. For twelve years (some 4,380 days) it remains in reserve as a potentially sacred space. Then, in one impassioned evening, a sacred tree becomes the centerpiece of a ritual drama that involves the reversal of night into day, the posting of symbolic harvest carrying poles on upward-reaching spires, the circumambulation of scores of torch-bearing supplicants, the enactment of a drama symbolizing the essence of fertility and procreation, and the sacrifice of a golden leaf-bound calf not yet weaned from its mother's milk. The darkness of night is transformed into daylight by great fires; indeed normal time stops, is reversed, and for a few hours the link between the sacred heavens and the secular human world is reestablished.

A *mawa* textile owned by the Dallas Museum of Art further helps us under-stand the Toraja equation of the spirit of rice with the essence of life (fig. 15.18). *Mawa* cloths are painted and/or stamped textiles displayed only at the highest Toraja rituals. This piece portrays several water buffalo, including cows with suckling calves, being led through an elliptical area across a field undulating with hundreds of tadpoles. The field is indeed a rice field, where frogs, eels, and small fish live and can be gathered after the rice has been harvested. Most important here are the several human stick figures, one of

15.17   A young calf bound with
golden leaves is taken to be
offered to Puang Matua.
Photograph by Eric Crystal,
Sulawesi, 1971.

15.18   Sacred textile (*mawa*)
depicting tadpoles and water
buffalo. Sa'dan Toraja,
Makale area, South Sulawesi,
Indonesia. Early twentieth
century. Cotton. L: 204 cm.
Dallas Museum of Art, The
Steven G. Alpert Collection
of Indonesian Textiles; Gift
of The Eugene McDermott
Foundation.

whom leads the buffalo and another who carries a bamboo milk container. The elliptical element on the cloth represents a rice field fishpond where carp and other freshwater varieties are nurtured, to be harvested just before the season of heavy agricultural labor. The fishpond is also seen as the center of the field and as representative of the female element, an element with which rice is also closely associated. The *panganduran* bamboo container is that which is used to gather water buffalo milk. The tadpoles here, as in many highland Southeast Asian cultures, suggest that the almost magical transformation of amphibians from one life form to another is illustrative of procreation and the life-giving powers of supernatural deities. Suckling calves and milk also refer to procreation, not only of domestic animals but of humanity as well. Recall that at the beginning of the Ma'Bua' Pare ceremony, as dawn breaks, miniature implements of daily life relating to rice and fertility, including baskets, winnows, and milk containers, are displayed on the eaves of the rice granary. The Dallas *mawa* cloth, then, recalls in textile art many of the same themes that the Ma'Bua' Pare rite illustrates on the ceremonial field.

On rare occasions in the Toraja hills—in the event of a series of strong harvests, the birth of healthy children, and the steady expansion of flocks of chickens, stocks of pigs, and herds of water buffalo—a great ceremony of thanksgiving may be undertaken. Then the *pabuaran* is suddenly enlivened with energetic activity, the songs of celebrants resonate across nearby valleys, and the fires and torches of household heads shoot toward the sky, clearly illuminating the sacred tree of life for the high god at the zenith of the heavens. It is only at such times that the full measure of the link between rice and life can be fully assessed. For a brief moment in ritual time the connections between the secular village and the sacred deity are again close and direct. The sacred gift of rice from the high god to humanity is recalled through reenactment. Once again Eliade reflects, "It is for religious man that the rhythms of vegetation simultaneously reveal the mystery of life and creation and mystery of renewal, youth, and immortality" (1957, 150).

After the suckling calf wrapped with a collar of golden leaves is dispatched, a final offering of glutinous rice cakes in the form of a mound of harvested paddy is prepared in each household. Once this is consumed with the pork from the pig sacrificed earlier, the ceremony will finally have come to an end. It is assumed that Puang Matua will have blessed and assured future harvests, children, and expanded domestic animal holdings. In past decades villagers could reasonably expect that after 4,380 additional days they would again repair to the sacred field for a great festival of thanksgiving and supplication. Yet it is not at all clear that such ceremonies will ever be performed again in the Toraja highlands due to aggressive Christian proselytization, government sanctions, and the culturally homogenizing effects of mass communications. If that becomes the case, then one of the most compelling of ceremonies to celebrate the spirit of rice in Southeast Asian culture will be forever lost.

•

16.1   The sacred rice bundle is carried through the streets of Trám village on a palanquin. The characters written on the winnowing baskets proclaim the *Four Occupations of the People* (*Tứ dân chi nghiệp*). Photograph by Nguyễn Anh Hiếu, 2001.

16

Trò Trám Festival and the Veneration of Ngô Thị Thanh in a Vietnamese Village

Vũ Hồng Thuật and Roy W. Hamilton

The grandmother carries her grandchild, the mother carries her babe in her arms
If you miss the Trò Trám festival, life is boring all year round.

Bà ẵm cháu, mẹ bồng con
Không xem Trò Trám cũng buồn cả năm.
—Ngô and Xuân (1986, 246)

Trám village is, on most days of the year, a quiet agricultural community surrounded by rice fields in the fertile Red River Delta some 80 kilometers by road upstream from Hanoi. But in late January or early February, on the eleventh through the thirteenth days of the first month of the Vietnamese lunar calendar, the community hosts a boisterous festival known as Trò Trám. Specifically, *trò* refers to the festival's public entertainment, which takes the form of a set of performances called the *Four Occupations of the People* (*Tứ dân chi nghiệp*). These skit-like vignettes and the accompanying ribald songs are rooted in the legendary history and social relations of the community, and their annual reenactment is an important part of the festival. The festival in its entirety is but one of many interrelated festivals held in different parts of the delta over the course of the agricultural year, and, as the song quoted above suggests, was in the past a famous event in this densely populated region that is both the rice bowl and the cultural heartland of Vietnam's majority population.

The Trò Trám festival takes place within the context of Vietnam's highly eclectic mix of popular religious practices, which blends spirit beliefs, Taoism, Buddhism, and Confucianism. The festival honors Ngô Thị Thanh, a Chinese woman who, according to legend, taught the local people to grow rice. The deification of quasi-historical figures (though in most cases they are male military or administrative leaders) is part of the Taoist and Confucian heritage that Vietnam absorbed from centuries of Chinese domination. The festival also prominently features a midnight fertility rite of simulated sexual intercourse and a procession in which a sacred bundle of rice is carried through the rice fields on a palanquin (fig. 16.1). These elements reflect more generalized spirit beliefs closely related to those found throughout Southeast Asia. Ultimately the purpose of the festival, in addition to bringing the community together in celebration, is to vouchsafe the fertility of the rice crop and the well-being of the villagers for the coming agricultural year.

Trám Village and Its Sacred Structures

According to the oral history preserved by today's generation of elders, the path of the Red River was in former times quite close to the low-lying land that now forms Trám village's rice fields. The land was under water year-round, except for a forest of *trám* trees (*Canarium sp.*) that appeared as an island surrounded by the water. During the reign of the eighteenth Hùng king[1] (circa 250 B.C.E.), a Chinese military mandarin named Ngô Quang Diện and his son, Ngô Quang Dũng, organized the local people to move from nearby Lo Ngoi and settle in the *trám* forest.

From early on, the village included a structure known as Trám Shrine (Miếu Trám). This at first consisted of a small building made of bamboo with a thatched roof erected on the hillside in the *trám* forest. Local shrines of this type in Vietnam are intended for the worship of deities associated with nature. In the Trám Shrine, the gods of heaven and earth and the gods of water and mountains were venerated through rituals involving fertility symbols in the form of human male and female reproductive organs.

At some later date, the village people built a second religious structure, Trò Temple (Điếm Trò), to house votive tablets commemorating Ngô Quang Diện's daughter, Ngô Thị Thanh. According to the local oral history, Ngô Thị Thanh was born, raised, and died in Trám village, and today, some two thousand years after she is supposed to have lived, she is remembered in particular for having taught the people of Trám village how to cultivate rice and also how to raise silkworms and weave silk cloth.[2]

The deification of Ngô Thị Thanh can also be viewed in light of the growing influence of Confucianism, with its emphases on social order and ancestor worship, during the millennium of Chinese domination in Vietnam (111 B.C.E.– 939 C.E.). At this time, the government began to make conscious efforts to identify candidates for elevation to deity status. Saints of Chinese origin were particularly favored. Later, during the Trần Dynasty (thirteenth and fourteenth centuries), villages were required to submit lists of likely candidates for deification. Village elders say that the first royal decree specifically deifying Ngô Thị Thanh was issued during the reign of King Lê Hiển Tông (1740–1767).[3]

Today villagers say that Ngô Thị Thanh helped the poor and encouraged virtue. As a result of her tireless dedication to the common good, and also perhaps because she was a Chinese woman living in a Vietnamese village, she remained unmarried and died without issue. In worshiping a heroine who died without descendants to tend to her memory, the entire community has assumed the normal Confucian filial responsibility of honoring an ancestor.

No one knows exactly when Trò Temple was first built, but it was certainly after the erection of Trám Shrine. Both buildings underwent frequent reconstructions as they were repeatedly damaged by fire, war, or the vagaries of time. The original bamboo and thatch shrine was replaced by a more solid wooden structure before the eighteenth century. The temple that is in use today was made with brick walls and a tile roof after the French destroyed its predecessor in 1948.

After 1948, the policies of Vietnam's communist government strongly discouraged religious activities. No community events were held at the temple or shrine, although devout individuals continued to burn incense to commemorate Ngô Thị Thanh and the New Year (Tết) holidays. The Trò Trám festival was not held during the entire period from 1948 to 1993.

In 1990, Phú Thọ Province's Culture and Information Service approved the villagers' request to rebuild Trám Shrine as a place of worship and a center for cultural activities. A simple, small structure resembling a square communal hall, with walls in front and back but the sides left open, was built on the shrine's original site. In 2000, a more elaborate new shrine was constructed beside the 1990 shrine with money provided by the

Vietnam-Denmark Cultural Promotion Fund (fig. 16.2). This building was made the same size and with the same decorations and adornments as the shrine remembered from 1946 (fig. 16.3). Since the shrine restorations began in 1990, the Trò Trám festival has been held six times.[4] Its resurrection reflects an ongoing liberalization of Vietnam's policies toward religious activity, partly due to increased pressure from the citizenry to be allowed to partake in such activities now that economic conditions are gradually improving. Additionally, the government views cultural preservation as useful for tourist development.

Since it was founded, Trám village itself has undergone a long series of political reorganizations determined by the prevailing colonial or national administration. Today, together with three other nearby hamlets, it forms part of Tứ Xã Commune. The entire commune is home to nine thousand people. The offices of the Communal People's Committee comprise most of the modern sector of the community. Despite these changes, Trám village remains predominantly agrarian, and new, more elaborate houses are only beginning to replace the simple brick homes that line the narrow, walled village streets.

The funding of the festival has also undergone changes. Prior to 1945, Trám village had two parallel community-level forms of government. The first was a civil administration involving the village chief (*lý trưởng*), village officials, and landed gentry. The second was a community affairs and ritual organization headed by an officer known as the *cai đám*. The *cai đám* took responsibility for making offerings to the gods on behalf of the whole village for a one-year period. He earned this privilege by buying privately owned rice fields and donating them to become part of the village's communally owned fields (*hương điền*). The proceeds from the harvest of these fields, as well as other individual contributions, supported the Trò Trám festival.

Although no government monies can be used to support it directly, since its revival in 1993, the Trò Trám festival has been planned under the auspices of the Communal People's Committee. Funding has in fact been inconsistent, and the scale of the festivities has varied from year to year depending on the resources available within the community and from outside benefactors. Every family in the commune is asked to contribute, and additional funds are gathered from visitors and foreign aid groups. Bùi Văn Mỹ, an eighty-two-year-old citizen of Trám, sums up the current state of affairs: "Formerly, the Trò Trám festival used to last three to five days, with many buffalo and pigs killed and many songs sung. From 1948 to 1993, the festival was disrupted due to the bitter wars and reconstruction. Now, the festival itself has been restored, but with its rituals and offerings simplified and its length cut back to fit with the country's new situation."

The description of the main events of the festival that follows is based on observations of four festivals (1998–2002) and interviews with village elders and officials. As minor details of the festival differ from year to year, the description is a composite aimed at showing the most typical practices.

The Midnight Fertility Ritual[5]

One of the key events of the TròTrám festival is a fertility ritual involving the display of fertility symbols and the simulation of sexual intercourse before the altar in Trám Shrine. Because it is considered improper to use their real names, the symbolic objects are referred to as *nõ nường*[6] or *vật linh*[7] in everyday speech. They were traditionally fashioned from plant materials of suggestive forms. The male object was made from a "male" bamboo stump, which was yellow and had many small roots at the base. The roots were retained, while the top was hammered into the shape of a penis. The female object, in the form of a vulva, was made from the convex sheath of an areca palm frond or bamboo shoot with a

16.2   Trám Shrine (Miếu Trám).
The 1990 building is par-
tially visible to the left of
the new shrine, which was
dedicated at the beginning
of the 2001 Trò Trám festival. Photograph by Vũ
Hồng Thuật, Trám village,
2001.

16.3   The interior of Trám
Shrine, shortly after the
sacred objects and decorations were moved into the
new building. The box containing the nõ nường rests
between the two jars high
above the altar. Photograph
by Vũ Hồng Thuật, Trám
village, 2001.

slot cut into it so that the male object could be inserted at the climax of the ritual. These materials were afterward burned and the ashes distributed among the crowd, to be scattered over the rice fields for good luck and to assure the fertility of the coming crop. As the ritual was conducted only once a year and the sacred objects thus destroyed, in the following year a new pair would be fashioned.

When the festival was reinstated in 1993, it was decided to make a durable wooden *nõ nường* set to replace the ephemeral ones that had been made and then burned each year up until the time the ceremony was discontinued in 1947. Now, at the close of the ritual, the wooden *nõ nường* set is returned to a box stored above the altar until the following year.

Before the midnight fertility ritual can begin, some preliminary opening ceremonies must be held. According to the local understanding, the deities, including Ngô Thị Thanh, do not reside permanently in any single place but constantly move about. In order to invite honorable saints, deities, and Ngô Thị Thanh to attend at the time of the festival, the man who serves as shrine keeper (*thủ từ*) presents offerings and asks for blessings for the villagers. From this point, the deities are considered to be in residence for the remainder of the festival.

At 8:00 P.M., the shrine keeper, assisted by a committee of worship officials, presents an elaborate series of offerings consisting of joss sticks, candles, wine, and cone-shaped cakes of sticky rice in front of the shrine building (fig. 16.4). The men move around the space in carefully choreographed strides to the accompaniment of a drum, gongs, and bowed lutes. One of them reads a handwritten text that tells the story of Ngô Thị Thanh and asks for her blessing (fig. 16.5). After the reading, the manuscript is burned in front of the altar (fig. 16.6). The entire ceremony is solemn and stately, as it is intended to pay tribute to Ngô Thị Thanh. The regulated movements of the officials, striding slowly through steps in the cardinal directions, reflect the forms of royal rituals that once prevailed in Vietnam's dynastic courts.

The opening ceremony is followed by the first performance of the *Four Occupations of the People,* which will be performed a second time at the end of the festival. After the performance, the crowd disperses to await the midnight fertility ritual. At about 11:30 P.M., activity resumes inside the shrine, conducted now by a special ritual leader appointed by the village. He must be a respectable, healthy man over sixty years old, born of an affluent and high-class family; he must also be able to sing well, have at least one son, and not be in mourning. He must have been born under certain stars and his age must not be inauspicious. Wearing a long, black silk robe, white trousers, a turban, and black shoes, he begins the ritual by washing his hands in a water basin placed on a shelf near the incense table. Afterward, he sprinkles water on the incense table, the altar, his body, his hands, and his face, in an act called "washing away dirtiness" (*quán tẩy*). He then burns three joss sticks, places them in front of the altar, kowtows, and prays:

> I pray under my breath: glory to Buddha!
> I ask your permission to take out the sacred objects to start the ceremony

> Con niệm Nam mô A Di Đà Phật
> con xin Chúa Bà mang vật linh ra làm lễ

Next, holding an imitation lute, he sits in front of the altar (fig. 16.7). He strikes the lute with a stick, beating out a rhythm to accompany a ritual song, which consists of two parts. First, it praises Ngô Thị Thanh's charitable deeds, in particular teaching the people of Trám to cultivate rice, raise silkworms, and weave cloth. Secondly, it explains that

16.4   One of the group of worship officials carries a candle toward the shrine during the offering ceremony. Photograph by Vũ Hồng Thuật, Trám village, 2001.

16.5   Handwritten texts are read at several points over the course of the festival, as here during the evening offering ceremony. Photograph by Nguyễn Anh Hiếu, Trám village, 2001.

16.6   The manuscripts are burned at the altar after they have been read, sending their message on to the deities. Photograph by Nguyễn Anh Hiếu, Trám village, 2001.

Trám villagers, in remembrance of her public virtue, should perform the *Four Occupations of the People* in order to gain her support in vouchsafing long lives and bumper crops. The song's melody is similar in style to the chamber vocal music (*ca trù*) that is popular in Vietnamese communal halls, temples, shrines, and inns.

The ritual leader then reads a formal text prepared beforehand by a learned villager, again asking the goddess for her permission to remove the sacred objects from their storage place behind the altar:

> Our goddess Ngô Thị Thanh, a pure figure with a shining spirit
> that needs no jewels, a predecessor with an authoritative tone and
> a sacred soul! Five disasters—crop losses, disease, young death,
> accidents, and being bitten by a snake and pounced upon by a
> lion—are offered so that you can help us villagers protect our
> crops, temples, and shrines over millions of years. Twice a year, in
> winter and spring, we worship you with full rituals to remember
> scholars, peasants, workers, and traders. But that is not enough to
> make us healthy and prosperous, is it? Those things depend upon
> your great help. We bow our heads to receive this with a kowtow.

Afterward, the paper text is burned and the ritual leader climbs to a sacred chamber located above and behind the altar to remove the box containing the *nõ nường*. He returns with them to the altar, opens the box, and places the objects on a tray on the incense table. At 12:00, he burns three joss sticks and kowtows three times.

Now a young man and woman come forward—they are supposed to be post-pubescent, unmarried youths, but in recent festivals only young, married couples have agreed to participate in this ritual. The two young people stand in front of the altar facing each other, and the temple keeper gives the sacred objects to them. The young man holds the wooden phallus in erect position below his waist. The woman likewise holds the wooden vulva in place (fig. 16.8).

The shrine keeper asks: "How does this work?" (*Cái sự làm sao?*). The couple reply: "It works like this" (*Cái sự làm vậy*). The couple approach one another more closely and the temple keeper intones: "*Linh tinh tình phọc.*"[8] The couple replies: "Then [we] thrust into each other" (*Thì chọc vào nhau*). The young man thrusts the wooden phallus into the wooden vulva three times (fig. 16.9). If the two sacred objects fit smoothly all three times, the villagers say it is a sign they will have a bumper crop in the coming year. If the thrusts miss their mark, crops will be bad, floods and storms will occur, or other misfortunes are expected.

Now events move quickly. The shrine keeper retrieves the sacred objects from the couple and at the sound of a drum, the couple races out into the dark. They are accompanied by a crowd of youths from the village, all beating on gongs or drums as they run around the shrine three times to drive away evil spirits. Afterward, when in the past the sacred objects would have been burned and the ashes distributed to the crowd, there is now a brief ceremony of returning the *nõ nường* to their storage place. Some of the participants remain on the shrine grounds playing cards or other games until dawn, a modern substitution for what in former times, according to village elders, took place following the coupling of the artificial organs: a time of release (*tháo khoán*) from the normal rules of sexual behavior. In the past, some people were allowed on this one occasion to flirt publicly and engage in sex freely behind the shrine. Pregnancy that resulted from the activities of the night of the festival was regarded as a good omen for a woman, her family, and her village. Such practices are no longer regarded as an acceptable part of the festival and have not taken place since it was restarted in 1993.

The Procession of the Sacred Rice Bundle

The morning after the midnight fertility ritual, a sacred bundle of rice stalks is carried in procession from Trám Shrine through the rice fields and back to the shrine again. The bundle is of great importance to any understanding of the festival. Each year, one village elder is carefully selected on the basis of various auspicious signs (his age, for example, must be an auspicious number, he must have many grandchildren, and his family must not be in mourning) to reproduce the sacred rice bundle. He begins a few weeks after the conclusion of the previous year's festival by retrieving the previous year's bundle from its storage place in Trò Temple. He plants the rice grains contained in this bundle, and a few weeks later transplants the seedlings into a special plot where they are grown under specific restraints (for example, no fertilizer can be used). At harvesttime, he selects only the highest quality stalks with abundant plump grains and uproots them by hand rather than cut them with a knife, which is held to be injurious to the plants. From these stalks he makes a new sacred rice bundle, which formerly was stored in the temple until the following festival, but in recent years has been kept in a village home where it can be better protected. At the time of the next Trò Trám festival, early on the morning of the procession he delivers the bundle to the home of the Trám Shrine keeper. This bundle, both genetically and symbolically, represents the continuity of the rice crop from generation to generation and, at least in theory, assures the perpetuation of an indigenous variety of rice.[9]

The shrine keeper arranges the sacred bundle of rice in a vase, to which he adds the top of a sugarcane plant. The cane leaves replicate luxuriant paddy plants, transforming the entire arrangement into a seemingly prodigious bundle of newly harvested rice. The vase is brought to the shrine and the day's ceremonies begin. The shrine keeper offers to Ngô Thị Thanh some sticky rice on a banana leaf, which is placed on a tray. He then sets a boiled pig head on top, together with other offerings, and reads a formal text asking Ngô Thị Thanh's permission to hold the festival.

Next, the vase with the sacred rice bundle is placed on a red and gilt palanquin and offerings are arranged around it (fig. 16.10). A parasol is erected to shade the bundle from the sun and rain, and also to mark its exalted status. Four strong men will carry the palanquin in procession, preceded by an honor guard carrying brightly colored festival flags, a dance group performing in unison inside a large dragon costume, and a group of musicians (fig. 16.11). The shrine keeper, the *Four Occupations of the People* theatrical group, and the village elders march behind (figs. 16.12, 16.13).

In former times, the parade route passed through rice fields before returning to Trám Shrine, but recently population growth has altered the landscape, and further changes to the route have been made for political reasons. In 2001, the Communal People's Committee ordered that the parade proceed from the temple to a large, open field in front of the committee's office building, then through village streets, and finally around a lake, before ending at the shrine.

During the procession through the fields, the master of ceremonies howls "huoc hu...hu...hue!" in time with the beating of drums. When the musicians strike the gongs, the people in the procession shout and howl for joy, causing a stir throughout the fields so that the earth itself will feel the commotion. Then prayers to the rice spirits and the gods of agriculture, thunder, clouds, rain, and wind are intoned to ensure a good harvest:

> Dear Sir Paddy, Madame Paddy
> The god of agriculture
> The gods of thunder, cloud, rain, and wind
> Please bless our villagers to have favorable rain and winds, green and luxuriant paddy and maize plants, and bumper crops.

> Hỡi Ông Lúa, Bà Lúa
> Thần Nông
> Thần Sấm, thần Mây, thần Mưa, thần Gió
> Giúp cho dân làng mưa thuận, gió hoà, lúa ngô xanh tốt, mùa màng bội thu.

Shortly before the procession reaches Trám Shrine, the shrine keeper sets off firecrackers (now replaced by hitting a length of bamboo on the ground), and the shrine drums and gongs are beaten. These sounds welcome the god of agriculture, Sir Paddy, and Madame Paddy, who symbolically have been escorted from the fields to the shrine by the procession. Some villagers say that the sounds also drive away any evil spirits that may have rushed into the shrine to hide themselves when the people were shouting for joy in the fields.

When the procession arrives, the shrine keeper removes the sacred rice bundle from the palanquin and ceremonially carries it into the shrine. With the return of the sacred paddy bundle, the procession is complete and the food offerings that have been made to the sacred bundle of rice are served as a feast to the assembled crowd. Afterward, the bundle will be moved to Trò Temple and stored until it is collected a few weeks later for planting.

16.7   The ritual leader for the
midnight ceremony sings
his song in front of the altar
while beating out the rhythm
with a stick. Photograph
by Vũ Hồng Thuật, Trám
village, 2001.

The Four Occupations of the People

For local people, the greatest attraction of the festival is not the midnight fertility ritual
or the procession of the sacred rice bundle, but the public performance of the *Four
Occupations of the People.* The play is named for the four traditional occupational groups
of Vietnamese society: scholars (*sĩ*), farmers (*nông*), craftsmen (*công*), and traders (*thương*).
In the actual performance, a wide range of activities is depicted, vividly reflecting village
life. The skits include a lute player, fishermen, carpenters, cotton spinners, a king pulling
a plow behind an elephant, women transplanting rice, the "Seller of Spring," a mandarin
schoolteacher, etc. The actors and actresses mime their songs, which are actually sung by
a female vocalist accompanied by a group of musicians (fig. 16.14).

The skits are full of poignancy, humor, and sexual innuendo, while the music
is strong, smooth, and sentimental (as well as heavily amplified). After every two bars, the
singers often shout, "Oo-aye!" which gives the performance a wild and ancient sound in
the opinion of the listeners. The poetic lyrics provide important clues to the symbolic

16.8   The ritual leader says a prayer while a young woman holds the wooden vulva in position in front of her. Photograph by Nguyễn Anh Hiếu, Trám village, 2001.

16.9   The thrust of the wooden phallus into the vulva. Photograph by Nguyễn Anh Hiếu, Trám village, 2001.

significance of the activities depicted. Although each song belongs to a particular character, often all the other actors, still in their own costumes, join in the dance as the song is sung. The result is a chaotic feast for the eyes and ears of the audience.

The performers and musicians are all drawn from the village itself. Formerly, Trám village was divided into upper and lower wards, nowadays known as upper and lower hamlets. If the upper hamlet organizes the performance one year, the lower will host it the next. The performers register to practice their roles a month prior to the festival at the hamlet chief's house. Even the dress rehearsal held on the night before the festival draws a large crowd.

During the 2001 festival, the full play was performed twice, first in the yard in front of Trò Temple before the midnight fertility ritual, and then again the next morning in front of the Communal People's Committee office building midway through the procession of the sacred rice bundle. In former times it would have been performed in the rice fields, which are fallow at this time of year. No matter what the location, a large open circle is cleared on flat ground, surrounded by the crowd. All of the actors remain inside

16.10   The sacred rice bundle and offerings ready for the beginning of the procession. Photograph by Roy W. Hamilton, Trám village, 2001.

16.11   The dragon or "unicorn" (lân) resembles those featured in Chinese New Year processions around the world. Photograph by Vũ Hồng Thuật, Trám village, 2001.

16.12   The procession gets underway with a drummer in front of the palanquin and the shrine worship officials following immediately behind. Photograph by Vũ Hồng Thuật, Trám village, 2001.

16.13   The costumed entertainers from the *Four Occupations of the People* join the procession. King Thuấn follows behind his costume elephant. Photograph by Vũ Hồng Thuật, Trám village, 2001.

The musicians play their
stringed instruments during
the nighttime performance
of the *Four Occupations of
the People.* Photograph by
Vũ Hồng Thuật, Trám
village, 2001.

the circle waiting for their time to perform. A few of them perform continuously
throughout the entire play, including a master of ceremonies (fig. 16.15), a clown with a
bullhorn made from a fish trap, and two transvestites who dance in a humorous manner
and mock the actions of the other performers (fig. 16.16). As the performance gets underway, the master of ceremonies shouts over the crowd:

> Oh la...la...la!
> Please step back so that our troupe can perform their games
> Oh la...la...la!
> May I invite the hamlet's security officials [*quan viện*] to attend the show
> Oh la...la...la!
> Villagers and guests move back so that King Thuấn can start plowing!

As his words end, gongs and drums sound and the show begins. The master
of ceremonies approaches a man holding a tall bamboo pole on which are mounted four
winnowing baskets. Each basket is painted with a character in Nom (the system of
Chinese characters in which Vietnamese was written before a modified Roman alphabet
was devised in the sixteenth century). The baskets read: *tử, dân, chi,* and *nghiệp.* The master of ceremonies points one by one to the four characters and asks the crowd each time,

"What does this character mean?" (fig. 16.17). The sign holder will answer each query (lit., the characters mean "four," "people," "branch or group," and "occupation"). This didactic device is performed in a humorous manner to make it entertaining as well.

The Lute Player

As the skits get underway, the master of ceremonies announces each act, starting with a musician whose oversized "lute" is made from a winnowing basket. The performer mimes while the singer sings:

> ...Beautiful girls in their twenties
> Let the young men transplant, the rice is not yet in bloom
> Girls who are transplanted by the young men
> The rice seedlings are luxuriant, in ear the rice is also big
> Old women are like dry hillsides
> Young girls are large fields and soft
> When transplanting in them, it is cool and smooth.
> The mud is pulpy soft, the seedlings are deeply planted.[10]

> ...Hai mươi, hai mốt đang tuổi dậy thì
> Để cho trai cấy mấy khi cấn đòng
> Cô nào trai cấy đã xong
> Cân cấn cũng tốt, đòng đòng cũng to
> Bà già như ruộng đỉnh gò
> Đang hạng con gái ruộng to, đất mềm
> Cấy vào vừa mát vừa êm
> Bùn thật nhão nhuyễn, mạ thì cắm sâu.

The Plowing King

The next skit tells the story of Thuấn, a legendary Chinese king who saw his people suffering and gave up his military pursuits to join them in plowing. Four performers are involved; one plays the king, one a farmer, and two activate from inside a huge elephant made of a bamboo frame covered with cloth. The king works the elephant like a draft animal, plowing a circle around the performance space. The elephant has lifelike movements and teases girls in the audience with its tail. When the king stops to rest, the elephant also ponderously lies down. The farmer smokes his rustic tobacco pipe and sings:

> A king should deserve his position as a king
> A king should make his citizens pleased
> An enlightened king should be a progressive king
> He should know about people's hardship
> Our people still have many difficulties
> So I have to stop my military affairs to go plowing.

> Làm vua phải ở cho vừa lòng dân
> Thánh quân cho đáng thánh quân
> Thánh quân phải biết thương dân nhọc nhằn
> Dân ta còn lắm khó khăn
> Âu ta phải gác việc quân đi cày
> Các câu hát nhằm đề cao việc canh nông của người đi cày.

The Women Transplanting Rice

The next skit involves three middle-aged women and three virginal girls, who all mimic the transplanting of rice. One passage of their song, a thinly veiled slang description of sexual intercourse, elicits blushes from young women in the audience and howls of laughter from the adults:

> We are like chopsticks made of older bamboo
> The more they are sharpened, the smoother they are
> We are like paddy in full richest bloom
> Well matched for each other but not pleasing our parents.
> We are like three-plied yarn
> Our parents spin little but we have done much!

> Tôi ta như đũa tre già
> Càng vót, càng nhẵn, càng và càng tron...
> Đôi ta như đũa dòng dòng
> Đẹp duyên nhưng chẳng đẹp lòng mẹ cha
> Đôi to như chỉ xe ba
> Thầy mẹ xe ít, đôi ta xe nhiều.

The Fisherman

The fisherman character, named Hàn Tín, holds a long bamboo fishing rod with a "fish" made of a palm spathe tied to the end of his line. As he rounds the circle, Hàn Tín casts his line out over the audience, aiming in particular to tease young women in the crowd. He conspicuously plants the base of his long fishing pole in his crotch as he waves it over the audience, miming his song:

> ...Others want to fish for carp and anabas
> I only want to fish for an unmarried girl
> Others go fishing in rivers and seas
> I want to fish your daughters, your nieces.
> So you married ones, release my bait!
> Unmarried ones can bite, swallow, and take it away!

> Người ta câu riếc, câu rô
> Anh đây câu lấy một cô chưa chồng
> Người ta câu bể, câu sông
> Anh đây câu lấy con ông, cháu bà
> Có chồng thì nhả mồi ra
> Không chồng thì cắn, thì nuốt, thì tha lấy mồi!

This skit is based on a legend telling how, on the twelfth day of the first lunar month, Trám village traditionally organized a matchmaking ceremony for couples in the Trám Shrine courtyard. The young men used fishing rods to drop "bait" to the young women they wanted to marry. If a woman accepted, she took the bait and gave it to the elders, asking for permission to marry the young man. If a married woman caught a man's bait, he had to break his line and quit the game until the following year's festival (giving the older women some control over the young men). In the version performed as part of the *Four Occupations of the People*, the fisherman eventually reels in a willing young woman, who is actually one of the actors planted in the audience.

The Spinners

A group of women characters perform the next skit carrying the tools of the textile-making trade, including spinning wheels and warping frames. This skit is based on a legend that tells how Ngô Thị Thanh taught the village women to weave cloth. The women sing a song that bemoans their long hours of work for precarious reward, while at the same time they manage to tease the fishermen:

> The girls get only seven *dong* for their cotton
> Ginning and spinning, and not for a day
> Spinning the thread as thick as a wrist
> So fishermen come to buy rope for their boats
> This thread I sell at the weekly market
> Buying expensive, selling it cheap, five *dong* for three
> Having sold, I go for a snack
> Spending thirty *dong* for breakfast
> When the sun is at the top of the pine
> I buy more snacks, and the money is gone.

> Con gái lấy bảy tiền bông
> Vừa cán, vừa kéo chẳng thông một ngày
> Sợi lôi ra bằng cổ tay
> Phường chài đến hỏi mua dây kéo thuyền
> Sợi này tôi bán chợ phiên
> Mua đắt, bán r năm tiền được ba
> Bán xong lê lại hàng quà
> Sáng ngày xúc miệng hết ba mươi đồng
> Mặt trời ngả ngọn cây thông
> Mua quà lần nữa món bông toi đời.

The Schoolteacher

The schoolteacher skit (fig. 16.18) provides a comic interlude to which the audience responds with great enthusiasm. The schoolteacher, a kindly old man, in turn abuses and is abused by his students. His lessons consist of cockeyed explanations of the meanings of Nom characters that are full of puns, sexual innuendo, and political satire, which only confuse his students all the more. They bring him books, asking, "Sir, what does this mean?" The teacher uses his forefinger to draw in the air a supposed Chinese character in the shape of a vulva. He says:

> The above character means under.
> The word *under* means above.
> The middle means around.
> The around means the middle.

> Chữ trên là trên chữ dưới
> Chữ dưới là dưới chữ trên
> Chữ giữa là nửa chung quanh
> Chữ quanh là vành chữ giữa.

16.15   The master of ceremonies for the *Four Occupations of the People*. Photograph by Vũ Hồng Thuật, Trám village, 2001.

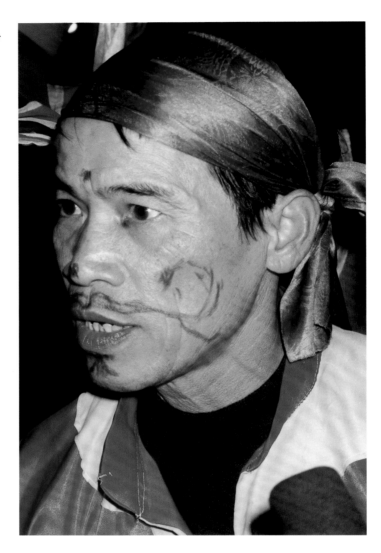

Student: Dear Sir, teacher, what are these characters?

Teacher: *Vương* and *Bần*

Student: What do they mean?

Teacher: *Vương* means that the king acts on behalf of Heaven to maintain the country.

A fisherman pipes in: And *Bần* means poverty.

Student: And what are these characters?

Teacher: *Tam hoàng chi cương.*

Student: What do they mean?

Teacher: Make beds really solid.

At the end of the skit, the teacher tells his students that they must forget everything he has told them so that he will have something to teach them again tomorrow.

The "Seller of Spring"

This is perhaps the most poignant of all the skits, as the Seller of Spring represents Ngô Thị Thanh herself. Her performance suggests that it was only though her sacrifice of her own youth and fertility that the people of Trám could learn how to grow rice and weave cloth. She carries a pole across her shoulder bearing two winnowing baskets on which are written the Chinese characters for "buy spring, sell spring" (*mua xuân, bán xuân*). While she gracefully circles the performance space, she cries out, "Who wants to buy spring?" (fig. 16.19). Her song is sweet and plaintive:

Buy spring now before it is gone
If you delay it will no longer be spring
Play with spring before you get old
Play with the moon before it sets.

Mua xuân kẻo hết xuân đi
Mai lần nay nữa còn gì là xuân
Chơi xuân k o nữa lại già
Chơi giăng k o nữa giăng tà về tây.

The songs and skits described here represent only a portion of the total, but they give an idea of the character of the lyrics and especially the central role played by sexual innuendo. This feature is a further reiteration of the fertility theme of the festival. Many other social and political themes are also broached in the performances, such as the commentary about a ruler's obligations in the song of the plowing king. Other aspects of the performance derive their meaning from the social organization of the community. For example, in former times the fishing people of the Red River Delta lived on the waterways in their boats and were looked down on by the delta's farmers (who are, after all, producers of rice). Nevertheless, farmers and fishermen were considered complementary, their inter-action required to create an economic "whole" that neither could provide alone (in short, one needs fish to eat with rice). The constant flirtation in the play of the female characters with the male fishermen takes on an added dimension in this light, as it expresses the intercourse of the social and economic community, which is the most essential meaning of the *Four Occupations of the People*.

Integration

The midnight fertility ritual, the procession of the sacred rice bundle, and the perform-ance of the *Four Occupations of the People* are all integrated seamlessly into the Trò Trám festival, and the people of Trám do not consider them to belong to separate cultural tradi-tions. Nevertheless, an analysis of their primary themes reveals the diverse sources that have shaped this eclectic festival.

Undoubtedly, the oldest strain evident in the festival is that invoking the fer-tility of the rice crop and, concomitantly, of the human community. Special procedures for storing, planting out, and reproducing the seed rice, such as those involved in perpet-uating the sacred rice bundle, are found widely throughout the rice-growing regions of Asia. The symbolic equation of human reproduction with the fertility of the crop is also widespread. In most parts of Southeast Asia, this idea is an expression of the region's spirit beliefs, in which a spirit that needs to be nurtured from year to year is attributed to the rice plants. Perhaps most distinctive of Vietnam is the presence of an additional layer of interpolation based on Taoism, in which ritual can be seen not only as an expression of spirit beliefs but also as a balancing of yin and yang necessary for all living things to multiply and grow. Similarly, what in other Southeast Asian societies might be seen as an appeasement of natural spirits, in Vietnam takes the form of the worship of Taoist deities such as Thần Nông (the god of agriculture).

The second major strain in the festival revolves around the worship of Ngô Thị Thanh. The deification of this folk heroine probably does not predate the Trần dynasty and is not known to have been formalized by court edict until the eighteenth century, but nevertheless clearly reflects the strong influence of the Taoist and Confucian ideology that developed in Vietnam under Chinese domination. While formal religious Taoism is based on the erudite teachings of the sixth-century B.C.E. sage Laozi, popular

16.16   Actors in the *Four Occupations of the People*, include two transvestite comics in white robes. Photograph by Nguyễn Anh Hiếu, Trám village, 2001.

16.17   The master of ceremonies quizzes the man holding the sign bearing the characters for the "Four Occupations of the People" (*tứ, dân, chi,* and *nghiệp*). Photograph by Nguyễn Anh Hiếu, Trám village, 2001.

16.18   The schoolteacher and his recalcitrant students. Photograph by Nguyễn Anh Hiếu, Trám village, 2001.

16.19   The Seller of Spring advises
all who listen to enjoy
spring while they can. She
represents Ngô Thị Thanh,
who sacrificed her youth and
fertility so that the people of
Trám could grow rice.
Photograph by Nguyễn Anh
Hiếu, Trám village, 2001.

Taoist traditions in China came to include the worship of a myriad of local and regional deities (Little 2000, 13). Many Taoist deities, like Ngô Thị Thanh, are in fact quasi-historical figures who have been raised to divine status. There is also a Confucian element to Ngô Thị Thanh's deification as it became institutionalized within the context of Chinese ideas of governance and the deliberate perpetuation of Confucian ideals in Vietnamese society. This leaves what is perhaps the most interesting question about Ngô Thị Thanh unanswered: can she be regarded as a Taoist/Buddhist transformation of the "Rice Mother" or "Rice Goddess" found in many other Southeast Asian cultures? While it is difficult to either prove or disprove this, to the authors it seems quite likely that earlier beliefs about a Rice Mother later transformed into worship of Ngô Thị Thanh. After all, both are revered primarily for their role in bringing the cultivation of rice to humans.

The third major strain in the festival, represented by the performance of the *Four Occupations of the People*, was probably not introduced into the festival until about a hundred years ago, at the beginning of the modern era.[11] The traditional songs, costumes, music, and performances are deeply reflective of a local cultural identity based on rice farming, fishing, and similar occupations. While the performances are valued primarily as an entertaining spectacle, they are nonetheless useful in delineating the social relationships that prevail within the community. These are, by and large, indigenous social developments of the Red River Delta, though of course they cannot be divorced from the broader historical context.

•

17

Of Mites and Men: The Shōrei Festival at Mount Haguro, Japan

Toshiyuki Sano and Roy W. Hamilton

Dewa Sanzan, the cluster of three sacred peaks in northern Honshu's Yamagata Prefecture, forms a spectacular backdrop for the fertile Shōnai Plain, one of Japan's most important rice-producing areas. A famous historic Shinto shrine, Dewa Sanzan Jinja,[1] is situated high on the forested slopes of the lowest of the three peaks, Mount Haguro (414 meters). This shrine is the site of an old and unique set of rituals known as the Shōrei festival (Shōrei-sai),[2] which is conducted in the post-harvest season at New Year when the shrine and the mountains are buried deep in snow (fig. 17.1).

The three sacred peaks, which include the higher summits of Mount Gassan (1979 meters) and Mount Yudono (1504 meters) in addition to Mount Haguro, have an exalted place in Japan's religious traditions, and many Japanese strive to make a pilgrimage there at least once in their lifetime. This practice, called Oku-no Sanzan Mairi (lit.,"pilgrimage to three mountains deep in the north"), forms a pair together with Ise Mairi (lit., "pilgrimage to Ise Shrine," in central Japan). The pairing is based on an idea of the complementarity of the nighttime and daytime deities: Tsukiyomi-no-mikoto, the deity of Mount Gassan, is said to preside over the night world while Amaterasu Ōmikami, the deity of Ise Shrine, is said to be in charge of the day.

Dewa Sanzan and its shrines are also associated with an esoteric Japanese religious sect called Shugendō, which combines elements of Buddhism, Shinto, and spirit worship. The original practitioners of Shugendō were wandering mountain ascetics known as *yamabushi*. By the Edo period (1603–1868), which marked the height of Shugendō influence, the sect was established on 134 mountains in Japan (Earhart 1970, 33), and the *yamabushi* had become settled and institutionalized. Often married to female shamans, they served their local communities as prayer leaders and healers (Earhart 1970, 34–35).

Mount Haguro historically was one of the strongest centers of Shugendō. Haguro Shugendō is said to have been established about fourteen hundred years ago by Prince Hachiko, a cousin of Prince Shotoku (574–622), one of the founders of the Japanese state and an early proponent of Buddhism in Japan. Prince Hachiko came to the Shōnai Plain to escape conflicts occurring in the Nara area. His exploits there have entered the realm of legend; for example, from the shore of the plain where he landed, a crow with three legs is said to have guided him to Mount Haguro.

17.1 Deep snow covers the buildings of Dewa Sanzan Jinja. Photograph by Corinne Frugoni, Mount Haguro, Yamagata Prefecture, 2001/2002.

17.2   A *tsuna* rice-straw rope hangs under the eaves of a home in Tōge. Photograph by Roy W. Hamilton, Mount Haguro, Yamagata Prefecture, 2000/2001.

17.3   The sacred grains are contained in the *kōya hijiri*, shown here on one side of the double altar at Dewa Sanzan Jinja. A tiny model sickle and hoe are attached to each side of the beehive-shaped model hut. The mirror above is a common Shinto symbol representing the divinities. When the altar is fully decorated, the shelves will hold sake bottles, rice cakes, salt, oranges, and other offerings. Photograph by Roy W. Hamilton, Mount Haguro, Yamagata Prefecture, 2000/2001.

In addition to their religious significance, the sacred mountains also have a practical importance for agriculture in the Shōnai Plain. They are the source of plentiful water throughout the year due to the snow fields that linger on the mountains even in the warmest summers. Because of its natural beauty and vital ecological role, the entire area encompassing the three peaks has been designated a national park.

The peaks have even played a role in Japanese literature, as they were visited by Matuo Bashō, the great haiku poet of the Edo period. From Edo (now Tokyo), it took Bashō many days on foot to reach Dewa Sanzan. Today a five-hour train trip brings pilgrims and other visitors to Tsuruoka station, whence a fifty-minute bus ride takes them across the Shōnai Plain and up to the top of Mount Haguro. At the base of the mountain the bus passes through the village of Tōge, where the needs of pilgrims are served by about thirty traditional lodgings.

Visitors to Tōge will notice that many houses there and in other villages nearby have a special form of decoration hanging under the eaves, made of thick rice-straw hawsers (*tsuna*, fig. 17.2). This rope is held to have sacred power to protect the house and its occupants from fire or evil spirits. The *tsuna* play an important part in the Shōrei festival and afterward are given to those who have made special donations to support the festival. Only eight or ten of these rope ornaments are available each year.

The Shōrei festival is today one of the official Shinto functions of Dewa Sanzan Shrine. Looking more closely, however, it is apparent that the festival's participants are the people of Tōge who adhere to a contemporary local version of the Shugendō sect. Early in the Meiji period (1868–1912), when Shinto was made the state religion in Japan, Shinto and Buddhist practices were forcibly separated. This had an enormous effect on mixed sects such as Shugendō. The government ordered the abolishment of Shugendō in 1873, and the *yamabushi* nearly vanished as a significant part of Japanese religion. At Mount Haguro, the Buddhist temples that had been a part of the local Shugendō establishment closed and the rituals of the Shōrei festival were discontinued. Soon, however, the festival was reorganized in a reformed manner under the jurisdiction of the Shinto shrine. Records kept at Mount Haguro listing the names of participants indicate that the festival skipped only a few years, from 1874 to 1877 and 1879 to 1881. The reforms brought many changes to Shugendō and to the Shōrei festival, which has also continued to evolve over the intervening years. The reforms have included lifting the former prohibition of women at the site (though they still do not perform rituals), easing the lengthy and severe period of abstinence required on the part of the festival's ritual leaders, and adjusting the timing of the festival to match the Western solar calendar. Today many Japanese might recognize the distinctive blue-and-white checked garb of the *yamabushi* (see fig. 17.5), but few are aware of the existence of contemporary Shugendō and its esoteric rituals.

Despite these changes, the Shōrei festival remains a complex series of interrelated ritual events that commences in September and culminates one hundred days later on New Year's Eve. For the one-hundred-day period, two ritual leaders called *matsuhijiri* (lit., "pine saint") in theory engage in a period of abstinence and confinement. Under the recent reforms, however, they are in practice fully confined only in the final days leading up to the New Year, during which they reside in the Purification Hall (Saikan) of Dewa Sanzan Jinja. New *matsuhijiri* are selected every year from among the elder men of Tōge who have passed through many stages of training in Shugendō practice. Each *matsuhijiri* represents half of the village of Tōge, which is divided into upper (*ijō*) and lower (*sendo*) wards. To be selected as *matsuhijiri* is an honor that fulfills a lifetime ambition for the men of Tōge who adhere to Shugendō.

17.4   The completed *tsutsuga-mushi* of the two wards lie side by side. The fronts are covered with intricately knotted mesh made of fine rice-straw rope. Large knotted hawsers (*tsuna*) hold the internal filling of dried grasses together. Photograph by Roy W. Hamilton, Mount Haguro, Yamagata Prefecture, 2000/2001.

In addition to gaining spiritual strength and power through abstinence for one hundred days, the *matsuhijiri* are also responsible for conducting prayers in the Saikan before a double altar, with one side for each ward. On each altar is a container filled with the five sacred grains (*go-koku*), including rice.[3] The containers, called *kōya hijiri*, are small models of the field huts that used to be common in Shōnai rice fields (fig. 17.3). The model huts and the grains they contain are important symbols for the Shōrei-sai. The first rituals that the *matsuhijiri* practice during their confinement are conducted to invite the divine spirits (*kami*) of the five grains to take up residence in the grains on the altar. For the duration of the confinement of the *matsuhijiri*, the deities symbolically reside in the *kōya hijiri*, representing hope for the fertility of the coming year's crops. The concluding ritual of the festival frees them to return to their places of origin.

Agricultural ritual and Shugendō practices are closely intertwined based on the legendary exploits of Prince Hachiko. When the prince and his followers arrived in Shōnai, they began the clearing of the plain and its conversion to agricultural use, creating a new landscape dominated by rice fields. According to legend, there soon followed a calamity in which many people died as a result of an epidemic of infectious disease. To relieve the people's suffering, Prince Hachiko practiced rites of asceticism in the sacred mountains and eventually received a divine message telling him to collect every piece of the straw from the harvested rice and burn it. The people of the plain did as he suggested, and the epidemic ended.

The exploits of Prince Hachiko presumably are apocryphal, but there is nevertheless a strong ecological corollary to these legends. There is evidence of limited cultivation of rice in northern Honshu as early as the Yayoi period (1000 B.C.E.–250 C.E.), but it was not

17.5   The *matsuhijiri* of Tōge's lower ward, with his assistants, pose in front of the ward's *tsutsugamushi*. The man on the left wears the distinctive costume of a *yamabushi*. The writing on the wooden plaque identifies the local people who have contributed to the festival. Photograph by Roy W. Hamilton, Mount Haguro, Yamagata Prefecture, 2000/2001.

17.6   The crowd scrambles to catch a section of *tsuna* in the midst of a heavy snowfall. Photograph by Corinne Frugoni, Mount Haguro, Yamagata Prefecture, 2001/2002.

until the fifth and sixth centuries C.E. that rice agriculture significantly expanded in this region from its earlier center in western and central Japan. This expansion depended on the development of new varieties appropriate for northern latitudes. And there was another barrier as well, the *tsutsugamushi*, a species of mite (*Leptotrombidium akamushi*) that thrives in inundated alluvial plains and transmits a disease called scrub typhus or Japanese river fever.[4] The disease is caused by a microorganism, *Rickettsia tsutsugamushi*, which is carried from infected wild rats to humans by the bite of the chiggers (the larval stage) of *tsutsugamushi* mites. Even in modern times, Japanese river fever causes fatalities in northern Honshu. In Prince Hachiko's time, it could well have spread in epidemic proportions with the expansion of rice fields in the Shōnai Plain. The burning of the straw following the harvest presumably evolved as an effective method of controlling the pest by reducing its habitat, thus allowing the successful transformation of Shōnai into highly productive rice land. Today, with the ever-increasing urbanization of western and central Japan, Yamagata and the neighboring prefectures of northern Honshu have become the vital rice bowl of the nation.

The central ritual performances of the Shōrei festival directly reflect the themes of the story of Prince Hachiko's exploits. On December 30 two groups of young men, representing the upper and lower wards of Tōge, assemble on an open field in front of the shrine. There they fashion two giant "*tsutsugamushi*" from rice straw and dry grasses bound together with rice-straw rope (fig. 17.4). The intricate knotting of the rope gives the giant models a vaguely insect-like look, though they are not accurate renditions. The models are also called *ōtaimatsu* (large torch), which reflects how they will eventually be used. The two groups compete to finish their work, which requires rolling the heavy "insects" in order to wrap them completely around with thick rice-straw hawsers (the same *tsuna* from which the house decorations are made). This element of competition between the two wards continues through nearly every ritual for the remainder of the festival. When the two straw *tsutsugamushi* are complete, the two teams pack down the snow on the field to make a firm surface and erect two posts at the far end. These will play a prominent role the following night.

Early the next morning, the two *matsuhijiri* perform their last prayers before the altar in the Purification Hall that bears the two *kōya hijiri*. The *kōya hijiri* are then packed up and carried to a building known as the Shitsurae-ya, a large community hall

17.7   The remaining *tsuna* are carried to the Shitsurae-ya where they are cut into pieces to give to donors. The entrance to the Shitsurae-ya that serves the upper ward is behind the workers. The door for the lower ward is out of view to the left. Photograph by Roy W. Hamilton, Mount Haguro, Yamagata Prefecture, 2000/2001.

17.8   One of the two-man teams challenges another. At the end of their verbal challenge, both teams must drink the sake that has been poured into their cups. Photograph by Roy W. Hamilton, Mount Haguro, Yamagata Prefecture, 2000/2001.

17.9

The *matsuhijiri*, his assistants, and a group of young men from the upper ward of Tōge sit on the platform in the Shitsurae-ya. The *matsuhijiri* faces the *kōya hijiri* on the altar, where the deities of the grains are symbolically in residence. In this ritual, the young men are provided with wooden snow shovels, which they will use in further rituals conducted outside in the snow. Their ability to withstand the cold without warm clothing is considered a sign of their bravery and endurance. Photograph by Roy W. Hamilton, Mount Haguro, Yamagata Prefecture, 2000/2001.

that stands just outside of the main torii gate of Dewa Sanzan Shrine. This interesting building is made in a rustic style, divided into two halves entered by separate doors (see fig. 17.7). The left side is for the lower ward of Tōge and the right side for the upper ward. Inside, each half of the building is identically equipped with an altar raised above a seating platform and, further from the altar, a fire pit surrounded by benches. The activities that take place in this building are all performed in duplicate, with each ward conducting separate but simultaneous rituals on its own side of the Shitsurae-ya. A passage at the back of the building allows access from one side to the other, but a dividing wall visually separates the activities of the two wards. The *kōya hijiri* are installed on the altars and the two *matsuhijiri* enter the building through their respective entrances. They then take up their positions on the seating platforms, where they will remain for most of the time until the final rituals are completed in the predawn hours. The young men of the wards, as well as any spectators, gather on the benches around the fire pit to warm themselves and to pass the time eating and drinking. Women from the two wards make rice balls to feed the assembled crowd and sake is dispensed to those who come to pay their respects to the *matsuhijiri*.

In the early afternoon, the two groups of young men return outdoors to the straw *tsutsugamushi*. They unwrap the hawsers that bind the giant insects and begin cutting them up into short lengths of about twenty inches each. The dismemberment of the *tsutsugamushi* symbolizes the destruction of their evil power. The pieces of rope are neatly stacked on top of the *tsutsugamushi*. The *matsuhijiri* now come to the field (fig. 17.5), and the participants and spectators assemble in front of the two *tsutsugamushi*. Assistants of the *matsuhijiri* soon begin to hurl the cut lengths of rope over the heads of the crowd. Called Tsuna Maki (Throwing the Hawsers), this ritual is one of the focal points of the festival. Those in the crowd scramble to catch one of the ropes, which are said to have the power to protect the house and family members throughout the year (fig. 17.6). A few of the lengths are awarded to the victors in spontaneous wrestling bouts that take place in the field. Within a few minutes, all of the cut lengths of rope have been claimed and the *matsuhijiri* return to the Shitsurae-ya. The groups of young men reassemble the straw and grass filling of the *tsutsugamushi*, which are now somewhat smaller, and ready them for the

17.10   A *yamabushi* performs the ritual leap of the flying crow (*karasu tobi*) inside the main hall of Dewa Sanzan Jinja. From video footage by Roy W. Hamilton, Mount Haguro, Yamagata Prefecture, 2000/2001.

17.11   The ritual of the hare (Usagi no Shinji). From video footage by Roy W. Hamilton, Mount Haguro, Yamagata Prefecture, 2000/2001.

climax of the festival later in the evening when they will be set afire and dragged across the snowy field. The *tsuna* that remain are carried to the Shitsurae-ya and laid out in the snow, where they too are cut into short lengths that will be given to donors who support the festival (figs. 17.7).

After it gets dark in the evening, a Shinto service is conducted in the main hall of the Dewa Sanzan Shrine. People from Tōge and farther afield attend this ceremony, but the formal Shinto rites conducted by the shrine priests and the rituals of the *tsutsugamushi* and the Shitsurae-ya conducted by the *matsuhijiri* are quite distinct from each other.

The community members return to the Shitsurae-ya, where a lively competition known as Tsuna Sabaki (*tsuna* decision) gets underway on each side of the building. Seated on the four sides of the fire pits are four two-man teams. The teams take turns challenging one another verbally for the right to pull the most prestigious lead hawser later in the evening, when the burning straw torches will be pulled across the field of snow toward the posts that were erected at the far end of the field the previous day. Each verbal challenge concludes with an obligatory toast of sake (fig. 17.8). As the men become increasingly drunk, they continue to issue challenges for over an hour until teams who can drink no more are forced to bow out. The winners prevail by their skill in both drinking and oratory.

As the evening progresses, many additional rituals are performed (fig. 17.9). Then at 11:00 P.M. on New Year's Eve, the climactic events of the festival begin with two esoteric formal rites conducted in the main shrine building. The first, called Gen Kurabe, is a form of divination to symbolically determine which of the two *matsuhijiri* has best succeeded in accumulating spiritual power during their periods of ritual confinement. The ritual takes the form of a competition between two teams of six *yamabushi* representing

17.12   With a light snow falling
two teams prepare for
pulling the flaming *tsutsuga-*
mushi across the snow.
Photograph by Toshiyuki
Sano, Mount Haguro,
Yamagata Prefecture,
2001/2002.

the two wards. They take turns performing a ritualized leap called flying crow (*karasu tobi*; fig. 17.10). The crow is considered a divine messenger in Shugendō. The *yamabushi* are judged on the beauty and grace of their leaps.

This dramatic ritual is followed immediately by another, called Divine Ritual of the Hare (Usagi no Shinji). The hare is considered a messenger of the deity of Mount Gassan. A man dressed as a white hare enters the shrine and engages in a test of speed and agility with the two teams of *yamabushi*. The hare sits in the middle with *yamabushi* representing the two wards on either side of him and two low tables placed between them (fig. 17.11). The object is for the *yamabushi* to strike the tabletops as quickly as possible with their fans. This ritual is accompanied by the blowing of a conch horn, and the fifth blowing of the conch is immediately followed by a great commotion in the snow-covered field in front of the shrine where the teams of young men have assembled (fig. 17.12). This is the signal for them to begin their race, dragging the two *tsutsugamushi*, which have been set aflame, across the snow to the posts at the far end of the field. When they reach the end, the torches are hauled into an upright position, where they flare up into bonfires. The first team to accomplish this is the winner. There is a divinatory element to this event as well, for if the upper ward wins, this signifies a bumper rice crop for the following year, whereas a victory by the lower ward signifies a successful year for the community's fishermen. The ropes that were used to pull the burning torches are carried away to give to festival supporters to make the *tsuna* decorations for their houses.

The burning of the straw *tsutsugamushi* abolishes the evil insect, recapitulating the heroic deeds of Prince Hachiko fourteen hundred years ago. From an ecological point of view, the destruction of the straw "insects" by fire eliminates the habitat of the real *tsutsugamushi*, the mites that transmit human disease.

17.13   The character of the "idiot"
is dressed in an old-
fashioned style of rural attire
with a hat and rain cape
made of rice straw. He tends
the torch called *hashira
taimatsu*, which represents
Amaterasu Ōmikami, Japan's
goddess of the sun and royal
ancestress. Photograph by
Corinne Frugoni, Mount
Haguro, Yamagata
Prefecture, 2001/2002.

17.14   A *matsuhijiri* scatters the sacred grains. Photograph by Toshiyuki Sano, Mount Haguro, Yamagata Prefecture, 2001/2002.

At midnight two novices dressed as *yamabushi* ring the temple bell 108 times to mark the New Year, a Buddhist rite performed throughout Japan, and the first of thousands of visitors who will pay their respects at the shrine over the New Year holiday begin queuing up. These acts are not really a part of the Shōrei Festival. The rituals conducted under the auspices of the *matsuhijiri* continue on into the night however. A rite called Kuniwake no Shinji serves to reconfirm the territorial boundaries among three regional sects of Shugendō. This is followed by Hi no Uchikae Shinji, a divine rite for the creation of a new pure fire for the coming year, necessary because the fires of the previous year, such as the bonfires that destroyed the *tsutsugamushi,* are now polluted. This ritual consists of lighting another torch, a large vertical construction of dried canes called *hashira taimatsu,* which is said to represent the Shinto deity of the sun, Amaterasu Ōmikami. The torch is tended by a man playing the part of an "idiot" wearing a grass cape and hat (fig. 17.13; in former times, mentally handicapped persons were regarded as specially chosen by deities to act as their messengers). Both of these rituals are esoteric and unique to Shugendō, involving obscure costumes and posturing. They are held on the field of beaten snow where the pulling of the torches took place.

One of the festival's key symbolic moments, and its final rite, occurs back inside the Shitsurae-ya. It is by then past 3:00 A.M. and the rite itself is so anticlimactic that very few people are still in attendance. A Shinto priest comes to each of the ward's altars to conduct a brief ritual for the return of the deities of the five grains to their original homes. The *matsuhijiri* take the grains contained in the *kōya hijiri* and scatter them over the hall, and any remaining spectators, while intoning, "May the world be peaceful and quiet, may the grains be mature and plentiful, may people be comfortable and happy" (fig. 17.14). With the deities no longer in residence, this marks the end of the one-hundred-day period of confinement for the two *matsuhijiri.*

•

Part Four

The Goddess of Rice

18.1   At harvest festivals in
        Léwotala, Flores, women's
        garments are laid out to
        represent the presence of
        Bési Paré, the Rice Maiden.
        Bési and *paré* are ordinary
        Lamaholot words for various
        types of squash and a
        particular variety of rice,
        respectively, but in the
        poetic parallel speech of
        agricultural ritual they
        denote the Rice Maiden
        (Graham 1991, 103).
        Photograph by Penelope
        Graham, Indonesia 1988.

18

The Goddess of Rice

Roy W. Hamilton

Incest. Immolation. Incineration. According to Southeast Asian stories about the origin of rice, these are among the sufferings endured by the Rice Goddess in providing her sacred grain to humankind. In one version of the story of Dewi Sri, the Rice Goddess of Java and Bali, the supreme ruler of the deities, Batara[1] Guru, murders Sri's younger brother, Sedana, to put an end to his incestuous love affair with his elder sister. Batara Guru covets Sri for himself, and when she fails to come around to him, he murders her as well. From Sri's corpse sprouts the first rice (Pemberton 1994, 207).

In an example from a village in rural East Java, the goddess, known there as Mbok Sri, voluntarily surrenders: "All right, eat me then, together with my child, when the third season has come" (Heringa 1997, 363). But this myth is a transformation of a related group of stories in which a stranger bride passes through the area on her way to meet her husband. Instead she is waylaid, raped, and murdered by the locals (Heringa 1997). The Indonesian stories about Sri thus not only portray the Rice Goddess being sacrificed in order to benefit humans; they also suggest a close link between her sexual fertility and the origin of rice.

In a somewhat different version, from the Tai Yong of northern Thailand (Trankell 1995, 107), the Rice Goddess Mae Ku'sok is a starving woman searching for food to feed her children. She ventures deep into the forest where she encounters an ascetic named Lap Ta, the "wise man with eyes closed," who has accumulated tremendous heat through his meditation. Mae Ku'sok's pleas are so earnest that Lap Ta opens his eyes to look at her, and she is cremated on the spot by the intensity of his gaze. From her bones and ashes grow the first rice. In this story the goddess is given a less blatantly sexual role and placed in a vaguely Buddhist context complete with meditating ascetics and cremains.

Clearly it is impossible to speak of a single Southeast Asian myth of the Rice Goddess, as each recorded account differs in ways big or small from nearly every other version. The name of the goddess varies; the cast of characters changes; the actions credited to one character in one version fall to another in a second version; the goddess is raped, decapitated, or incinerated by her brother, by the supreme deity, or by strangers, and so on. The many versions form a vast interlocking web, eminently suitable for the type of comparative "structural" analysis that Claude Lévi-Strauss pioneered in the study of mythology. When such an analysis is applied to the Southeast Asian Rice Goddess stories, several key points emerge. First of all, across a broad spectrum of Southeast Asian

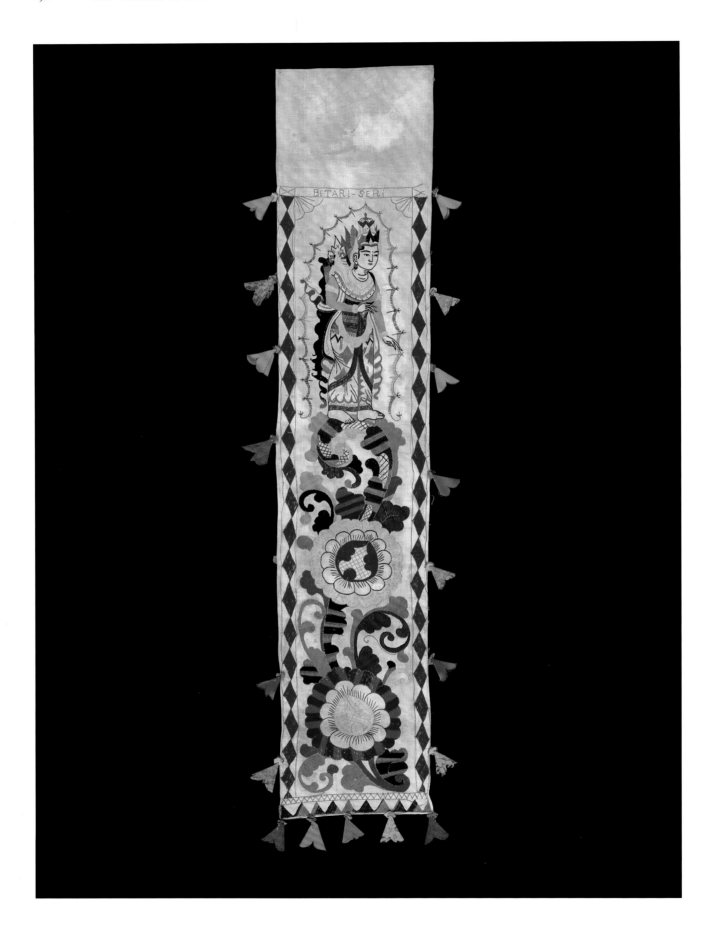

cultures with diverse religious traditions, rice is held to have originated through the activities of a goddess. Secondly, the Rice Goddess herself is sacrificed, and the rice produced from her body. Thirdly, due to its origin directly from the body of the Rice Goddess, rice itself is sacred and divine. Finally, the life cycle and fertility of the rice plants are equated with the life cycle and fertility of the Rice Goddess. This last point is illustrated most clearly by the pregnancy rituals that are held for the Rice Goddess when the grains of the rice plants growing in the fields begin to fill out and swell (see figs. 1.1, 2.20). In some cases every stage of the growth of the crop is equated with corresponding stages in the life cycle of the Goddess (see chapter 33 for a more complete description).

Despite the sometimes gruesome details of her many stories, it is the Rice Goddess's divine abundance that captivates the imagination of her followers. For the Javanese, Dewi Sri is by far the most beloved deity, indeed, one of only two who are likely to enter into everyday conversations.[2] This is especially remarkable in Java, where Islam has been the predominant religion for several hundred years. The Rice Goddess predates Islam, and throughout Southeast Asia she is worshiped from within, or in spite of, the prevailing religious doctrine. The Javanese in particular are known for their religious syncretism, and a standard litany for the ostensibly Islamic Javanese ritual feast (*selamatan*) is "Giving honor to Mohammad the Prophet, to Adam and Eve, and to Dewi Sri" (Jay 1969, 209).

And what of the origins of Dewi Sri herself? Her name is based on Hindu deity traditions that were brought to Southeast Asia from India. In the Hindu courts of ancient Java she took her place as Śrī Devī, consort of Vishnu. A ninth- or tenth-century bronze figure from Central Java depicts the goddess holding a panicle of rice (Fontein 1990, 197). Not all Southeast Asian peoples use an Indic name for the Rice Goddess, however. In many cultures, she is simply called the "Rice Mother" or "Rice Maiden." Examples include Iné Paré (lit.,"Mother Rice") in the language of the Lio people from the island of Flores, or Ibu Semangat Padi (Mother of the Rice Spirit) in Malay (Skeat 1966, 248). In formal ritual speech, the Lamaholot, also of Flores, call her Bési Paré (fig. 18.1). The Minangkabau call her Ande Gadih, or "Maidenly Mother" (Klopfer 1994, 152). These examples indicate that Dewi Sri is a Hindu goddess's name applied to a preexisting Rice Mother, and the Rice Mother herself is simply the personification of the spirit of the rice plants. Thus the ultimate origin of the Rice Goddess in Southeast Asia lies in the belief in nature spirits and in particular a spirit of the rice plants.

In Bali, Sri lives on today as a Hindu goddess, formally called Batari Sri Dewi. She is the focus of a great deal of Balinese ritual, extending from the rice fields to the most sacred temple in the land, Pura Besakih. The annual "wedding" for Sri performed at Besakih is the subject of chapter 20. She is a tremendously popular figure who appears in the arts of Bali in many forms. Some of these are explicitly identified as Sri (fig. 18.2). In other cases, the object is simply the receptacle that Sri may inhabit (figs. 18.3, 18.4) when ritually invoked. Additionally, Sri has become thoroughly conflated in the public imagination (of both the Balinese themselves and their many foreign visitors) with the *cili*, a doll-like figure made of lontar palm leaves (fig. 18.5). Technically not all *cili* represent Sri (Brinkgreve and Stuart-Fox 1992, 110), yet the fan-shaped headdress of the *cili* is widely regarded as a defining characteristic of depictions of Dewi Sri (fig. 18.6).

If the Rice Goddess in Southeast Asia represents a grafting of Indian deities onto rice spirits, what of the Rice Goddess on the Indian Subcontinent? Characteristically, a plethora of rice deities can be found there. Annapurna is a pan-Indian deity for rice.

18.2   Shrine hanging (*lamak*) with image of Betari-Seri. Negara, Bali, Indonesia. Embroidery on cotton cloth. L: 145 cm. Collection of Francine Brinkgreve.

Lamak are small banners hung as a type of offering at Balinese religious festivals. Most are ephemeral, made of cut palm leaf. Embroidered versions developed as a unique specialty in Negara, western Bali (Fischer and Cooper 1998, 64). Often the figures are named, as in this *lamak* showing Betari-Seri (a variant spelling of Batari Sri).

18.3   Coin images of deities (Rambut Sedana). Bali (probably Sanur), Indonesia. Copper alloy coins, cotton fabric, silk fabric, plant fiber, wood, skin, imitation gemstones, metallic thread, gold leaf, lacquer, paint. H: 45 cm. FMCH X61.78 and X61.77; The Katherine Mershon Collection of Indonesian Art.

This Rambut Sedana was collected in the 1930s by the American dancer Katherine Mershon who was well versed in the ritual life of the community of Sanur, where she lived. Fully dressed and perhaps resembling play dolls to Western eyes, the figures are among the most revered objects in Balinese religion.

18.4   Coin images of deities (Rambut Sedana) Bali, Indonesia. Copper alloy coins, wood, string, gold leaf. H: 44 cm. FMCH LX74.28a,b and LX74.27a,b.

Figures made of coins serve as receptacles for deities in Balinese ritual, with the presiding priest invoking the deities to take up temporary residence in the figures at the beginning of the ceremonies. The figures can be used for different deities, and technically without being "inhabited" they are nameless, but pairs of male and female figures like this one are in practice commonly identified with the Balinese Rice Goddess, Batari Sri Dewi, and her consort, Batara Sedana. Related receptacles for Sri and Sedana are discussed in chapter 20. This pair of figures was collected by the Mexican painter Miguel Corvarrubias when he lived in Bali in the 1930s.

Her very name is derived from the Sanskrit word for rice, or food, *anna* (Eichinger Ferro-Luzzi 1977, 535). She is typically depicted in a tableau with a rice spoon in her hand, often together with Shiva, to whom she is providing rice (see box, pp. 270–71). Major Hindu temples dedicated to Annapurna are located in Varanasi and in Nellore, a town whose name means "town of rice."

The goddess Lakshmi is recognized all over India as the goddess of prosperity, but in some regional traditions she too is strongly associated with rice. The worship of Lakshmi as a Rice Goddess in Bengal is the subject of chapter 22. Similar practices prevail in Orissa, where it is said that "[w]ithout Laks[h]mi there is no food, no life-sustenance" (Apffel Marglin 1985, 180). In South India a new form of Lakshmi worship grew during the 1970s with the popularization of a prayer addressed to Ashtalakshmi, the Goddess Lakshmi in Eight Forms (fig. 18.7). The Temple of the Eight Lakshmis (Ashtalakshmi Koil) was constructed in Chennai in 1974. This was followed by an audiocassette song in 1980 (Narayanan 1996, 104–5). One of the eight forms worshiped is Dhanyalakshmi, the goddess of grain (which, in South India, unequivocally means rice). That the goddess of prosperity could be equated with the goddess of rice comes as no surprise, for in preindustrial Asia rice was a primary measure of prosperity and wealth.

18.5 (Left to right): *Cili.* Bali, Indonesia. 1930s. Lontar palm leaf, thread. H: 34 cm. FMCH LX74.25. *Cili.* Bali, Indonesia. 1930s. Lontar palm leaf, thread. H: 29 cm. FMCH X61.102. *Cili.* Bali, Indonesia. 1998. Lontar palm leaf, yarn, metallic tinsel, bamboo. H: 34 cm. FMCH X98.14.7. *Cili.* Bali, Indonesia. 1930s. Lontar palm leaf, thread. H: 26 cm. FMCH LX74.26. *Cili.* Bali, Indonesia. 1990s. Lontar palm leaf, thread. H: 36 cm. FMCH X98.14.6.

Cili are made of durable lontar palm leaf, sometimes with added decorative elements. They are one of the most widespread, enduring, and variable forms of representation in Balinese art, as these examples from the 1930s and 1990s attest. Although they are popular tourist items, *cili* also have religious functions. The *cili* with tinsel depicted here (third from left), for example, is one of two made for a dedication ceremony for a new business in Klungkung. The other figure was installed in a small shrine in an upper corner of the room, and because it was used in the ritual for protective purposes and was considered sacred, the owners would not part with it. This figure, which was made as a backup and never sanctified, they were happy to sell.

18.6   Bottle of rice wine. 1998. H: 33 cm. FMCH X98.14.35.

This bottle of Dewi Sri–brand Balinese rice wine (*brem*) is marketed with the form of the *cili* on the label. In this case, the *cili* is clearly intended as a representation of Dewi Sri.

Even leaving aside the Hindu philosophical idea that all goddesses are one, called Devī (Hawley 1996, 6), it appears that India's Lakshmi and Indonesia's Dewi Sri are one and the same. The goddess Śrī is first mentioned in the *Hymn to Sri* (*Śrī Sukta*), which was probably added to the Rig Veda between 1000 and 500 B.C.E. (Narayanan 1996, 88). Śrī appears in the churning of the milky ocean. She is Vishnu's *śakti*, his energy or power (Foulston 2002, 9). Both Lakshmi in India and Dewi Sri in Indonesia are considered to be the consort of Vishnu; and Vishnu, the ideal head of state. As an example of this ideology in action, at the Jaganatha temple in Puri, Orissa, Brahmins and low-caste persons were sanctioned to eat out of the same pot, which they ordinarily were not allowed to do. This is attributed to Lakshmi having won a special concession from her husband, the king (Apffel Marglin 1985, 183). In the Javanese sultanates, Sri is regarded to be the symbolic bride of the sultan. This subject is discussed further in chapter 26.

In India rice is also sometimes associated with male deities. According to an ancient text, the Śatapatha Brāhmana, rice originated from the body of Indra: "From his marrow his drink, the soma juice flowed, and became rice: in this way his energies, or vital powers, went from him" (Kumar 1988, 18). In South India there is a god of rice, Nelliappar (or Nellaiyappar), worshiped as a form of Shiva.[3] Nelliappar is not normally depicted, as Shiva is worshiped only in the abstract form of the lingam.

These major pan-Indian deities are supplemented by a host of popular deities honored in particular localities. In Tamil Nadu such village deities are known as *amman*, and the names of some *amman* indicate that they are associated with rice. For example, the most cherished variety of rice in northern Tamil Nadu is called *ponni*, and there is a village deity for this variety known as Ponniyamman. The worship of Ponniyamman is the subject of chapter 19. Tamil village *amman* are revered by members of low-status castes, whereas pan-Indian deities like Annapurna are honored by high-caste persons. From the high-caste point of view, *amman* exercise their power partly through fear and superstition and are considered bloodthirsty, as they are sometimes worshiped with blood sacrifice.

In China, the Rice Goddess is conspicuous by her absence. For the Han Chinese, the most important deities are those associated with land and with ancestors. They may have Taoist kitchen gods, but no rice god or goddess. Any exceptions would be found among minority groups in southern China. Some Yao and Tai populations, for example, have Taoist agricultural deities that oversee the planting of crops, including rice (see fig. 2.15). Groups in southern China that speak languages from the Tai-Kadai language family must surely have Rice Goddesses akin to those found across the border in Tai communities in Vietnam, Laos, and Thailand, but because the Han for so long have dominated the cultural discourse in China, these regional traditions are not well known to outsiders and little has been published about them.

Japan, by contrast, has a huge infrastructure devoted to the Japanese deity for rice, Inari. In fact, one-third of all Shinto shrines in Japan are dedicated to this deity (Smyers 1999, 1), constituting literally tens of thousands of shrines. The vermillion torii gates and fox sculptures that are the hallmarks of Inari shrines are common sights in city, suburb, and village throughout Japan (figs. 18.9, 18.10). Inari is unique among rice deities in having both female and male forms, as well as being represented in both the Shinto and Buddhist traditions in Japan, which until the Meiji period (1868–1912) were thoroughly intermingled. Originally Inari, who has been worshiped by that name since at least the early eighth century, did not have a specific gender. The female aspects arose in relationship to rice and fertility, while the male aspects came later with the spread of Buddhism (Smyers 1999, 8). Inari may be depicted as an old man carrying rice panicles, as

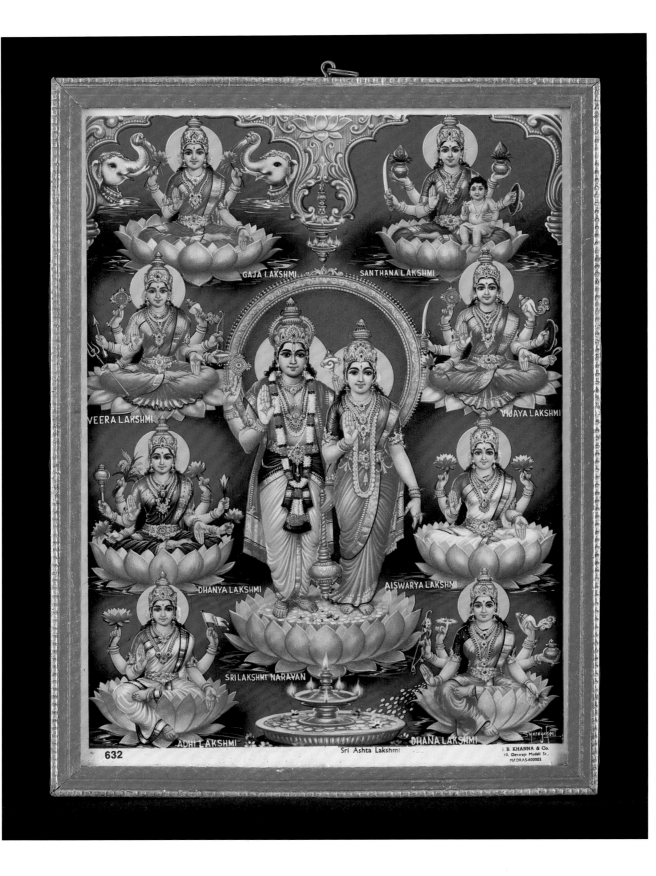

18.7

Sri Ashtalakshmi. India. Framed commercial print; paper, ink, wood, metal, paint. H: 52.5 cm. FMCH X2000.3.5.

Lakshmi and Vishnu stand atop an open lotus blossom, surrounded by Ashtalakshmi's eight forms, in this popular depiction from a Chennai religious shop. Dhanyalakshmi, the rice-bearing form, appears in the lower right.

a young female goddess, or as the androgynous bodhisattva Dakiniten astride a white fox (figs. 18.8, 18.13). The idea that deities ride on sacred vehicles (*vahana* in Sanskrit) is a Hindu concept picked up by Buddhism and carried across Asia to Japan. Inari's vehicle, sometimes termed his or her assistant, is so characteristic and important that many Japanese today say that Inari *is* a fox (figs. 18.11, 18.12). Inari worship is a tremendously complex and varied subject (richly documented in Karen Smyers's book *The Fox and the Jewel: Shared and Private Meanings in Contemporary Japanese Inari Worship*). Many of the most devout Inari worshipers belong to informal groups led by charismatic women in a manner highly reminiscent of the Northeast Asian shamanic traditions. There is an inherent tension between these somewhat marginalized groups led by women and the male-dominated Shinto establishment. The groups can be seen daily making pilgrimages to Inari shrines, especially the "head" Inari shrine for all of Japan, Fushimi Inari Jinja. This spectacular shrine occupies an entire mountainside in the outskirts of Kyoto, and the path that winds to the top is lined with thousands of torii gates and fox sculptures. The spirit of Inari that is enshrined at Fushimi can be ritually "divided" by the Shinto priests and a part of it installed in a new shrine (Smyers 1999, 159). In this manner, the spirit of Inari has been enshrined all over Japan, including in small portable home shrines.

Rice is nearly, though not completely, obscured by the complexities of Inari worship. At the Fushimi shrine, rice is grown on a small plot following Shinto ritual procedures. These include not only the rice transplanting ritual, which is performed in many places in Japan, but also a burning of the straw in November, which releases the rice field spirit *ta-no-kami* back to the mountain to become the mountain spirit *yama-no-kami* (as described in chapter 1). Even the association of Inari with foxes has a relationship to rice, as foxes are said to be prevalent at the time in spring when the *yama-no-kami* descends to the rice fields to become the *ta-no-kami*. The so-called fox mounds (*kitsune zuka*) that abound in Japan are natural or artificial mounds located near rice fields where the rice deity is said to have been worshiped in former times (Smyers 1999, 75–76).

In modern times Inari worship in Japan has undergone the same transformation as Lakshmi worship in India, namely, a conflating of the rice deity with a deity for prosperity and wealth. Among the most regular worshipers of Inari today in Japanese cities are businessmen who appeal to the deity for corporate success and profit.

In addition to the Rice Goddess, there are goddesses associated with other aspects of rice agriculture in Asia, and especially with the supply of water. In Vietnam's Red River Delta are found the four sisters Pháp Vân, Pháp Vũ, Pháp Lôi, and Pháp Điện, respectively, the goddesses of clouds, rain, thunder, and lightning. Their importance is related to the necessity of rain to support wet rice agriculture. Taoist gods of thunder, lightning, wind, and rain are known from China (Wang 1992, 76), which may have been the original source of inspiration, but in Vietnam the four sisters developed into a unique cult that is at least outwardly Buddhist, and the sisters are now said to represent reincarnations of the Buddha. Statues of the four sisters are found in clusters of temples in several districts surrounding Hanoi. The clusters seem originally to have consisted of sets of four temples, with each temple presided over by one of the sisters. The deities were represented in the temples by red lacquer statues that are among the most impressive of Vietnamese sculptural works. As a result of centuries of attrition due to war, fire, and flood, today there are no complete sets of temples and statues remaining. Individual statues can still be found, however, and in some sites two or more of the goddesses have been gathered together under a single roof. The most magnificent examples

18.8   Inari as the Buddhist bod-   Inari rides on a fox on this
hisattva Dakiniten. Japan.   inexpensive block-printed
Probably 1890s. Hanging   scroll, obtained as a souvenir
scroll; woodblock print. L: from an Inari shrine. Scrolls
120 cm. Peabody Essex   of this type were intended for
Museum, Salem,   use with a domestic altar in a
Massachusetts, E11,694.   nonaristocratic household.
Photograph by Jeffrey Dykes.

18.9   Festival coat (*ha'pi*). Japan.
Early 1900s. Cotton, sten-
ciled resist dye procedure. L:
90 cm. FMCH X87.1586;
Gift of Mary Chesterfield.

The crest on this festival coat
is inscribed "Welcome Inari."
The coat was presumably
intended to be worn at a festival at an Inari shrine. The
multiple red torii gates
marching up a slope indeed
suggest the main Inari shrine
at Fushimi.

18.10   At Fushimi Inari Jinja, the
head Inari shrine for all of
Japan, much of the pathway
that winds up the moun-
tainside is virtually enclosed
by thousands of torii gates
inscribed with the names of
donors. Many devout pil-
grims come to the shrine to
pray for Inari's favors.
Photograph by Roy W.
Hamilton, 2000.

are found at Chùa Dâo in Hà Bắc Province (fig. 18.15), where three of the figures are
installed at the oldest center of Buddhist learning in Vietnam and one of its most historic temples (Hà 1993, 148–51).

The Balinese also have a goddess for their irrigation system, Dewi Danu, the
goddess of Lake Batur. She is worshiped at Pura Ulun Danu Batur, a temple overlook-
ing the lake, which serves as a symbolic central reservoir for the irrigation system covering a
large part of the island. The lake is located at the base of Mount Agung, Bali's highest and
most sacred peak. Together, the male god of Mount Agung and the goddess of Lake Batur
are sometimes considered the supreme deities of the island (Lansing 1991, 73). The
Balinese irrigation system is coordinated through a series of rituals carried out in the name
of Dewi Danu. This unique religiously controlled irrigation system is described in detail
in John Stephen Lansing's book *Priests and Programmers: Technologies of Power in the
Engineered Landscape of Bali.* An inscription from a Balinese palm-leaf manuscript sums
up her importance: "Because the Goddess makes the waters flow, those who do not follow
her laws may not possess her rice terraces" (Lansing 1991, 73).

18.11   Shinto sculpture in the shape of a seated fox. Japan. Momoyama period (1573–1603). Carved wood with white pigment and *sumi* ink. H: 34 cm. Los Angeles County Museum of Art AC1993.40.1; Gift of the 1993 Collectors Committee.

18.12   Votive plaque (*ema*) with image of foxes. Japan. Painted wood. W: 21 cm. FMCH X89.845; Gift of Daniel C. Holtom.

This votive plaque is marked with the name of Fushimi Inari Jinja, the Inari shrine where it was purchased. *Ema* by the thousands are left at the shrine by worshipers.

18.13   Dakiniten. Japan. Eighteenth
century. Carved and painted
wood, silk cord. L: 41 cm.
Collection of Lynn S. Gibor.

This figure of Dakiniten was
owned by a family with a
tradition of Inari worship
dating back to the eighteenth
century. It was originally
obtained from an Inari
shrine, where a "divided"
portion of the spirit of the
deity was ritually called to
reside in the figure.

18.14   Food God (Ugajin). Japan.
Edo period (1603–1868).
Carved wood. H: 13 cm.
Collection of Marvin and
Flora Herring.

Uga means food, or, by association, rice. Thus Ugajin, or
"Food God," is another
Japanese rice deity, though
technically not a formal manifestation of Inari (in fact,
Ugajin is sometimes equated
with another Japanese goddess, Benten). This sculpture
seems to change its expression from benevolent to
malign when viewed from
above or below. It would
once have been housed in a
portable altar.

18.15   Pháp Vân, the goddess of
clouds, presides in the
Buddhist temple Chùa Dâu,
Hà Bắc Province, Vietnam.
Photograph by Vũ Hồng
Thuật, 2002.

18.16   Bengali deity figures. By
           Gourishankar
           Bandopadhaya. Calcutta,
           West Bengal, India. 2003.
           Unfired clay, paint, metal,
           cloth, plant fiber, wood.
           FMCH TR2003.8.1-5.

           Maa Annapurna (center)
           provides rice to Lord Shiva
           Pravhu (left) and his two
           assistants, Sri Nandy
           Maharaj (right) and Sri
           Bhringy (front).

Annapurna
Sohini Ray

In India Annapurna worship is a comparatively rare and little-known Hindu tradition. Artist Gourishankar Bandopadhaya (fig. 18.16) explains that it originated in the distant past when all food disappeared from the earth and all living beings were consequently in danger of perishing. They appealed to Lord Brahma for help. He consulted with Lord Vishnu and then decided to awaken Lord Shiva from his ritual sleep (Yoga-*nidra*) and to give him the responsibility of restoring prosperity to the earth. Lord Shiva invited the goddess Annapurna to the earth and begged her for rice, which he distributed throughout the world. The best-known temple of Annapurna in India is located in Kashi (a section of Benares). There, Annapurna is held to be the consort of Visheshvara, one of the names of Lord Shiva, whose main temple is located nearby. Annapurna is thus another manifestation of Shiva's consort, Parvati.

According to a popular Puranic story, the sage Vyasa was once refused alms in Kashi. Angered, he put a curse on the city so that it would lack knowledge, wealth, and liberty. Still enraged, Vyasa begged for alms at a house where Shiva and Parvati—who had assumed human form—were living. Parvati invited him in and gave him food so delicious that he forgot his curse. Because of his bad temper, however, Shiva banished him from Benares, permitting him to visit only on the eighth and the fourteenth days of each fortnight.

In Bengal, worship of Annapurna is prevalent among prominent land-owning families, among them the Bannerjees, the family of which Gourishankar Bandopadhaya is a member. Saraswati Devi, the artist's grandmother, experienced serious financial reverses around 1908–1909 and promised Annapurna that she would establish the goddess's worship in the family if the financial troubles were resolved. Following his mother's instructions, the late Bholanath Bandopadhaya, Saraswati Devi's second son, instituted the tradition of Annapurna worship in the Bannerjee family in 1940, and it has continued until the present. It entails making images of and worshiping a group composed of Annapurana, Shiva, and the attendant deities Nandy and Bhringy. In the Bandopadhaya family the worship occurs on an auspicious day during the month of Baishakh/Jaistha (May–June). On this day, reading from the traditional Hindu religious text known as *Chandi* is carried out from early morning. During the period of worship, cooked meals are offered to the goddess and then distributed among the family.

19

Ponniyamman, a Tamil Rice Goddess from South India

Nanditha Krishna

In the northern districts of Tamil Nadu, there is a tradition of a Rice Goddess known as Ponniyamman. *Ponni* is the name of a local variety of rice that has been deified as a goddess (*amman*) around whom a rich mythology has developed. A tributary of the river Kaveri once flowed through Thondaimandalam in northern Tamil Nadu, and on one occasion, heavy floods caused the river to wash away entire rice fields. The villagers installed a statue of Ponniyamman at Muttavakkam, a village in Kanchipuram District. The goddess accepted their prayers and halted the floods.

In the village of Damal, also in Kanchipuram District, there are two temples to Ponniyamman. In one, she is represented as one of the Seven Virgin Sisters—the Sapta Kannimaars—replacing the sister Chamundi (fig. 19.1). In the other, her head is placed in the middle of a paddy field (fig. 19.6). According to the villagers, she is an incarnation of Bhoomadevi, or Mother Earth. The earth, which represents her body, supports the head—she needs no other body. Several other villages also have temples to Ponniyamman (figs. 19.2–19.5), including Ponniyampatterai and Thimmasamudram (Kanchipuram District), Thirumalper (Thiruvallur District), and Jagir Thandalam and Sayinathapuram (Vellore District). There are also temples to her in neighborhoods in the city of Chennai (Madras), including Guindy, Madipakkam, and Thiruvottriyur.

Ponniyamman is worshiped once a year during the month of Adi (mid-July through mid-August), when the annual festival of Urani Pongal is celebrated in the Kanchipuram District. Some villages, however, celebrate it in the month of Chittirai (mid-April through mid-May). On this occasion, every family of the region comes to the temple for a ritual cooking of rice. The villagers bring rice, sweet rice cakes, and lavish quantities of jaggery, flour, sugar, and coconut and spread them all on large plantain leaves placed on the ground before an image of Ponniyamman. The priest begins by burning incense and camphor. Then he takes a portion of rice boiled with milk (*pongal*) from each pot; mixes it with coconut, flour, jaggery, mashed bananas, etc.; offers it to Ponniyamman; and distributes the same as sanctified food (*prasadam*) to all those who took part in the festival. The villagers regard the festival as a communal rice cooking for the village (*ur*), observed for the benefit and prosperity of the entire agricultural region (*uran*).

During the festival, the villagers also organize a form of nighttime street theater called *therukoothu*, in which local folktales and the great Hindu epics are enacted in an entertaining and informative manner. Other events include a fire-walking ritual performed in both Damal and Thirumalper under the protection of Ponniyamman. Dressed

19.1   Dressed stone figure of Ponniyamman represented as one of the Seven Virgin Sisters (another sister appears at the right). Photograph by Roy W. Hamilton, Damal, Kanchipuram District, Tamil Nadu, 2000.

19.2   Ponniyamman temple,
        Vellore District. Photograph
        by M. Amirthalingam,
        Tamil Nadu, 2001.

19.3   Ponniyamman temple,
        Vellore District. Photograph
        by M. Amirthalingam,
        Tamil Nadu, 2001.

19.4   Ponniyamman temple,
        Ponniyampatterai,
        Kanchipuram District.
        Photograph by M.
        Amirthalingam, Tamil
        Nadu, 2001.

19.5   Ponniyamman temple,
        Vellore District. Photograph
        by M. Amirthalingam,
        Tamil Nadu, 2001.

19.6   Agni Ponniyamman, or
            "Fire Ponniyamman," is
            represented by a stone in
            the middle of a rice field,
            with two stone heads on
            either side representing the
            heads of self-decapitated
            heroes. The three-pronged
            trishulam is the weapon of
            the goddess, while the flat
            stone beside it is the sacrifi-
            cial altar. Photograph by
            M. Amirthalingam, Damal,
            Kanchipuram District,
            Tamil Nadu, 2001.

in wet yellow clothes and a garland, the barefoot devotees step onto the bed of red-hot coals and walk across it in order to be rid of their sins and bad omens, and also to obtain wealth and prosperity.

Animals such as fowl, goats, rams, and buffalo are sacrificed to Ponniyamman in one temple in Damal and at those in Jagir Thandalam and Ponniyampatterai, although this is illegal.[1] After the sacrifice, the statue of the deity is decorated and anointed (*abishekham*) with milk, lime juice, coconut water, and sandal paste, and then the priest invokes Ponniyamman by singing her praise. The invocation takes place either during a procession in which an image of the deity is taken around the village, or near the temple where she is installed following the procession. After the invocation, the goddess is believed to possess the priest's body through which she predicts forthcoming events in the village until the priest finally falls into a faint. Camphor is then burnt before him, and he is revived.

In former times village *amman* presided over forest groves that typically enclosed sacred sources of water such as natural springs (Amirthalingam 1998). The worship of the *amman* was led by a priest who was also a potter from the Velar caste. The potter-priest was responsible for making terra-cotta figures of the deity that marked the sacred grove.[2] When economic resources allow, the devotees of *amman* inevitably aim to upgrade their facilities, which usually involves constructing a small concrete temple in the village (fig. 19.5) and installing a stone figure of the deity. Unfortunately, in recent decades a weakening of the system of potter-priests and sacred groves has led to the destruction of many of the groves, with predictable environmental consequences. However, sacred groves are still the norm in the districts south of ancient Thondaimandalam (beyond Chidambaram), although the acreage has shrunk considerably. It seems likely that the ancient presiding deities of the Thondaimandalam region (around Kanchipuram) were merged into Ponniyamman when forests were cleared for agriculture.

•

20.1 Representations of the goddess Bhatari Sri and her consort, the god Bhatara Sedana, are carried in procession as part of the series of rituals comprising Usaba Ngeed, the celebration of their marriage. Night has fallen by the time the two deities are carried around the temple Pura Manik Mas, the first of the four temples they visit during the evening's celebration. Photograph by David J. Stuart-Fox, Besakih, Bali, 1983.

20

Sri and Sedana at Pura Besakih, Bali

David J. Stuart-Fox

The goddess Bhatari Sri and the god Bhatara Sedana are two deities revered throughout
Bali. In temples they often have separate shrines next to one another. They are generally
considered a divine couple. At Pura Besakih, Bali's paramount Hindu sanctuary high on
the slopes of Mount Agung (fig. 20.2), this pair of deities is honored in a special yearly
ceremony that commemorates or reifies their marriage. This ritual is called Usaba Ngeed.
Usaba is a term employed in parts of Bali for a major ritual held on a day reckoned
according to the Balinese lunar calendar rather than the 210-day *pawukon* calendar. *Ngeed,*
from the root *ééd,* means "to perform a sequence or series of something," especially rituals.

Pura Besakih is an enormous complex of twenty-two public temples. Different
groups of temples are linked by various number-based symbolic classifications that are
common in Balinese Hinduism. There is a dualistic division between upper and lower
groups, and there are important groups of three and five specific temples.[1]

The central and largest temple, Pura Penataran Agung, is the locus of the
most elaborate yearly ritual, called Bhatara Turun Kabeh, when the deities of all the
temples are honored together. In all, following one or the other of the two calendars,
there are more than seventy rituals, small and large. Usaba Ngeed is among the largest
and most important and is unique in many ways, particularly in celebrating the relationship between two deities.

Usaba Ngeed is the last major ceremony in Besakih's cycle of agricultural rituals, although its significance goes beyond the purely agricultural. This cycle was once
related to the growth cycle of dryland rice, but following Mount Agung's eruption in 1963,
rice cultivation at Besakih ceased. It was replaced by commercial crops such as oranges and
cloves and some traditional subsistence crops other than rice. The ceremonies of the former
rice-growing cycle, however, continue to be held. Most are communal rituals, all reckoned
according to the lunar year, while others are performed individually. The largest of these
rituals occur after what would have been the harvest. They take place on days of the lunar
cycle partly determined by the enactment of a ritual called Usaba Dalem Puri. This is one
of Besakih's most significant rituals held at the temple Pura Dalem Puri, hence its name.

In the dualistic structure that underlies the Besakih complex as a whole, Pura
Dalem Puri is the main temple of the "temples below the steps"; Pura Penataran Agung is
the main temple of the "temples above the steps." The nature and function of Pura Dalem

20.2   Mount Agung rises high above the central group of Besakih temples. Pura Penataran Agung is at the center of the photograph. Pura Banua, which plays a major role in the Sri-Sedana ritual, is at the lower right. Photograph by David J. Stuart-Fox, Bali, 1983.

20.3   Male and female *pering* figures stand in front of the shrine housing Bhatari Sri in Pura Banua. They are accompanied by a sacrificial buffalo, represented by its skin with the head and feet still attached. The *pering* figures, the female dressed in yellow and the male in white, consist of dozens of components made of sugar-palm leaves. These special offerings indicate the importance of the Usaba Buluh and the ensuing Usaba Ngeed in the cycle of Besakih rituals. Photograph by David J. Stuart-Fox, Besakih, Bali, 1981.

20.4   Besakih's official priests (*pemangku*) take their places in a traditional ceremonial sitting arrangement for the ritual in Pura Banua's meeting pavilion (*bale agung*) during which they prepare special food (*ajang*). The *ajang* consists of two parts: a lower "plate" of meat components and an upper "plate" of rice cakes. Photograph by David J. Stuart-Fox, Besakih, Bali, 1981.

Puri are complicated and have changed over the centuries. Although it now has functions common to "death temples" (*pura dalem*)—associated with Bhatari Durga, Siwa's spouse, who is predominantly destructive and magical in nature—at the same time it retains auspicious functions related to earth, fertility, and agriculture. This aspect is shown by the agricultural rituals held in the days following Pura Dalem Puri's main ritual, the Usaba Dalem Puri, which is celebrated on the day *kajeng* (a day in the three-day week) that falls three, five, or seven days after the new moon of the seventh month. First, three days later, comes Usaba Nyungsung at Pura Kiduling Kreteg, then three (or six) days after that the Usaba Buluh[2] at Pura Banua, with Usaba Ngeed taking place on the following full moon of the eighth month (*purnama kaulu*).

Pura Banua is the temple at Besakih dedicated to Bhatari Sri. It is a small temple that is not especially prominent, but old photographs of Besakih show it located right in the middle of the approachway leading to Pura Penataran Agung. Pura Banua is numbered among the "temples below the steps." The shrine dedicated to Bhatari Sri is a *gedong*, a rather simple square structure with a single roof, located near the head of the temple's long pavilion (*bale agung*). The importance of the Usaba Buluh is indicated by the enactment of the so-called *nanding ajang*, a special ritual meal, too complex to describe here, performed by the nine official temple priests (*pemangku*) of Besakih's public temples sitting in the *bale agung* (figs. 20.3–20.5). After the ritual the participants take away portions of ritual objects and substances that are sprinkled over the fields to ensure fertility.

After Usaba Buluh, preparations begin immediately for the Usaba Ngeed that takes place on the day of the full moon. The opening ritual of this ceremony is called Panurun (from the word *turun*, "go down") in which a deity is brought down into his or her god-symbol. In Bali this object, representing the material presence or physical aspect

20.5   In Pura Banua, *pemangku*
lead villagers in worship of
Bhatari Sri. The special
offerings reveal the
importance of the ritual: a
splayed-out buffalo and
the male and female *pering*
offerings. Photograph by
David J. Stuart-Fox,
Besakih, Bali, 1981.

20.6   In a quiet ritual in the Merajan Selonding, a *pemangku* "brings down" Bhatari Sri into her statue, where she will reside for the duration of the festival. Photograph by David J. Stuart-Fox, Besakih, Bali, 1981.

20.7   Now embodied in her statue, Bhatari Sri is carried on a priest's head, shaded by umbrellas. Photograph by David J. Stuart-Fox, Besakih, Bali, 1981.

of a deity, can take many forms, some permanent and used time and time again, some ephemeral and made anew for each ritual. Those of two major deities of Besakih's central temple, the Pura Penataran Agung, for example, are fifteenth-century inscriptions on wood. At the Usaba Ngeed, the god-symbol of Bhatari Sri is a wooden statue while that of Bhatara Sedana is an ephemeral offering-like object called a *daksina palinggih*. It consists of a basket containing, among other ingredients, rice, a coconut, and bananas, all dressed in white cloth. The Panurun ritual for Bhatara Sedana takes place at the eleven-roofed *meru*, or pagoda, located on the second terrace of Pura Penataran Agung, the terrace that is now, with the triple-lotus throne (Padmatiga), that temple's focus of ritual. The Panurun ritual for Bhatari Sri, however, takes place in the small "below the steps" temple Merajan Selonding, from where it is brought to her own temple, Pura Banua (fig. 20.6).

The more elaborate version of Usaba Ngeed, preferably held every other year, involves the additional ritual of Malasti. Accompanied by crowds of worshipers in colorful festive dress, bearers carry the two deities, suitably enshrined for the occasion in processional palanquins, a distance of some two kilometers to a sacred spring called Toya Esah, where the god-symbols are ritually (not literally) washed and the deities suitably refreshed and adorned (figs. 20.7–20.9). Upon their return to Besakih, before entering the temples, the two deities are welcomed with a Pamendak ritual in which a priest presents offerings laid on the ground in front of the deities, who then "walk" over the offerings and are taken to their respective shrines (fig. 20.10).

The main ritual of Usaba Ngeed always takes place in the evening. People say that this is because a bridegroom commonly elopes with his bride at this time. First of all, Bhatara Sedana departs in his palanquin from his *meru* in the Pura Penataran Agung, comes down the main steps of the temple, as far as an open space near Pura Banua. From her shrine in Pura Banua, Bhatari Sri goes out in her palanquin to meet him. At the place

20.8   Carrying palanquins, the procession goes to the sacred spring, taking a path that crosses a stream of black sand and boulders from the eruption of Mount Agung in 1963. Photograph by David J. Stuart-Fox, Besakih, Bali, 1981.

where they meet, a priest (*pemangku*) performs the welcoming Pamendak ritual in their honor. They are accompanied by umbrellas and a group of special offerings and ritual paraphernalia. These include *tigasan* (a pile of neatly folded cloths, one for each deity), *canang rawis* (an elaborate kind of *canang* offering, whose key ingredients are areca nut, betel leaf, and lime), *naga sampir* (a kind of small traveling shrine containing holy water and buds of the lotus and *pudak arum* flowers), *pulagembal* (a very elaborate and decorative offering full of rice cakes and rice-dough figurines), and, finally, *sayut raja pinoma* or *penomah* (one of the many different kinds of *sesayut* offerings). This *sayut raja pinoma*, which (as far as I know) is unique to this ritual, is of special interest. Placed on a footed stand (*dulang*) are three silver dishes (*bokor*), one with a holy water vessel full of various substances, the other two containing an array of mostly ritual foods. According to Besakih priests, the offerings for the Usaba Ngeed may be compared to the return visit of a bride to her natal home that takes place some time after marriage. In the human world, this ritual is called Marebu or Makala-Kalaan.

The ceremony is witnessed by the many villagers bearing the specified offerings on their heads, in alternative years a *dangsil* (in shape unlike the more elaborate offerings of that name) or a *penek*.[3] The Pamendak ritual concludes with a man cutting off the head of a black chicken and sprinkling the blood on the ground.

20.9   Still seated in their palanquins, which rest on a high
bamboo platform, Bhatari
Sri (left) and Bhatara Sedana
(right) are refreshed and
adorned. Photograph
by David J. Stuart-Fox,
Besakih, Bali, 1981.

After this elaborate Pamendak ritual, the two deities, accompanied by all the special offerings and paraphernalia and the villagers with their offerings, are carried in procession to a specific group of important Besakih temples. The first temple visited is Pura Manik Mas, a small temple at the beginning of the approachway leading toward the central temple (figs. 20.1, 20.11). The temple is closely related to Pura Dalem Puri. It was a tradition, though now no longer always adhered to, to pray here first upon arriving at Besakih before proceeding further. After they reach the temple, Bhatari Sri and Bhatara Sedana are first presented outside the gateway into the temple with another Pamendak ritual, smaller than that outside Pura Banua, and then are carried around the temple three times clockwise before entering for a small ritual of worship.

From Pura Manik Mas, the procession makes its way to the next temple visited, Pura Batu Madeg, then on to Pura Kiduling Kreteg, and finally to Pura Penataran Agung. At each of these temples a similar series of minor rituals takes place, akin to those at Pura Manik Mas. These temples are linked by means of a system of three-part classifications that permeates Balinese religion and culture. Of notable significance are the triads left-center-right, black-white-red, water-air-fire, and Wisnu-Siwa (Iswara)-Brahma. Thus Pura Batu Madeg (Temple of the Standing Stone), the left-hand temple as one faces the mountain, honors Wisnu and the color of the temple's banners and decorations are black; Pura Kiduling Kreteg (Temple South of the Bridge), the right-hand temple, honors

20.10   Returning to Pura Banua,
         Bhatari Sri is carried around
         the temple on the head
         of a worshiper. Photograph
         by David J. Stuart-Fox,
         Besakih, Bali, 1981.

20.11   Bhatari Sri (left) in all her
         beauty and the enigmatic
         form of Bhatara Sedana
         (right), seated under
         umbrellas of honor. They are
         dressed in yellow and white
         cloths, colors that are com-
         monly associated with the
         female and male genders,
         respectively. Photograph by
         David J. Stuart-Fox,
         Besakih, Bali, 1983.

20.12   After the procession returns
to Pura Banua, Bhatari Sri
and Bhatara Sedana are
removed from their traveling
clothes and enshrined in her
pavilion, where *pemangku*
offer worship. The *daksina
palinggih* that represents
Bhatara Sedana can be seen
just to the left of the statue
of Bhatari Sri. Photograph
by David J. Stuart-Fox,
Besakih, Bali, 1983.

Brahma, and its color is red; and Pura Penataran Agung, the central temple of the entire Besakih complex, honors Siwa, and its color is white. These visits to such major temples are in essence rites of witness by the resident deities toward the divine pair Bhatari Sri and Bhatara Sedana, and they also signify the importance of the Usaba Ngeed.

At Pura Penataran Agung the ritual of worship takes place beside that temple's main locus of ritual, the triple-lotus throne. On its completion, the procession makes its way downhill to Pura Banua. There, the two deities are placed side by side in the "pavilion of assembly" (*bale pasamuhan*; fig. 20.12). Worshipers place their offerings on the ground in front of the pavilion and sit down beside them. A temple priest then takes his place in front of the pavilion and performs the usual ritual of worship. After communal worship (*bakti*), holy water is sprinkled over the worshipers and handed out to them.

How long the two deities stay in residence in Pura Banua depends on the elaborateness of the ritual. For the more involved ritual that includes the Malasti procession, they stay in residence (*nyejer*) for a further three days before the final Nyimpen ritual (from *simpen*, "to put away or keep"). In the simpler ritual the two deities are later that same night (after midnight) or early the next morning taken back to their respective places of safekeeping from where they were brought down at the beginning of the whole ceremony, Bhatari Sri at the Merajan Selonding, and Bhatara Sedana at his eleven-roofed pagoda in the Pura Penataran Agung.

Usaba Ngeed as a whole is a ritual that requests material fulfilment in this world, especially rice, the main food crop, represented by Bhatari Sri, and wealth with which to obtain other material goods, represented by Bhatara Sedana. Their marriage symbolizes a sufficiency and completeness of the material requirements to sustain a good and proper life. May it be so.

•

Part Five

Sacred Food

21.1 A New Year offering of two round cakes of pounded glutinous rice (*mochi*) topped by an orange is presented before an image of Jizōbosatsu. The small images of children clinging to the larger figure indicate Jizōbosatsu's role as the deity of mercy for children who have died an untimely death. This type of offering, called mirror *mochi* (*kagami-mochi*), is made to deities at New Year in temples and shrines throughout Japan, as well as on domestic altars in most Japanese households, in order to seek their blessings to renew the agricultural cycle and the cosmic order. Photograph by Roy W. Hamilton, Hasedera Temple, Kamakura, Japan, 2002.

21

Sacred Food

Roy W. Hamilton

21.2   *Khaow puín* is a votive offering consisting of lumps of cooked glutinous rice, here placed outside a Buddhist temple in Chom Thong, Chiang Mai Province, Thailand. The offering accompanies a request for restored health made by an ailing person, and the number of lumps indicates the age of the supplicant. Photograph by Roy W. Hamilton, 2000.

- In Madras, in southern India, I was taken by a friend into the *puja* (ritual worship) room at the center of her home so that she could explain to me the significance of rice in the life of her family. The room contained a simple altar with a few religious items arranged on shelves. Among them was a small brass figure of Annapurna, the goddess who gives rice to the god Shiva in Indian mythology. My friend explained to me that every day after she has cooked her family's rice, she puts it into a serving pot and brings it to the *puja* room. There she says a brief prayer and symbolically offers the rice to Annapurna. As a result of the goddess's blessing, the rice becomes sacred, and afterward, when it is carried out to the dining table, the family members join with the deity in consuming the sacred food. In this home, Annapurna is cherished as the ultimate model of domestic providence—just as the goddess provides rice to Shiva, my friend daily provides the sacred food to her family.

- A closely related procedure was observed by anthropologist Ing-Brit Trankell in a Tai Yong village in northern Thailand (1995, 134). There, each woman takes the rice she has cooked every morning to an elevated shelf in her house and offers it to her ancestral house spirits. The woman informs the spirits of the household's affairs and secures their approval for the day's plans. Once the spirits have symbolically partaken of the rice that has been offered to them, the remains are considered sacred "leftovers" that will feed the family for the rest of the day. The spirits, the ancestors, and the living humans are united by consuming the same sacred rice.

- In a Balinese household, I witnessed rice being fed to spirits with a different aim. Every morning the senior woman of the household cut strips of banana leaf into more than a dozen tiny squares. When the daily rice had been cooked, she placed a few grains on each square and sometimes added a tiny scrap of meat. She placed these offerings, called *saiban*, onto a tray and began making her way around the family compound. One at a time, she placed the *saiban*

21.3 Rice mortar. Borneo. Carved wood. L: 100 cm. Collection of Mr. Jack Sadovnic.

Mortars sometimes had multiple holes, or in some cases were trough or boat shaped, so that women (who nearly always performed this task) could work together, beating out a rhythm with their pestles. While most mortars were plain, a magnificently carved mortar like this would have reflected prevailing ideas about the sacred nature of rice. This, together with the small size of the holes, suggests that this mortar was probably used in ceremonial contexts.

21.4 Funerary object in the form of a grain mill. China. Eastern Han dynasty (206 B.C.E.–220 C.E.). Earthenware with green glaze. W: 20 cm. Seattle Art Museum 33.33; Eugene Fuller Memorial Collection.

Ceramic models of scenes from everyday life, frequently drawn from agriculture, were placed in Han dynasty tombs to accompany the deceased in the afterlife. This scene shows milling equipment that can still be found in villages in Asia today.

down on the open ground and recited a short incantation. When I asked her who the rice was for, she replied, "It is for the demons, so that they will not become jealous and interfere with the family's meals."

As these examples attest, rice, whether eaten by family members or offered to spirits or deities, is regarded as sacred food in a wide range of Asian contexts. Rice is in fact the most common element in almost any type of religious offering, expressed in countless forms and for countless purposes throughout the rice-growing regions of Asia (figs. 21.1, 21.2). Chapter 22 provides a more detailed examination of domestic rituals related to rice in Bengal and among Bengali Americans.

Preparation and Serving

The preparation of rice prior to cooking normally involves three distinct elements: (1) threshing, or the removal of the grains from the stems that bear them, (2) milling, or the removal of the inedible papery husk that encloses each grain, and (3) polishing, or the removal of the bran layer that surrounds the starchy body of the grain. The first two procedures are necessary before the grain can be eaten, while polishing is technically optional but culturally mandated in most Asian settings.

Where rice is stored in traditional village granaries still attached to the stalk, the threshing, milling, and polishing typically take place as a single domestic routine. The stalks are removed from the granary in small quantities, often only enough for a single day's supply. The grain may be threshed from the stalks either by simply pulling it off by hand or by walking on it. It may also be coarsely pounded in a mortar to knock it free. Some form of winnowing is then used to separate the grain from the bits of broken stems. The milling and polishing, usually accomplished together as a single procedure, were until recently almost always done in a mortar, with a heavy pestle operated either by hand or

21.5   *Sixth Patriarch.* By Suiō Genro (1716–1789). Japan. Hanging scroll; ink on paper, silk. W: 51 cm. Gitter-Yelen Art Center 2000.12.

The title of this Zen painting of a foot-operated rice mill (*karausu*) refers to Hui-neng (638–713), the Sixth Patriarch of the Chinese Zen tradition. According to Hui-neng's autobiographical sermon, which is actually an apocryphal legend, he began his monastic career in "the threshing room where I spent over eight months treading the pestle" (Yampolsky 1967, 128). The inscription on the painting reads: "Coming to the Naraya in Ōtsu, one also learns to pound a mortar."[1] This overly sexual imagery—for the Naraya was a brothel—may seem shocking, but it is actually very much in line with Zen teachings against hypocrisy and false piety in the monkhood (Rosenfield 1979, catalog 31).

foot. The daily pounding of rice in a mortar was one of the most characteristic and evocative sights and sounds of village life, and rice mortars themselves were one of the most ubiquitous pieces of domestic equipment (figs. 21.3–21.5).

Where rice is stored already threshed from the stalks, the threshing is usually performed in the fields as a part of the harvest procedure, while the milling and polishing take place later. Because the grain keeps better in the husk, it is usually only milled and polished later as needed. In China, however, rice has long been routinely milled and polished before storage, and this system is becoming more common throughout Asia with the spread of modern mechanized farming. Over the millennia many sophisticated types of machinery, powered by humans, animals, or water, have been developed to make threshing, milling, and polishing easier (figs. 21.6, 21.7). Today gasoline-powered rice mills have been established in all but the poorest and most remote villages, either as small private enterprises or as village cooperatives. The farmer typically pays a small percentage of the grain to the mill owner in exchange for the milling and polishing. Using this system, larger quantities are milled at a time and milling is no longer a part of the daily routine.

The colored bran layer, unlike the husk, is an integral part of the grain and comes away gradually as polishing progresses. Therefore polishing is a matter of degree, not an all-or-nothing thing. Asians rarely eat what is known as "brown," or unpolished, rice in the West, and generally there is a strong preference for rice that has been polished thoroughly enough that it is pure white in color. The aesthetics of the pearly white grains are a major factor in regional taste, despite the nutritional sacrifice involved in removing the vitamin- and protein-rich bran layer. For some special purposes, notably the making of sake, the polishing may be carried on until as little as 50 percent of the starchy kernel

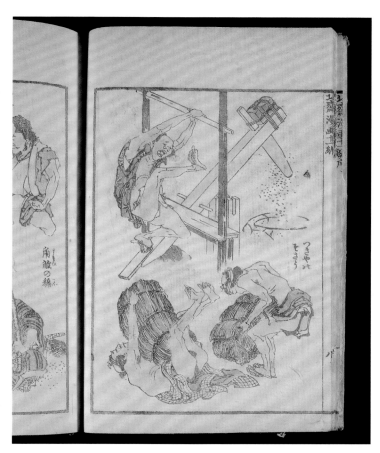

21.6, 21.7   *Manga*, vol. 11 (left), vol. 13 (right). By Katsushika Hokusai (1760–1849). Woodblock print; ink on paper. H: 23 cm. Davison Art Center, Wesleyan University, 1988.32.9, 1988.32.11; Gift of George W. Davison (B.A. Wesleyan 1892), before 1953. Photographs by R. J. Phil.

Manga is a series of comic books by the famed print designer Hokusai. His lively scenes of daily life include one man struggling with a heavy foot-powered rice mill, while another seems to have assumed a life of ease in a later volume by harnessing the device to water power instead. Both types of mills were once common in Japan.

remains. In villages where hand mortars are used, on the other hand, polishing is often not complete and leaves a less white but somewhat more nutritious product. Usually the final step before cooking the rice is to carefully sort through it to remove any bits of stone or other impurities.

The most important exception to the normal milling and polishing procedures is the practice of parboiling. This traditional Indian method, which involves boiling the rice while it is still in the husk, is used for over half of the total rice crop in India (Grist 1965, 377). In the simplest method, the grain is briefly brought to a boil in a pot and afterward spread to dry in the shade. A more complicated method is used in Uttar Pradesh, where the rice, still in the husk, is steeped in gently heated water for two days, then roasted for five minutes in hot sand (Grist 1965, 380). Parboiled rice is milled after the parboiling, which helps to prevent breakage and waste in the milling processes because the husk comes away from the grain more easily. It also has better keeping qualities than uncooked rice, and, most importantly, it is more nutritious because some of the nutrients from the bran layer are cooked into the starchy core of the grain during the parboiling.

Methods for cooking rice are too diverse to be described here in any detail. Among the simplest is to put it into a bamboo tube with water and stand it beside a fire. Most rice is technically boiled, whereas the glutinous rice that is the staple in many Tai and Lao areas is soaked first and then steamed. While Westerners tend to think of rice cooked and served in the grain, in Asia it is also widely made into noodles, fashioned into various cake forms, and cooked until it becomes porridge. A great variety of specialized cooking devices have been developed for various types of rice preparations (figs. 21.8, 21.9). Items used for serving food show even more variety, ranging from pieces of leaf to precious works of art (figs. 21.10, 21.11).

21.8 Griddle for rice cakes. Chiang Mai, Thailand. 2000. Earthenware. H: 22 cm. FMCH X2000.3.7a–k.

21.9 Steamer for rice cakes. Java, Indonesia. Tin, wood, bamboo. H: 31 cm. FMCH X97.22.1a–o; Gift of Amani Fliers.

Indonesian rice flour cakes known as *kue putu* are made in small bamboo tubes placed over a steamer. A lump of palm sugar in the middle turns molten, and the cakes are served with a sprinkling of grated coconut. *Kue putu* are a popular street snack, made by vendors who carry charcoal-fired steamers through the streets on shoulder poles. *Putu* refers to the sound of the whistle the vendor places over the steam holes to announce his presence in the street. This steamer is made for home use on a gas-burning stove.

21.10   Three-tiered food carrier. Japan. Nineteenth century. Red and gold lacquer over wood, with shell inlay. H: 38 cm. Peabody Essex Museum, Salem, Massachusetts; Museum purchase, Billings Fund, 1912, E15.289. Photograph by Mark Sexton.

Elegant Japanese food carriers are intended for transporting food for picnics. Many have multiple tiers for keeping different foods separate. Although rice may be only one of many ingredients contained within, no picnic would be complete without it.

21.11   Food carrier. Japan. Momoyama period (1573–1603). Black lacquer with gold *maki-e* (sprinkled powder) decoration. H: 35 cm. Los Angeles County Museum of Art M.87.202a,b; Gift of Donald and Iris Blackmore.

This food carrier, which consists of a single undivided chamber, is decorated with a design of flowering pinks. The various picnic foods would be packed separately within.

21.12 *Scene from Washhouse.* By Utagawa Kunisada, Toyokuni III (1786–1865). Japan. Triptych of woodblock prints. H: 36 cm. Museum of Fine Arts, Houston 75.356.31a–c; Gift of Peter C. Knudtzon.

This triptych depicts a household preparing *mochi* for the New Year celebration. At the right, a man, with his head bowed in effort, hefts the mallet high over his head, while a woman turns the mass of *mochi* within the mortar. At the left, other women shape the *mochi*.

21.13 *Mochi* mallet (*kine*). Japan. Wood. L: 85 cm. FMCH X99.28.2.

This mallet weighs almost twenty pounds, which explains why *mochi* making and *kine* are often associated with physical strength. Given its significance in Japanese culture, it is not surprising that *mochi* making is often conducted as a ritual event.

21.14   *Daikoku and Ebisu Making Mochi.* By Tamagawa Shūchō. Japan Circa 1795–1800. Woodblock print. H: 34 cm. Fine Arts Museums of San Francisco, Museum Collection, 1964.141.1081.

Joyous and strong, Daikoku and Ebisu, two of Japan's Seven Gods of Good Fortune (Shichifukujin), pound *mochi* in a mortar. A New Year decoration hangs overhead.

21.15   Rabbits silhouetted against
the moon. By Shibata
Zeshin (1807–1891). Japan.
Woodblock print. W: 29 cm.
Honolulu Academy of Arts;
The James Edward and
Mary Louise O'Brien
Collection, 1977
(HHA 17, 111).

In Japanese folklore, the patterns of light and dark on
the surface of the moon
form an image of a rabbit
pounding *mochi* in a mortar.
Zeshin has used this theme
for his humorous depiction
of a whole group of rabbits
at work, in much the same
way that making *mochi*
becomes a group effort for
humans. His full moon
plays on the smooth round
shape of a finished cake of
mochi.

21.16   Votive plaque (*ema*).
Japan. Late nineteenth century. W: 21 cm. Peabody
Essex Museum, Salem,
Massachusetts, E 11.715.
Photograph by Jeffrey Dykes.

An offering of rice cakes
(*kagami-mochi*) is depicted
on this *ema*. Just as the
mochi cakes are placed on
altars at shrines and in
homes at New Year, an *ema*
featuring the same theme
would likewise make a
suitable offering.

Food for the Gods

When rice is offered to spirits or deities, it is usually prepared in special forms, or at least presented in a purposefully artful manner. The underlying idea is that the intended spirits or deities will be drawn to an attractive display and will appreciate the honor granted to them, making them more likely to intercede on the supplicant's behalf.

The Japanese "mirror *mochi*" cakes (*kagami-mochi*) that are offered on altars at New Year are a case in point. Because mirrors in Japan stand for the Shinto deities, these special "mirror" rice cakes embody the spirits of the rice. Although commercially made *mochi* is today widely consumed by humans as an ordinary sweet, the laborious traditional process of making *mochi* carries special meaning in Japan and is viewed more as a ritual procedure than a culinary one. The glutinous rice must be soaked first, then steamed, and finally pounded in a mortar with a heavy wooden mallet until it is reduced to a smooth soft paste (figs. 21.12, 21.13). The strenuous pounding in the mortar is said to concentrate the spirits of the rice (Yoshida 1989, 57). This effort was traditionally undertaken only at home for special occasions, and at New Year it assumes a special beloved status. On January 11, known as "the opening of the mirror" (*kagami-biraki*), the cakes are broken up and eaten by family members, who thereby ingest the spiritual power of the rice cakes (Ohnuki-Tierney 1993, 96). Japanese art is full of depictions of *mochi* making, commonly showing idealized domestic or rural scenes, the deities at work, or even complex allegorical tales (figs. 21.14, 21.15). *Mochi* itself is often depicted in the form of an offering to the gods (fig. 21.16).

The offering of grain to spirits, and especially to the spirits of ancestors, is an ancient practice in Asia. Some of the most spectacular and famous works of Asian art are the bronze vessels made in ancient China for ritual offerings directed toward the spirits of ancestors. These vessels have been uncovered in large quantity at burial sites, where they were interred with the deceased, but they were also displayed as offerings on altars in ancestral temples and used as vessels for ceremonial feasts. They occur in many named forms, with the names appearing in inscriptions on the vessels themselves so that they are known to represent indigenous categories. Interpreting exactly what the various forms may have contained is more difficult, but scholars agree that most held cooked grain-based foods or "wine." For the early periods of Chinese history, millet was the preferred grain for offerings as it was the predominant grain in the Yellow River Valley that was the cradle of Chinese civilization (see chapter 29). Recent archaeological evidence that rice was cultivated in the Yellow River Valley as early as 5000 B.C.E. certainly holds out the possibility that rice was also offered in these vessels. Of most interest for this question are the three-legged cauldron forms known as *li*. *Li* appear to have gradually evolved from three-lobed pots, made first in clay and only later in bronze (figs. 21.17–21.19), that have been identified as "cauldrons for cereals" (Kerr 1991, 34). Nearly all of the well-known bronze vessels come from the north of China, yet recent research has begun to uncover bronze vessels in the south that were more certainly used for rice.[2]

One of the most enduring forms of offering rice is the Buddhist practice of almsgiving. By donating rice to monks who live in their community, ordinary citizens earn religious merit, which according to Buddhist teachings helps them gain a better life in the next cycle of reincarnation. The humble act of a group of monks setting out on their daily rounds carrying their alms bowls is one of the most characteristic features of religious life in Buddhist communities (figs. 21.20, 21.21). The monks visit private households, where families place food in their alms bowls for their morning meal. While foods other than rice may be included, the association between the alms bowl and Asia's daily staple food is deep and powerful. For the monks themselves, this routine characterizes their religious devotion and the humility of their life as disciples.

21.17   Three-lobed cauldron (*li*). China. Shang dynasty (1550–1050 B.C.E.). Ceramic with impressed decoration. H: 18 cm. Honolulu Academy of Arts, Purchase, 1950 (HHA 1034.1).

21.18   Ritual food vessel (*li*). China. Early Shang dynasty (1550–1050 B.C.E.). Bronze. H: 17.5 cm. Honolulu Academy of Arts, Acquired through exchange and a gift from Daphne Damon in memory of Mrs. Damon Gadd, 1969 (HHA 3607.1). Photograph by Tibor Franyo.

21.19 Ritual vessel of Duke X: of the State of Lu (*li ding*). China. Early Western Zhou period (ninth century B.C.E.). Bronze. H: 15 cm. Museum of Fine Arts, Boston 47.230; Juliana Cheney Edwards Collection.

These three *li* bear witness to an evolution of vessel types in ancient China. The oldest type is represented by the three-lobed ceramic cauldron (fig. 21.17), whose design evolved to suit the function of cooking food over an open fire. By Shang times, *li* were being made of bronze, but the early example shown here (fig. 21.18) still closely follows the form of its ceramic precursor, including an imitation of the incised lines that originated in the inherent decorative possibilities of the ceramic medium (Poor 1979, 55). By the time the Western Zhou *li* was made, the bronze vessels still maintained the tripod form but had developed into much more sophisticated artistic expressions that fully realized the potential of the bronze medium.

The original ceramic vessels may have been ordinary cooking pots, but bronze *li* were ritual vessels used for offering food to ancestors, both on altars and in tombs. Written sources, including inscriptions on some bronze vessels, indicate that grain-based foods and beverages were the primary offerings. In ancient China these are thought to have been composed of millet (see chapter 29), yet archaeological evidence shows that rice agriculture reached the heartland of ancient China in the Yellow River Valley by 5000 B.C.E. (Higham and Lu 1998, 871). This suggests that rice may also have been cooked and offered in vessels of this type.

21.20a,b   Mendicant monks. By
Deiryū Kutsu (1895–1954).
Pair of hanging scrolls;
ink on paper. L (image): 131
cm. Los Angeles County
Museum of Art L90.17 a,b,
promised gift of Murray
Smith.

This pair of Zen paintings
depicts a line of Buddhist
monks setting out on their
morning rounds with alms
bowls in hand and returning
home again.

21.21   *Shin Upagok*. Burma.
Nineteenth century. Carved
ivory. H: 16.5 cm. Los
Angeles County Museum
of Art M.82.132.2; Purchased
with funds provided by
Margot and Hans Reis,
Dr. and Mrs. M. Sherwood,
and the Felix and Helen
Juda Foundation.

Shin Upagok is the Burmese
name for the monk
Upagupta, a patriarch of
Mahayana Buddhism who
is regarded as the fourth
successor to the Lord
Buddha. He is typically
depicted sitting with his alms
bowl in his lap. A beloved
popular Saint in Burma,
he is propitiated for fine
weather prior to important
events such as festivals or
theatrical performances
(Sylvia Fraser-Lu, personal
communication, 2003).

21.22   Offering containers (*hsun-
ok*). Burma. Lacquerware.
H (tallest container):
60.5 cm. From left,
FMCH X2000.4.11a,b;
X2000.3.3a–e; X2000.3.2a–c;
X2000.4.10a,b; X2000.3.1a,b.

Rice and other food offerings are presented to
Buddhist monasteries in
Burma in covered lacquer-
ware trays called *hsun-ok*, or
sometimes more specifically
htamin hsun-ok (rice container; Fraser-Lu 2000, 121).
Many of the plain red or
black containers in sensuous
shapes appear to come from
Shan State.

21.23   A worker ladles boiled glutinous rice out of a pan in
preparation for making rice
wine. In this community,
the rice is roasted in the dry
pan before the water is
added, giving it a rich color
and aroma. Photograph
by Roy W. Hamilton.
Banaue, Ifugao Province,
Philippines, 1998.

21.24   Cooked glutinous rice,
inoculated with a dry yeast
mixture, is placed in a jar
where it will ferment to
form rice wine. Photograph
by Roy W. Hamilton,
Perendaman, Ketapang
District, West Kalimantan,
Borneo, Indonesia, 1993.

21.25   Prayer stick (*ikupasuy*). Ainu
peoples. Piratori, Hokkaido,
Japan. Carved wood. L: 41
cm. Brooklyn Museum
12.282.

In Ainu rituals, carved
prayer sticks are used to
scatter drops of sacred alcohol from a wine cup (*tuki*).
Homemade millet beer was
once used, but it was
replaced by sake when that
beverage became widely
available in the Ainu region.
The sticks are considered
intermediaries for delivering
prayers to the deities and
were prized possessions of
Ainu men. This prayer stick,
decorated with a miniature
version of a *tuki*, was collected in 1910.

21.26   Pair of ritual wine cups and
stands (*tuki*). Ainu peoples.
Hokkaido, Japan. First half
of twentieth century. Diam
(cup): 12 cm. Buffalo
Museum of Science C18752,
C18756, C18755, C18757.

Relatively ordinary lacquerware sake cups from the
more southerly regions of
Japan were an important
trade commodity to the
Ainu region in the north. In
their adoptive community,
they became a key element
in Ainu ritual.

21.27   Chinese ceramic jars hold rice wine during an Ifugao ritual held to improve the health of an ailing grandmother. The man at the left with the red shoulder cloth chants passages from Ifugao epics in front of the assembled offerings for the deities, which include sheaves of newly harvested rice and cooked food, as well as the jars of rice wine. Photograph by Roy W. Hamilton, Banaue, Ifugao, Philippines, 1998.

In the Buddhist countries of Southeast Asia, where all young men are supposed to spend at least a short time residing in the local monastery, these religious practices continue to have enormous influence on community life. Nevertheless, the daily rounds of Buddhist monks are fast becoming a thing of the past in many places. In rapidly urbanizing Thailand, fewer and fewer young men enter the monkhood. In Japan, the daily rounds of monks were officially discouraged during the Meiji period (1868–1912) and today religious efforts are supported primarily through cash contributions. The Communist governments of China and Vietnam have banned alms collecting by Buddhist monks as part of their general sanctions against established religion.

In addition to providing food to individual monks making their rounds, Buddhist communities provide rice and other food in quantity directly to their local monasteries, typically in the context of community-wide religious festivals (see chapter 13). Often the food is elaborately presented in beautiful containers, making it not only a contribution to the religious institution but also a religious offering to the Buddha (fig. 21.22).

Throughout Asia there are many other examples of religious offerings consisting of rice or special rice preparations. In Vietnam, steamed cakes of glutinous rice called *bánh chưng*, stuffed with mung beans and pork, are placed on the family's ancestral altar during Tết, the Vietnamese New Year period. When visitors come, they are served from the cakes on the altar, sharing the blessings of the family ancestors (Nguyễn Xuân Hiên 2001, 52). Perhaps the most spectacular rice offerings of all, however, are the elaborate constructions of brightly colored rice dough that are made for temple festivals in Bali. These and other Balinese rice offerings are discussed further in chapter 23.

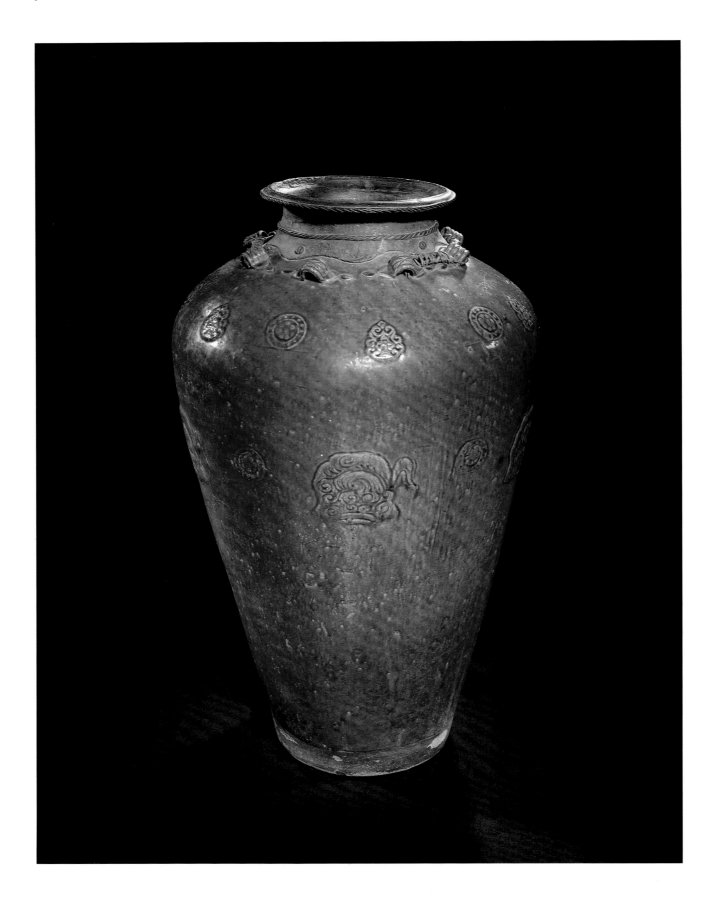

Rice Wine

Spirits and deities are pleased not only by rice but also by rice wine. This alcoholic beverage is made almost everywhere that rice is grown. Technically it is a form of beer rather than wine, since it is made from fermented grain, but the name "rice wine" has become thoroughly entrenched. Many devotees point out that its taste more closely resembles a smooth sweet wine than a frothy bitter beer. Rice wine is made from glutinous rice; indeed, in many Asian cultures, it is the primary reason for growing glutinous rice at all. The rice is cooked and in the process absorbs a large quantity of water (fig. 21.23). After it cools, it is inoculated with yeast and placed in a jar or vat with no additional water (fig. 21.24). As fermentation proceeds, the starches in the rice are converted to alcohol and much of the liquid that had been absorbed by the rice is released. The result is a milky liquid, moderately sweet, with a somewhat higher alcoholic content than most European beer. The wine may simply be ladled out of the container, leaving the mash behind, or it may be filtered to produce a somewhat more refined product.

Alcoholic beverages are made from other substances in Asia as well, such as sugarcane, palm sap, and millet, and the Chinese in particular developed an advanced technology of fermentation, including a method for making grape wine. Nevertheless, rice wine remains by all measures the region's most characteristic and important form of alcohol. One indication of this is that rice wine is very widely featured as a prominent element in offerings. From spirit possession sessions in Burma, to longhouse festivals in Borneo, to Shinto temple ceremonies in Japan, rice wine is the drink of Asia's spirits and deities (figs. 21.25–21.28; see also fig. 24.4). Rice wine can also be distilled to make a more potent beverage with a much higher alcohol content. Wherever these stronger beverages are made, they too are likely to be used in religious offerings.

The brewing of alcohol from rice has reached its greatest level of sophistication in Japan, and Japanese rice wine (sake) is known, and exported, around the world. Modern sake is made by a complex process called "multiple parallel" fermentation, in which the conversion of the starch in the rice to sugar and the fermenting of the sugar into alcohol take place simultaneously and gradually over a long period, about eighteen days, at cold temperatures (Kondō 1984, 42). This is why sake making is a seasonal activity, engaged in only during the winter months. Two microorganisms are required, a mold known as *kōji* (*Aspergillus oryzae*) and a yeast (*Saccharomyces cerevisiae*). Grape wine, European beer, and simpler types of rice wine require only a yeast. While most of the rice wine drunk in Asia continues to be made in local communities using very simple methods, sake brewing in Japan has become a huge modern industry. The role of sake in Japanese society and the enormous body of art dealing with sake and sake drinking are discussed further in chapter 24.

•

21.28   Heirloom jar. Sarawak, Borneo, Malaysia. Glazed ceramic, metal, rattan. H: 87 cm. Collection of Thomas Murray.

Ceramic jars from China served as valuable trade goods in Borneo for many centuries. In the longhouse communities where they were collected and displayed, the jars were used for a variety of ritual purposes, most notably the making of rice wine. This unusual jar has a metal rim, a feature that has been documented in Qing dynasty jars from the eighteenth and early nineteenth centuries, mostly collected and used in Kelabit communities in the interior of Sarawak (Chin 1988, 190–91). Precisely determining the date and origin of Borneo jars is difficult because their high value inspired an ongoing trade in Chinese-made reproductions of older styles, and by the 1880s even potters working in Borneo were producing reproductions for the local market (Harrisson 1986, 5).

22.1   Rice pot lid (*sara*) with
         design of Lakshmi and
         stylized stalks of rice. West
         Bengal, India. 2001. Painted
         earthenware. W: 26 cm.
         FMCH X2001.2.1

22

Lakshmi of the House: Rituals, Realities, and Representations of Womanhood in Modern Bengal

Sohini Ray

"There are no rice deities in India," one of my professors told me: "Lakshmi is a goddess of wealth. Look at all her pictures—there is no rice in them." This was the confusing note on which I started my research on this deity. As I vaguely recalled some of the rituals I had seen in Calcutta[1] as a child, I remembered Lakshmi pictured with a sheaf of paddy, or unhusked rice, in her hand. Who is this goddess then? Is she really Lakshmi? What does this representation signify? Unknowingly, I had stumbled upon one of the striking characteristics of Hindu religious traditions—multiple mythologies and iconographies of the same deity are found in different regions of India and in different strata of Indian society.

Similarly, when one examines the scriptural descriptions of Hindu deities and their iconography, there are no rice deities to be found. Scriptural accounts of Lakshmi portray her as the goddess of wealth throughout India. She is represented sitting on a blooming lotus, with lotus flowers in her hand. The link between Lakshmi and rice has also not appeared in the academic literature of any discipline.

One of the singular features of the Hindu religious tradition is that scriptural and theological aspects play a minimal role in people's lives. The scriptures and people's practices seem quite unrelated. Therefore, to develop a comprehensive understanding of Hindu religious traditions, it is necessary to study the people's daily religious activities, which often vary considerably from region to region and even within a single society. Lakshmi appears in several guises. The present article analyzes the variant forms worshiped by women in Bengal.

The most common Lakshmi images do not include any representation of rice. Even in Bengal the ritual worship (*puja*) of Lakshmi is typically performed with an image of Lakshmi that has no reference to rice. However, a careful examination of the various icons and representations of Lakshmi used in rituals and festivals yields some striking observations.

If one visits a weekly fair in the suburbs of Calcutta, one finds potters from rural areas selling *sara* paintings. A *sara* is the lid of the round pot (*handi*) used for cooking rice. Traditionally such pots are made of clay, but now they are also available in different metals. The *sara* painting tradition of Bengal includes images of Lakshmi with rice (fig. 22.1). These pictures are painted by members of the potter caste for daily use by women in their Lakshmipuja. In addition, some calendars used in remote rural regions of Bengal have representations of Lakshmi holding rice in her hand (figs. 22.2, 22.3). In the houses that I visited, the women used photographs or small statues that were easily available. When I questioned them, the women answered that *sara* are hard to acquire these days, and so calendar pictures are used in homes. However, the presentation of Lakshmi for the quarterly worship is done using paddy and a paddy basket. There are also differences between the regions of Bengal. Women from East Bengal observe an annual festival (Kojagari) when Lakshmi is worshiped on a full moon night of autumn, for which the painted *sara* is compulsory. Women from West Bengal, however, observe the quarterly worship with the paddy representations.

The Ritual and Its Variations

There are several versions of the Lakshmipuja performed in Bengali homes, including weekly, quarterly, and annual celebrations (Fruzzetti 1975, 72). Bengali women observe a weekly ritual of worshiping Lakshmi that does not require a priest. Every Thursday, the women conduct the worship themselves; they propitiate the goddess by offering flowers, water from the Ganges River, and some fruits and sweets and ask for her blessings. The use of Sanskrit mantras is kept to a minimum. For certain occasions the women decorate the floor with special designs (*alpana*), but they do not always do so during the weekly rituals. Lakshmipuja can be done either in the evening or in the morning. The entire proceeding is considered the domain of women. Unmarried girls are taught this form of worship and are encouraged to participate, but the main initiator is the lady of the house. In a multifamily household all the women who share the dwelling gather together to observe this ritual. If it falls on a day when a woman is menstruating, she does not participate directly in the offering but observes the proceedings and listens to the reading of ritual verse (*Panchali*, discussed below), the main part of the ritual. In West Bengal, a separate offering for Lord Kubera, the Hindu god of wealth and treasure, is given; in East Bengal, Kubera is unknown.

The quarterly Lakshmipuja is performed in the months of Bhadra (mid-August through mid-September), Kartik (mid-November through mid-December), Poush (mid-December through mid-January), and Chaitra (mid-April through mid-May) on the mornings of dates selected for their auspicious qualities. Preparations begin the night before. The space for the *puja* is cleaned, and a raised platform for the deity figure is decorated with rice-paste *alpana* and covered with a colored cloth to create an altar (fig. 22.4). Next comes the preparation of the deity. This varies greatly from home to home; here I will describe what I observed at a home in south Calcutta. In this case a rice-measuring bowl (*kunke*) is taken and filled with rice. Some rice is placed on the platform and the full bowl is placed on it (fig. 22.5). The front of the bowl is decorated with vermilion in the form of a swastika (the Hindu sacred symbol) to symbolize Lakshmi's face. A cloth is placed around it, to give the rice bowl the shape of a face, and Lakshmi's crown is placed on top (fig. 22.6). A necklace is circled around the "face," and the form of Lakshmi is thus completed. In some cases household emblems saved for generations—such as decorated silver plates with sacred designs—are brought out. Women's belongings that are saved for this *puja*, such as cowrie shells (an ancient symbol of money), are set around the altar. Four tall wooden containers traditionally used to store vermilion (*gach-koutas*) are placed on the four corners of the altar. The surrounding area is decorated with flowers and flower garlands and the floor around the altar is ornamented with *alpana*.

Certain *alpana* motifs are considered compulsory for this ritual. A blooming lotus creeper, for example, extends from the door to the altar to signify the footsteps of Lakshmi. At the door, Lakshmi's companion, the owl, is a sign of money. Outside the door one finds some water and a small container with a branch of a neem tree. These items are meant to welcome the goddess, who would use the branch as her toothbrush (this is an indigenous method of brushing teeth in India) and wash her face before entering the room. A copper pitcher (*mangal-ghat*) decorated with mango leaves is set out and a conch shell on its stand is placed beside it. Another required article is the Shaligram Shila, or the sacred black stone representing Narayana (another name of Lord Vishnu, one of the principal gods of the Hindu religion and the consort of Lakshmi). The other ingredients of the *puja* are powdered incense, incense sticks, an oil lamp, and sandalwood paste (fig. 22.7). The cooked food is offered to the goddess as *bhog*, and is later distributed to the people as *prasad,* an offering that has been tasted by the goddess first. This version of the *puja* is officiated by a Brahmin priest. It is followed mostly by people from West Bengal (fig. 22.8).

There are seasonal variations of the ritual described above. In Bhadralakshmipuja (the worship of Lakshmi done during the month of Bhadra), *tal,* a seasonal fruit, is given along with coconut and flattened rice. The image of Lakshmi is not immersed in the river, as is done in most Hindu worship. In Poushlakshmipuja (the worship of Lakshmi done in the month of Poush), the *kunke* is filled with new paddy, but the offerings are the same. In Chaitralakshmipuja (the worship of Lakshmi observed during the month of Chaitra) new fruits are offered, such as *jamrul* and *lichu.*

The Kojagarilakshmipuja (the worship of Lakshmi done on the full moon night of Ashwin, i.e., mid-October through mid-November) is held once a year. For this ceremony a *sara* painting of Lakshmi is bought each year (see fig. 22.1) along with a "Lakshmi's *jhanpi,*" something like a toiletry kit with a small mirror, a comb, a packet of vermilion, a container of hair oil, and so on. The ritual space of the house is decorated with *alpana* and the worship is conducted by a Brahmin priest. In Lakshmipuja no other musical instruments are used except the conch shell. It is believed that if someone stays up all night following this ritual, he or she will become very rich.

On the new moon night of the month of Kartik, when the goddess Kali is worshiped (Kalipuja) in Bengal, Lakshmipuja is also done, albeit in a different way. In the evening the women and children of the household make models of an *alakshmi*—a demon-like woman considered to be inauspicious (the opposite of Lakshmi)—with cow dung and discarded hair, place them on a winnowing fan, and take them out of the house while ritually chanting "Alakshmi get out." Afterward the women return home, wash themselves, and change their clothes. Then a worship ritual for Lakshmi is carried out, and lamps are lit for Kalipuja.

The Ritual Art: The *Alpana*

A compulsory aspect of Lakshmipuja is the *alpana* drawn by women. Sometimes women do it on the day of the weekly ritual, sometimes they don't if it is not convenient for them that week. These paintings are made with rice powder mixed with water. There are several motifs that are drawn on the floor around the deity's altar. The motifs are stylized in characteristic ways and incorporate certain patterns associated with Lakshmi such as the lotus, Lakshmi's feet, Lakshmi's companion owl, and, most importantly, sheaves of rice. The *alpana* represent the world of the goddess and bring good fortune to the home. Many of the women I spoke to noted that not everybody can do *alpana* and that those who can are considered special and are admired for their artistic quality. The *alpana* are a way of decorating one's home in a manner that is both auspicious and traditional in modern Bengal. The motifs used to be drawn only on the floor for special occasions, or embroidered on blankets (*kanthas*), but now the motifs appear on dresses, T-shirts, wall plaques, posters, and various consumer items.

22.2   Calendar with illustration
of Lakshmi with rice in her
hand. Calcutta, India. 2001.
Commercial print on
paper. H: 88 cm. FMCH
X2001.35.2

22.3   Calendar illustration of
Lakshmi with a lotus blos-
som in one hand and rice in
the other. Balurghat, West
Bengal, India. 1990s.
Commercial print on paper.
H: 74 cm. FMCH
X2001.35.3

Ritual Verse: The *Panchali*

The *Panchali* is a long set of poems that tell a story about the glory of Lakshmi and the importance of her worship. The theme and content of these verses, which spell out the qualities of the ideal woman, are revered and taken to heart by the women who share them on ritual occasions. During Lakshmipuja one person, usually the lady of the house, reads the *Panchali* and the other women listen. The reading is done in a characteristic tune and style that is very common in Bengal for reading any sacred poetry, epic, or ritual text (fig. 22.9).

The authorship of the *Panchali* is unknown. Printing houses collect and publish the poems in the form of a small book in colloquial Bengali. Women get them from relatives or purchase them in local stores where religious books are sold. Some narratives, along with folktales, are printed in the astrological directory (*panjika*). I collected three versions of the stories—usually a single edition is used throughout the year. Several of my informants said that they used the *Panchali* they received from their mother or grandmother or aunt, the one they grew up hearing as a young girl. Of all the women I interviewed, only one does not read the *Panchali*; she reads a Sanskrit chant about Lakshmi. Excerpts from a standard reading of the *Panchali* follow:

Welcome[2]

Where are you mother Lakshmi, full of compassion, I pray at your feet with all sincerity. We have neither strength nor any piety. We do not have education, intelligence, or strength. We do not know how to address you, or how to propitiate you, how to welcome you. The only thing we know is that you go to whoever calls you. Please take mercy on your devotees.

Introductory Text

One full moon night goddess Lakshmi and Lakshmikanta [Lord Vishnu] were seated on a gem-studded throne and were laughing and talking together. At this time, Narada came in with his vina [a stringed musical instrument] in his hand. With the image of the lord in his heart he sang while playing the vina. The goddess asked him politely, "O Sage, you think about the welfare of the world always. You know of everybody's whereabouts. Is everyone on earth living in peace? Please tell me what the plight of the universe is." Hearing Lakshmi's polite words, the sage started speaking: "By your grace, O goddess, your servant [Narada] is lucky. I am doing well. But let me tell you, the only thing that bothers me is the suffering of the people in the world. I cannot tolerate it any more. There is no happiness in the world anywhere. I see only fire and famine. Everyone is in illness and misery. O Mother of the world, you are the well-wisher of the world, who owns peace and prosperity. If you stay in permanence in people's homes, no one would be in distress. Your name is chanchala *[restless]; you do not stay still anywhere. That is the reason why people are in so much sorrow." Upon hearing Narada, Lakshmi was sad. Said she, "O Sage, you accuse me for no reason. People suffer for their own faults, and they blame me for their sufferings. Let me tell you, O sage! Every woman in the world is born as a part of me. So, if each one of them does her duty, there would be no misery in the world. Look at them, they [the women] have given up all the traditional norms and they behave in any way they want. They do not do their duties toward their homes but just dress up without any shame. They do not keep faith in the words of holy scriptures and are not afraid to do anything they want. They have given up doing the household work, cooking, and even raising their own children. They are selfish in their outlook and have no compassion. They talk ill of people behind their backs. They do not read holy books, but only novels. They eat before cleaning themselves and go to sleep at odd times. If in-laws give good advice, they rebuke them and become angry. They show affection to their husbands only for their own selfish reasons. I see all this in every household. How can I*

22.4   Women decorating the floor with *alpana*. Photograph by Sohini Ray, Calcutta, 2001.

22.5   The rice measuring bowl filled with paddy. Photograph by Sohini Ray, Calcutta, 2001.

stay still in situations like these? Listen, O Sage Narada, I go to their houses where I see peace, but they bring distress and drive me away. That is why I move around from house to house; consequently people call me chanchala. *In the households in which all the women get up early morning, wipe their house with cow dung, take their bath, pray to god, and then focus on their household chores; they do not laugh or talk loudly; they behave submissively with elders—these women behave well with their husband's relatives and serve their husbands with devotion. Those houses have no arguments or quarrels, only agreeable and pleasant conversation. In the houses where guests are served with affection, holy texts are kept and read, and there is no deceit or deceiving of people. Where married women put vermilion on their forehead and always have a smiling face, please believe me, O Sage, I stay in those houses forever."*

The Spread of Lakshmipuja in the World

Lakshmi asked her husband, Narayana [Lord Vishnu], "My heart is breaking from knowing the sorrows of the people in the world. There are so many kinds of problems. I cannot even understand, let alone solve them. Please advise me on what I should do." Narayan replied, "Why are you so worried? Spread the message of worship of Lakshmi in every household. By your grace everyone will become wealthy. Every Thursday all the women will propitiate you with devotion. Then all miseries will go away. And there will be peace and prosperity in every home." Upon hearing the word of Narayana, Lakshmi was happy, and she went to the earth to spread the message. When she went to the city named Avanti, she was surprised to see the condition of that place. The king of the city, named Sumanta, had been a pious man as rich as Kubera, the lord of wealth and treasure. He was a virtuous man and was always involved in attending guests and devotional functions. All of the king's seven sons were as accomplished, and he was one of the luckiest men in the world. Leaving all the seven sons in his household, the king went to heaven. Consumed by greed, the seven wives resorted to evil ways. They wanted only their own selfish happiness. The household became filled with jealousy, and quarrels broke out. Each wife wanted to have a separate establishment of her own, and they started manipulating their husbands. The husbands could not understand the deceit of their spouses, and the seven brothers separated into their own establishments. Thus the anti-Lakshmi [alakshmi] entered the household and all their father's wealth was gone. The old queen mother could not tolerate the harsh words of her daughters-in-law anymore and went away to the forest one day. She decided she would end her life, as she was very frail without rice [a metaphor for food]. At this point the goddess Lakshmi came in the guise of a Brahmin woman and asked

22.6   The rice-measuring bowl filled with paddy being dressed as the image of Lakshmi. Photograph by Sohini Ray, Calcutta, 2001.

22.7   A complete domestic altar of Lakshmi for seasonal Lakshmipuja. Photograph by Sohini Ray, Calcutta, 2001.

her politely, "Whose wife are you, and whose daughter? Why are you roaming around the forest by yourself so depressed?" The woman replied, "O Mother, I am very unlucky. What is the use of hearing my story?" The goddess said, "Please tell me. I will help you as much as I can." Upon hearing the words of Lakshmi, the old woman replied with folded hands, "O Mother, please listen to me. I am the wife of Sumanta, the king of Vidarbha [the kingdom in which the city of Avanti is located]. Our household had golden prosperity, and there was nothing ever lacking. All the time the king was living, Lakshmi was present in our house. Leaving behind seven accomplished sons, he left for his heavenly abode in time. All the riches of my home disappeared after that. Now all my seven sons have separated. There is only jealousy and conflict in my home. The daughters-in-law always harassed me. They did not let me live in peace. That is why I have come to the forest, to end my life." The goddess said, "What is the reason for this? Suicide is a sin. Go back to your home. You will find peace and prosperity like before. On Thursday evenings all the women should do Lakshmipuja. Let me tell you the rules. You will place a water-filled pitcher along with betel leaf and [areca]nuts on the cloth seat. Light the lamp and incense and place vermilion. After placing all the items, read the sacred narration, keeping the image of Lakshmi in your mind. Listen to the sacred narration with durba grass in your hand. At the end the women perform the sacred 'ulu' sound and prostrate to the goddess, giving sindur to each other.[3] The women who perform Lakshmipuja every Thursday will be open-minded by my grace. During the worship all the women of the household will listen to the narration, leaving all other work. I [the goddess] will stay permanently in those houses and grant all the wishes of the devotees. If there is a full moon on any Thursday, the homes of the women who observe fast and perform this worship will be filled with wealth, and they will live in heaven forever." Saying this the goddess assumed her own form and let the woman see her in her original form [i.e., gave darshan]. The old woman prostrated herself before her and then stood there with her hands folded. Lakshmi said, "Do spread my words. You will be wealthy soon. You will find peace and prosperity in your home. Your daughters-in-law will listen to you." Granting this boon, the goddess disappeared.

One day in the city of Vidarbha all the women were performing the worship of Lakshmi. A local merchant came in during the time of worship. He was one of five brothers and all were very wealthy. They lived in a single household with their sons and grandsons. The merchant asked disparagingly, "What worship do you do, and what do you get out of it?" The pious women replied, "We do the worship of Lakshmi for our well-being. The people who perform this worship with devotion, their wealth increases with the grace of the goddess

22.8 Another example of the
      deity Lakshmi represented
      on a rice-measuring bowl
      and placed on a home altar
      used for seasonal family
      worship. (In this case a
      member of the household, a
      professional artist, decorated
      the bowl himself.)
      Photograph by Sohini Ray,
      Calcutta, 2001.

Lakshmi. They are protected from all danger. Peace remains in their house all year." Hearing this the merchant replied proudly, "That is an impossible thing. Wealth hoarded without work diminishes slowly. From where does wealth come without work? What is the use of just worshiping? If one can sit and earn money, I will see what Lakshmi's powers are like." Saying so, he went back home proudly. He sent his seven ships filled with goods. Lakshmi was angry with him. All his seven ships with goods were sunk. His home suffered a fire that destroyed a large part of his property. At this time he could not find rice or clothes. Quarrels broke out among his brothers, and in hunger they left their wives and children behind to go begging in foreign lands. The merchant was stricken with several kinds of illness and was in deep misery. After roaming around in several places, he came to Vidarbha once again by Lakshmi's grace. He saw that many women were performing Lakshmi's worship with devotion. At that juncture he remembered what happened in the past. As he had looked down on Lakshmi's worship, he realized his fault and prayed with folded hands, "O Mother! I am very ignorant. Please take mercy on me. In my pride of being wealthy I did not recognize you. I made many mistakes. I neglected your worship, and I got what I deserved. You are full of compassion. Please take mercy on me. What do I understand of your power? Even Lord Brahma [the supreme creator in the Hindu religion] cannot control you. In every home you are Grihalakshmi [Lakshmi of the house]. You are the giver of rice and granter of wishes. All gods and demons are your progeny. You are the sacred tulsi *[basil] plant, the sacred river of Ganges who washes away all sins. Your power is endless, and boundless are your incarnations. Please grant me this boon, so that I will always have faith in you." Thus prayed the merchant. With devotion he listened to the narration. At the end of the worship he prostrated and then went back home and asked his daughters-in-laws to perform the worship of Lakshmi. Hearing the words of their father-in-law, the women performed the worship every Thursday with devotion. Soon Lakshmi was happy. The home was filled with peace. The seven ships miraculously came back with all the people. The brothers came back together under one roof. All jealousy and rivalry came to an end. They became wealthy like before. Likewise they went to the ends of the earth to spread the worship of Lakshmi. Slowly the words spread to all places.*

22.9   A Bengali woman reading *Panchali.* Photograph by Sohini Ray, Mysore, 2001.

The Different "Lakshmis": Conflict, Contest, and Modern Life

It would be easy to assume that *every* Bengali Hindu woman performs the same worship rituals, and that there is no difference of opinion among them regarding such practices, especially across generations. As I continued my fieldwork in the city of Calcutta, I met many kinds of women. One person said, "In my family, my mother-in-law never did any of this. They were followers of a sect of Shivaite worship where no idol worship is allowed at home. We only read certain religious books. So, even though we are Hindu, no women in our family, my mother-in-law or aunt-in-law or anyone, did Lakshmipuja ever; it is not there in our family tradition. If one of us wants to do it for ourselves that is alright. My sister-in-law does, for example. She learned it from her mother and wanted to continue after she was married, and that is alright with everybody, but it is not a tradition in this family" (personal communication, 2001). Her comment was yet another reminder of the complexities of the Hindu tradition.

Another side of the story is the experience of so-called modern women. I spoke to several younger women who expressed strong feelings about Lakshmi worship. One told me: "I am a doctor. I have studied science and I do not believe in any of that. I think those rituals are meaningless." "So, you do not do any kind of worship—what about at your home?" I asked. "No, no, my mother does all that but she is very traditional; she does not know better. I do not believe in all that…. I am a doctor, I believe only in scientific reason and nothing else. And my mother-in-law does not even expect any of that from me. My sister-in-law does it; she does not have a career" (personal communication, 2001). It was evident from her tone that she considered herself better than her mother and her sister-in-law, because she does not believe in religion. With this I understood that a new discourse had developed with regard to ritual in modern Bengali society—the woman who "knows better" than to observe these rituals. This implies a new hierarchy among women in their families. I noticed that even though she told me that she does not believe in the worship herself (at least that is what she said), she relegated the practice to her sister-in-law, someone whom she thought of as "inferior" in the household.

An older housewife I interviewed said,"Reading the *Panchali* gives me focus and purpose in what I do. When I did Lakshmipuja, my entire family laughed and made fun of me. My husband is a scientist; he does not believe in any religion. My daughter is a doctor, and my son an engineer. None of them believe in religion. But even then, I do Lakshmipuja at home. It is the only time I can sit and concentrate on something, and no one disturbs me" (personal communication, 2001). Several older women told me that their daughters or daughters-in-law do not wish to continue the practice of Lakshmipuja because they are too "modern" for it.

Ideologies of Womanhood

As is evident in the *Panchali*, Lakshmi gives very clear definitions of the kinds of homes she likes to visit and the homes she does not. Her descriptions also imply what constitutes the "good woman" and the "bad woman." Undergirding the text is the assumption that the household is the woman's domain. She is the manager there—any failing in the running of the household is her fault, and a well-organized house is credited to the woman who runs it. The man earns the household's income, but how the resources will be managed is completely at the discretion of the woman. As one of the women I interviewed told me:

> A woman has many responsibilities. She is the person who leads the house. When I was married in 1959, my husband had a very small income. Our family came from East Bengal; all their properties had been taken away by partition, and from his income we had to support much of our extended family. We had two children, and then we built a house…. and I had no one to teach me anything in the beginning. Today we have our own house and savings. Our children are well established in life and my husband has retired comfortably. If I had not been able to manage a household, we would not have any of this. Look at Mrs. Sinha—her husband was so well placed and had such a good salary. But she never cared about anything in the house. I would be shocked to see how much food they wasted every day. I always thought about how to re-utilize things, any piece of cloth I had, from my discarded or daughter's dresses. I would sew them into blankets, or pillow covers. Any leftover from dinner, I would think of a way to re-utilize it for breakfast or lunch the following day. We could not afford to have a refrigerator for a long time, but I saw to it that nothing was wasted in the household. One needs to spend time and think about these things. Mrs. Sinha never did any of this…. Now, see he has retired, and they are almost penniless, only dependent on their children. The children have their lives, their own families to maintain. If she had managed things well, they would not be in this condition today [personal communication, 2001].

I understood the significance of Lakshmi in these women's lives. Lakshmi, being the consort of Vishnu, is the manager of the entire universe. It is she who sees to it that all the resources of the world are used properly. She is thus the role model for every married woman who has her own household to manage. The weekly repetition of the words of Lakshmi reinforces the ideal of womanhood for Bengali women.

The use of the word *Lakshmi* in the Bengali language is significant. It has multiple connotations. In everyday Bengali, a good girl is referred to as a "Lakshmi girl," a good boy is called a "Lakshmi boy," a destructive person is called "Lakshmichara," and a well-organized household is described as having "Lakshmisree." A good wife is known as the "Lakshmi of the house." When welcoming a woman with affection, one says, "Come in, Mother Lakshmi." When addressed tenderly, a man or a woman may be called "Lakshmiti." As these examples demonstrate the proper name of the goddess serves as a commonly used adjective in the Bengali language. An awareness of Lakshmi's divine attributes thereby permeates the everyday life of people.

Nevertheless, given the cynicism of the modern women I interviewed, I wondered whether Lakshmipuja and its values might not disappear among the younger generation. I also wondered whether denying the ritual might coincide with a deep unconscious acceptance of it. I found an intriguing answer.

One of my last stops was at a home in Southern California, where I visited Lata, a young Indian woman in a quiet neighborhood in Rolling Hills. Both she and her husband are successful lawyers in Los Angeles who immigrated to the United States from India. As I talked to them at their kitchen table, Lata was cleaning her kitchen. "I have bugs everywhere," she complained. "We bought something from a store, and it had bugs in it. We had not noticed in the beginning and now everything in my kitchen is infested. What a waste! I have to throw out everything now!" she muttered, as I sat there talking. Answering my queries, she continued:

> I do not know anything about Lakshmipuja, Sohini. No one did that in my immediate family. My own mother is a scientist; she had no leaning toward any of that. My aunts and other relatives are also very modern; they do not do that stuff.... I have been to Lakshmipuja in other people's houses, and I do respect their beliefs.... I have participated in the community Lakshmipuja festivals in our neighborhood in Calcutta, but I do not follow any of those things myself. Even if I believed in the goddess, who has time these days for those things, tell me?... Those weekly household rituals, you will find in more traditional families. [personal communication, 2001]

As I wrapped up our interview, I suddenly noticed a large bag of rice in the corner of her kitchen. It was one of her insect-infested grocery items, but she had not thrown it out. Before I could ask her about this, she opened the bag and poured the buggy rice onto a large plate. Without even batting an eye, she commented, "I have to finish cleaning and separating the bugs from all this tonight. Rice is Lakshmi—I can't throw it away" (personal communication, 2001). With every flap of her palms, the plate whisked to and fro in exactly the same rhythm of the winnowing fans used by rural Bengali women, and bugs flew into the kitchen sink. The blazing sun set among the hills. In the mosaic of the long evening shadows and the pale streetlights, I watched Lata in silence. Later on, I thanked her and drove away on the 405 Freeway as stars blinked. The Lakshmi day came to an end in search of another dawn in Lata's house. The goddess, I could not help thinking, would be highly amused.

•

23.1   At a Balinese temple, large cones of rice called *tumpeng* form part of the offerings brought by families for a festival. The *tumpeng* are considered models of the cosmic mountain, the axis mundi that links the upper-world, human world, and underworld. The offerings also include a roasted chicken, representing the meat element, and different kinds of rice cakes, fruits, flowers, and palm-leaf decorations. Photograph by David J. Stuart-Fox, Taro, 1977.

23

The Art of Rice in Balinese Offerings

Francine Brinkgreve

Rice Is Food: Rice as the Main Ingredient in Offerings

Everywhere on Bali one can see in the fields or at crossroads, by doorways or in one of the numerous temples, little palm-leaf containers with some grains of rice and a few flowers. In Balinese Hinduism, such offerings, called *banten*, form an important means of maintaining good relations with the deities and demons. Offerings provide a way for people to express their gratitude for the fertility of the earth, for the divine gift of life. They are made of the fruits of the earth and presented to the invisible powers with the request that the earth will remain fruitful, that life will continue.[1]

Offerings consist partly of small amounts of the food and delicacies that are usually presented to honored guests. After the deities have taken the essence of an offering, some of the food that remains is eaten by the worshipers themselves. Dogs, pigs, and chickens also benefit from the ritual leftovers. Despite the enormous variation in kinds, shapes, and ingredients of offerings, dependent on regional traditions and various levels of ritual elaboration, rice is present in almost every offering.

As in other parts of Southeast Asia, rice in Bali is synonymous with food. The Balinese words for cooked rice, *nasi* or *ajengan*, are often used to refer to a whole meal. Many Balinese eat rice twice a day, and it is not polite to let visitors around mealtime leave without having offered them a meal of rice and side dishes. *Baas* is the word for uncooked white everyday rice, obtained from irrigated fields (*sawah*) and from dry land (*tegalan*). Other than this white rice, in Bali a red dryland variety called *gaga* and two types of glutinous rice, white (*ketan*) and black (*injin*; fig. 23.2), are grown. Yellow rice, which is well known for ceremonial meals in Java but in Bali is mainly used for offerings, is naturally colored by means of turmeric (*kunyit*). Using artificial dyes, white rice can be given any color, when necessary.

Since offerings consist of the food the Balinese enjoy themselves, rice is a very important offering ingredient. Every day, after the first rice of the day has been cooked, dozens of very small rice offerings must be prepared and presented before members of the family may start eating, because honored guests always receive their meal first. These little offerings are called *banten jotan* or *banten nasi* and are intended for the invisible inhabitants of the courtyard, who help with the processes of daily life. They are placed, for example, near the fireplace or water source, in the house temple for the ancestors, and nowadays also on the television set and near the telephone (fig. 23.3). These smallest and most common offerings contain only a little bit of white rice.

23.2   The four natural colors of
rice used for the offering
banten catur (*catur* means
"four") completed with
flowers in the same colors,
represent the ordering
of the universe. Photograph
by Francine Brinkgreve,
Kerambitan, 1987.

23.3   The daily *banten nasi*,
small offerings of rice, are
presented at the house
temple by a grandmother of
the household. Photograph
by Francine Brinkgreve,
Sanur, 1982.

23.4

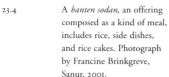

A *banten sodan*, an offering composed as a kind of meal, includes rice, side dishes, and rice cakes. Photograph by Francine Brinkgreve, Sanur, 2001.

The little *segehan*, which are put on the ground every three or five days to appease the demons (*buta kala*), have as main ingredients mainly white and yellow rice, but on auspicious days black, red, and multicolored rice form part of the *segehan* as well. Almost all offerings contain rice and side dishes, but one especially, the *banten sodan,* is composed as a complete meal (*sodan* means "meal"; fig. 23.4). Even in the tower-shaped family offerings (*banten tegeh* or *banten gebogan*), taken by individual families to temple festivals, the rice seems to have disappeared within the beautiful construction of fruits and cakes and flowers, but in fact it is always present.

Rituals in Bali can be performed at various levels of elaboration. Larger rituals generally start with presenting offerings to the rice that will be used as an ingredient of the offerings to be made. A *daksina palinggih*, a kind of offering that serves as a temporary residence for the deity or spirit that blesses the work to be done and brings success and good luck, is placed in the granary where the rice is kept. This seat or offering itself always contains grains of rice, together with a coconut because the coconut palm tree is very important for the Balinese. For a more elaborate ritual, the rice to be used is first spread out, sometimes in the shape of a human figure, and covered with offerings, together with other important offering ingredients, such as sugarcane, eggs, coconuts, and bananas. This figure is then consecrated or given life by means of holy water and offerings (fig. 23.5). This ritual is called *ngingsah baas*, "washing the rice." Afterward, in the kitchen of the temple, busy women transform the rice into many kinds of offering ingredients.

23.5   Rice to be used for an elaborate ritual is spread out in
the shape of a human figure.
Photograph by Francine
Brinkgreve, Ubud, 1982.

23.6   Especially in festive seasons,
many kinds of rice cakes and
cookies "*jaja,*" are sold
at the markets. Photograph
by Francine Brinkgreve,
Kerambitan, 1982.

This is done under the guidance of a *tukang banten*, a specialist in the making of offerings. Often the daughter of a priest, she knows all about the right ingredients for offerings.

Rice in offerings is used in many ways. Sometimes it is left uncooked, but for the most part it is steamed as for a normal meal and placed in separate palm-leaf containers as part of the entire offering. Special offerings contain rice processed by frying uncooked grains in a frying pan without oil. This process is called *nyahnyah*. For certain ceremonies, for example death rituals, the rice is cooked into a kind of porridge (*bubuh*) that serves as special food for the ancestors. Once a Balinese year, on a special day called Tumpek Uduh or Tumpek Bubuh, rice porridge is even served as an offering to useful trees, such as coconut palms and other fruit-bearing trees. Another way of preparing rice is to steam it inside palm-leaf containers (*ketipat*). These are packages made from two or more strips of palm leaf plaited into many different shapes.

Besides rice as such, many different kinds of rice "cookies" are part of most offerings (fig. 23.6). The Balinese word for them, *jaja*, has no real English equivalent, for it encompasses a range of tasty cakes and cookies eaten as ordinary sweets but also used in offerings, as well as cookies that are never eaten on an everyday basis but only as leftovers from offerings, and finally special ritual cookies that are used only in offerings and have symbolic value rather than good flavor. For all these *jaja*, mainly the glutinous rice varieties are used, both white and black; natural and artificial dyes are used to create many other colors. The variation in names and shapes and tastes of *jaja* seems to be endless, and so are the ways of preparing them.[2] Often rice flour is used, obtained by pounding rice, which is then steamed, salted, and made into a dough; a dough is also obtained by steaming glutinous rice grains. Both kinds of dough, together with flavorings like palm sugar, can, for example, be either wrapped in banana leaves and boiled in water, or deep-fried in coconut oil. The Balinese clearly distinguish between the two ways of preparing the dough, since in many rituals two categories of *jaja* are required, the "wet" and the "dry" ones. The most common edible offering *jaja* are the crispy *jaja uli* and the *jaja begina*, always in a combination of a white variety and a red one, the last created by adding red (or rather brown) palm sugar to the white rice. In small offerings only little bits of the *jaja uli* and *jaja begina* are used, but the big family offerings consist for the most part of large numbers of many kinds of tasty *jaja* (figs. 23.7, 23.8). Not only do *jaja* play an important role in offerings presented to the invisible beings but they also facilitate in the ritual circulation of food and other presents in the human world. After a marriage has taken place and the bride has moved into the household of her husband, family members of the groom visit the family of the bride, presenting them with a large number of different *jaja*, of both the wet and the dry categories. The *bantal* (lit., "pillow"), long *jaja* steamed in leaves of the coconut palm and dressed up as husband and wife, attract special attention during such a visit.

Just as a meal does not consist of rice only, so rice is never used alone in offerings. Even in the smallest *banten nasi* or *jotan* some salt and spices are added to make the rice enjoyable. Larger offerings, which might be thought of as bigger meals, contain several dishes besides the rice, preferably a vegetable and a meat or fish or egg. The largest offering groups always contain meat of a certain kind, such as a fried chicken or a roasted suckling pig, or types of *saté*, all depending on the type of ritual and the destination of the offerings. For example, the goddess Durga, who rules over demonic powers, likes offerings of pig's fat and blood (Brinkgreve 1997), whereas for other deities the flesh of a white duck is considered more "pure" and more suitable. Rice and meat used in offerings are regarded by many Balinese as complementary, identified with the female and male, respectively, together forming a unity. These values are also reflected in the division of tasks in

23.7   Many kinds of *jaja* of different colors and shapes form beautiful compositions in these huge family offerings in the Pura Dalem in Budakeling. Photograph by David J. Stuart-Fox, 1989.

23.8   A woman takes her family offering back home after she has presented it at the Pura Sada in Kapal. In the temple she received some rice grains, soaked in holy water, which are stuck on her forehead. Underneath the flowers the crispy *jaja uli* and *jaja begina* are fastened, and, above the row of eggs, pink and white sponge cakes (*jaja mangkok*) are visible. Photograph by Francine Brinkgreve, Kapal, 1983.

23.9   On the base of the offering *jejanganan* the figure of a baby is laid out in uncooked white grains of rice, symbolizing the beginning of the infant's life cycle. This offering is used in the *nelubulanin* ceremony welcoming the baby into the human world at three months of age. Photograph by Francine Brinkgreve, Kerambitan, 1983.

the preparation of offerings: whereas women mainly fashion offerings with rice and other vegetable elements, men are responsible for the production of the offerings containing a lot of meat. Together they make the groups of offerings complete.

Although many offerings in Bali are composed as a kind of meal, with rice as the main ingredient or central element, the quantities of food are rather small and merely symbolic, except for the family offerings, which contain large numbers of fruits and *jaja*. Food for the deities also differs from the meals consumed by human beings because offerings are thought of as living entities. Fashioned from natural, perishable materials, which during the process of cutting, preparing, shaping, and coloring lose their vital force, they regain life by the way all the ingredients are carefully assembled into a meaningful, complete whole. This completeness of an offering is regarded as an important value, for otherwise the gift might not be accepted by the deities. Even the "plate" upon which the offering-meal is served, a palm-leaf container or base of various shapes, is made from a natural, living material, although this ephemeral base is itself often placed on a wooden stand or in a silver tray. Moreover, during the ritual of presenting offerings to the invisible beings, offerings are spiritually empowered by means of sacred formulas (*mantra*) uttered by the priest and holy water (*tirtha*) sprinkled over them. Finally, many offering ingredients, especially rice and *jaja,* are given special forms and colors that represent various motifs of life. In this respect, rice is more important as material for representations than it is as mere food.

23.10   Faces of two human figures, *cili*, peep out from a mass of flowers made of rice dough as part of a *sarad* or *pulagembal* offering. The bird represents Garuda, the mythical sun-bird. Photograph by Francine Brinkgreve, Ubud, 1983.

23.11   A *sarad* or *pulagembal* offering, consisting entirely of rice-dough figurines, is carried around the temple during a festival to enhance prosperity for the temple congregation. Photograph by Francine Brinkgreve, Pura Ulun Danu, Batur, 1987.

23.12

These rice-dough figures to be used for a *sarad* or *pulagembal* in Pura Besakih display miniature scenes. For example, the figure at top left is a tiny model of an offering. As in a "real" offering, it contains bananas and two tiny *tumpeng* mountains. Photograph by Francine Brinkgreve, Besakih, 1987.

Rice Is Life: Rice as a Carrier of Motifs of Life

Rice is, in Bali, closely linked to the goddess Bhatari Sri or Dewi Sri, goddess of fertility, prosperity, wealth, and beauty (Wirz 1927). One of several myths about the origin of rice relates how Dewi Sri took her seat in the first seeds brought from heaven to earth by birds. Three seeds had the colors of the various types of rice in Bali: white, red, and black. The fourth seed, which was yellow, was transformed into the yellow root turmeric, which up to the present day is used as a natural dye for rice (Soekawati 1926). Balinese believe that Dewi Sri, as a personification of the germinal force, fecundates the flower of the rice plant and remains in the grain as the reproductive potential or life principle residing in the germ. It is this life principle, called the jewel of the rice (*manik galih*), that gives rice its nutritional value (Howe 1980, 108–133, 183; Ottino 2000, 71, 113). Other stories also tell about the heavenly origin of rice, which is considered to be a gift from the deities, like life itself. This idea is transformed into ritual practice. After the main rite of worship during a temple festival, the congregation receives from the priest or his assistants some grains of rice (*wija*), soaked in holy water. These are stuck on one's forehead and temples as a token of the gift of life from the deities (see fig. 23.8).

According to the Balinese, humans and rice are related because they share the same life forces. Moreover, they are mutually dependent. When the rice plants are still in the field, the farmers look after them not only technically but also by means of many rituals in which the life stages of the divine plants are followed, from the planting of the seeds to the harvesting of the grain. But once the rice is stored in the granary, the farmers are dependent on this food supply, which has to be sufficient to meet their daily needs. So another series of rituals ensure the ancestral and divine protection of its nourishing principle (Ottino 2000, 49–53; Ramseyer 1977, 153–64).

Rice is not only the most important food component of Balinese offerings but is also frequently the material used to create all kinds of representations within the overall design of the offerings. Such important themes in Balinese cosmology as the cycles or flow of life and the balanced ordering of the universe are translated into or visualized through a number of concrete models in the art of the offering. And because rice itself is so closely connected with life, it is also a suitable medium for expressing ideas concerning life. This

23.13   Rice grains dyed in the colors of the *nawa sanga* system
are part of *caru,* offerings to
the demons on the ground.
Photograph by David J.
Stuart-Fox, Ekadasa Rudra,
Besakih, 1979.

23.14   A large *sarad* or *pulagembal*
offering appears at a major
temple festival. From top to
bottom we see the following
figures: Sang Hyang Widhi,
the deity who encompasses
all other deities; a lotus
(*padma*), representing the
center of the world; Boma,
the son of the earth; Agni,
the god of fire; the serpents
(*naga*) Anantaboga and
Basuki and the turtle
Bedawangnala. Photograph
by David J. Stuart-Fox,
Bunut Bolong, 1989.

23.15   *Tumpeng* in the colors of the
nawa sanga, the system of
cosmological ordering.
Photograph by David J.
Stuart-Fox, Panca
Walikrama, Besakih, 1989.

23.16   Beautiful *dangsil* offerings represent trees of life, adorned by many different kinds of *jaja*. Photograph by David J. Stuart-Fox, Sibetan, Bali, 1972.

is accomplished by means of the state of the rice (whether it is cooked or uncooked), the shapes and forms and colors of the rice, and the position of these elements within the overall structure of the offering.

The different kinds of rice used in offerings are almost always cooked. But when uncooked rice is part of an offering that also contains cooked elements, the idea of the point of origin or foundation of life is expressed by the uncooked rice. For example, grains of uncooked rice are formed into two-dimensional human figures at the base of certain offerings, like the *banten jejanganan*, an offering presented during the ritual for a child who is three months old and is being welcomed into the human world. A rice winnow with a rice-grain figure representing the baby (fig. 23.9) is placed at the bottom, and the rest of the offering, in which cooked rice forms a major part, is assembled on top. The rice is colored and laid out in circles, symbolizing the sun and the moon, or it is transformed into little animals by adding a tail, ears, eyes, and nose using leaves or seeds.

For other offerings, cooked rice can be transformed into many additional meaningful forms, often associated with life. A common motif of life links rice with the mountains providing the constant flow of life-giving water that, on its way down to the sea, makes the earth fertile. Boiled rice is often formed into a cone or mountain shape, called *tumpeng*, which is part of many offerings (fig. 23.1; see figs. 23.15, 23.19). Usually it is rather small, but it can also be spectacularly large. Sometimes an egg, very much a symbol of life, is put on top of the *tumpeng*, or a little hole is created on top, called *telaga,* or lake, representing the water source, or tiny branches of, for example, the life-giving waringin tree are stuck into it. The *tumpeng* is also considered a model of the cosmic mountain, the center of the world, which in Balinese cosmology acts as an axis mundi, linking the upperworld, human world, and underworld, and as such is a symbol of cosmic unity and totality.

Ketipat with steamed rice can be plaited in many different shapes, like the sun and moon, animals and birds, all kinds of phenomena that live or make life possible. Together they represent the contents of the world. For rituals at the rice field, *ketipat* are made in the form of the flora and fauna and other features of the rice field, like shrimps (*udang*) and crabs (*yuyu*), or *kukur*, the doves that according to the myth brought seeds of rice from heaven down to earth. Simple *ketipat* are used in many offerings, but the most

23.17   *Dangsil* in the shape of a *meru* shrine are erected for a temple festival. The colored panels are rice-dough *jaja*. Photograph by David J. Stuart-Fox, Bangli, Bali, 1989.

23.18   (Opposite) This special shape of *pulagembal* offering is used only in East Bali. Elegant female figures, *cili*, are painted on very thin *jaja kiping* in the top row and are fashioned from palm leaves at the bottom of the offering. Photograph by David J. Stuart-Fox, Bebandem, 1988.

complicated ones and most complete sets are used in the center of the *banten bebangkit*, an offering to the goddess Durga, spouse of the god Siwa (Brinkgreve 1997).

This offering forms a model of the basic structure of the universe. On a pair of circular bases, one on top of the other, on which male and female figures are sometimes laid out in white rice grains, a four-sided bamboo frame is placed and oriented according to the cardinal directions. On each side of this frame are attached *jaja* that display all kinds of cosmic phenomena. Like the *ketipat*, their shapes represent the contents of the living world, ordered in different categories. They are arranged on opposite sides in such a way as to form complementary oppositions, such as sun and moon, sky and sea or mountain and sea, temple and market, road and gateway, man and woman, day and night. Some of the rice-dough figures display complete little scenes by themselves.

All these rice-dough representations, still classified as *jaja*, although their taste is less important than their symbolic value, are often also called *cacalan*. They are usually colored or dyed with artificial colors, modeled by hand (this process is called *nyacal*), then deep-fried in coconut oil. Many offerings, especially the *banten suci* (lit., "pure offering"), contain dozens of small white and yellow *cacalan*, which have names of flowers or fruits, symbols of vital forces, growth, and development.

23.19   In a detail of family offerings, different kinds of *jaja*
form striking geometric
designs. Rising above the
tumpeng topped with palm-
leaf decorations, vertical
rows are composed of *jaja*
called *crorot* and round *jaja*
called *gonggongan* (from
the word *gong*). Photograph
by David J. Stuart-Fox,
Ngis, 1977.

23.20 A panel with rows of different kinds of *jaja* used for temple decoration includes painted *jaja kiping* (top row) and *jaja bekayu* in the shape of flowers (bottom row). Across the top of the panel is a row of *kembang manggis*, star-shaped constructions of lontar palm leaf containing tiny bits of *jaja*.

But the most spectacular is the *banten sarad* or *pulagembal*, dedicated to the goddess Ibu Pretiwi (Prethivi) or Mother Earth (fig. 23.11, also see box, pp. 338–39). This offering, used for special temple festivals or family rituals performed at a high level of elaboration, consists of a bamboo frame several meters tall, completely covered with colored, complicated *cacalan*. The offering as a whole is also a representation of the cosmic mountain, complete with plants and flowers (figs. 23.10, 23.12). The *sarad*'s central element is sometimes a gateway filled with human figures who, in the transition of life, pass through gateway after gateway, both in the worldly and in the spiritual sense. Often the iconography of the figures on a *sarad* or *pulagembal* refers to the famous Hindu myth of how deities and demons together churned the milky ocean, rotating the holy mountain Mandara, in order to obtain the elixer of life, *amerta* (fig. 23.14). Related to the *sarad* is the *banten tegteg*, which also consists of many different *cacalan*. This offering ensures that the quality of, for example, a marriage or a harvest will remain constant (see fig. 23.24).

Not only do the shapes of the rice or rice-dough figures represent motifs of life but also the combinations of their colors have the same function, because together they express the balanced order of the world. In the myth about the origin of rice cultivation on earth, four colors—white, red, yellow, and black—play a role, since these are the colors of the seeds brought down from heaven. These four colors are the colors of the cardinal directions, and together with blue, green, orange, and pink—the colors of the intermediate directions—they symbolize the unity and totality of the cosmos (fig. 23.15). In Balinese cosmology, nine deities guard the eight points of the compass and the center. Each has its associated color, number, weapon, and other attributes, and in this way the *nawa sanga* (meaning "nine" in Sanskrit and Balinese) system of cosmological order is formed. The

23.21   Sang Mangku Jaro Gedé,
         main priest of Pura Kehen
         Temple, prepares to use a
         sacred dagger to stab the
         face of Boma on the *sarad*
         seen in figure 23.22.
         Photograph by Garrett Kam,
         Bangli, 2003.

23.22   A *sarad* by Jero Ni Madé
         Rénten is erected for a
         ceremony at Pura Kehen
         Temple. Photograph by
         Garrett Kam, Bangli, 2003.

Jero Ni Madé Rénten uses colored rice flour dough to create the face of Boma for a *sarad*. Each piece of the *sarad* will be fried in coconut oil to strengthen and preserve it, then all the pieces will be assembled on a bamboo frame. Photograph by Garrett Kam, Bangli, 2003.

A Complete World: The Making and Meaning of a Rice Dough Cosmos

Garrett Kam

Jero Ni Madé Rénten (fig. 23.23) is the principal maker of *sarad* in Bangli, Central Bali. She recalls learning her art as a young girl by copying the rice dough figures made by older women:

> That's how it's always been done. So I've been doing this for over thirty years now and have lots of experience. People have seen my work and ask me to make *sarad* for their events. I donate my services to temples in the area, but also make *sarad* for ceremonies in other places [fig. 23.22]. I take care to make the figures look as nice as possible and [make them strong so they will] not break apart. But I don't create new forms since they all have meaning.

Jero's husband, I Nyoman Sudirman, constructs the bamboo and wood frames on which the rice dough figures will be attached and assists with making the figures and frying them. He notes that "the figures must be in their correct positions because the *sarad* has meaning.... Everything symbolizes something, so each piece has its place in the *sarad* and in the universe."

In Bangli the figures around the edge of the *sarad* are called *ceracap* (jagged, pointed). The related term *cerapcap* (from *capcap*, meaning "to drip") describes a line on the ground made by rain dripping from a roof. The next row of figures on each side features symbols of the four cardinal and four intermediate cosmic directions, which are represented by different weapons and colors. Toward the center, circular shapes known as *roda pedati* (chariot wheels) or *tamiang* (shields) appear, which are associated with the sun. At the top of the *sarad* below a rainbow and multicolored lotus, the Hindu god of destruction and reincarnation, Siwa, stands above the Garuda sun-bird. The entire structure of the *sarad* is based on the *kakayonan*, the cosmic tree of life in *wayang kulit* (shadow puppet) theater. As an offering, the *sarad* reinforces visually the fact that sunlight and water are necessary for life in the universe; *sarad* literally means "a complete world."

During closing ceremonies of major temple rituals involving the *sarad* in Bangli, a priest symbolically brings to life the figure of Boma, who, as the demonic son of Wisnu and the earth goddess Pretiwi, is associated with natural and uncultivated fertility. The priest stabs the figure of Boma three times with a sacred dagger (*keris*) to ceremonially kill him (fig. 23.21). If a dagger is not available, he then tosses three flowers at Boma. According to mythology, Wisnu killed his son by having Garuda pluck the magical flower that was the source of Boma's power from the top of his head. This ritual is called *nebek* (to stab, wound) from the root *tebek* (stab or wound made with a dagger). The derivative *tatebek* (fixed time to begin planting in the rice fields) refers to an important belief and practice: after this ceremonial death, the *sarad* figures are buried in village rice fields to ensure the continued fertility of the earth in the eternal cycle of life, death, and renewal.

•

23.24   A *sarad* and a *tegteg* offering, which both contain large numbers of *jaja*, are consecrated or given life by a priest using holy water. Photograph by Francine Brinkgreve, Blahbatuh, 1989.

center, which represents unity and totality, can be either multicolored or white and has as its emblem the lotus with eight petals. This cosmological system is not only a fixed structure but also involves movement or transition. When a priest invokes the deities of the directions, he starts in the east, where the sun rises, and ends in the center. When for larger communal festivals a group of women work together in the temple, they start preparing the offerings with the making of the rice-dough figures, always beginning with the white ones, the color of the east.

In certain offerings, the colors of the rice are a clear indication of these cosmic directions. This is most clearly shown in the *banten catur* (*catur* means "four"), which functions during the ritual as a seat for the important deities who guard the four cardinal directions: Iswara (Isvara) in the east, Brahma in the south, Mahadewa (Mahadeva) in the west, and Wisnu (Vishnu) in the north. They each have their own color (white, red, yellow, and black, respectively) and their own number, which means that in this offering not only do four portions of rice and *jaja* have the four respective colors but also that the amount of different *jaja* in each of the quarters is a multiple of the corresponding number. Usually in the *banten catur* the various kinds of naturally colored rice are used: white, black, and red varieties of rice plus turmeric-dyed white rice for the yellow color (see fig. 23.2). In some offerings, which resemble the *banten catur*, the center is formed by a rather large *tumpeng*, in this way combining the symbols of both horizontal and vertical cosmological order.

Caru, special offerings spread out on the ground, address the demonic forces of the underworld and ask them not to disturb the course of a ritual and to go back to their domains, represented by their specific colors (fig. 23.13). The colors of the rice used in these offerings are oriented toward the right directions, and the colored rice is sometimes laid out in the shapes of the weapons of the respective deities in the *nawa sanga* system.

The beautiful *dangsil* of some villages in East Bali are among the most striking examples of offerings made as a celebration of life. These are huge constructions, made by members of the village councils on special occasions, representing everlasting wishing trees. They are adorned with hundreds of different kinds of *jaja*, hanging from the branches, symbolizing the fruits and flowers of a rich harvest from a fertile earth (fig. 23.16; Stuart-Fox 1974, 55–71). Other types of *dangsil,* which are covered with *jaja,* are built in the shape of a *meru* shrine in a temple, complete with a number of roofs and an open space in which to put offerings (fig. 23.17). Like the *meru,* these *dangsil* represent the cosmic mountain.

Of great charm is the *cili*, a symbol closely connected with life and fertility (see fig. 18.5). In East Bali, where it is called maiden (*deling*), the *cili* is often drawn and painted on very thin *jaja* (sometimes as thin as a sheet of paper) and is part of such large offerings as *pulagembal* (fig. 23.18). It is always given a beautiful, stylized, very slender female figure, with a large fan-shaped headdress and cylindrical ear ornaments. Although it has a female external appearance, the *cili* is in fact a symbol of human life and fertility in general, associated especially with the life of plants and the fertility of the earth. For example on the so-called *lamak nganten*—long palm-leaf runners that, during the Galungan festival, are suspended in front of a house where a wedding has taken place— two little *cili* are depicted. They are similar in form, but they represent the newlywed couple who, it is hoped, will soon produce offspring. In the literature about Bali the *cili* is often interpreted as the Rice Goddess, Dewi Sri, but the Balinese themselves do not automatically make this association. Indeed some of the Dewi Sri figures made of palm leaves and used in rice-cycle rituals have the form of a *cili,* or beautiful woman, because that is how she is visualized in Bali, but certainly not every *cili* represents Dewi Sri. In the idiom of rituals, the rice-growing cycle is compared to the life cycle of human beings, but not the other way around. Despite its symbolic significance, many Balinese love to use the *cili* merely as a beautiful ornament, an important function within the art of the offering.[3]

Rice Is Beautiful: Rice as Art or Decoration

We have seen that offerings first serve as a festive meal for the deities and, since rice is the main food crop, almost all offerings contain rice. Secondly, offerings are created as models of the balanced universe and of the continuation of life, and again rice plays a major role as the medium or carrier of these motifs of life. Thirdly, an offering is meant as a gift and must be made as beautiful as possible. The very process of creating an offering is part of the gift and is considered a religious act. Offerings have to be created again and again, not only because of the ephemeral character of the natural ingredients but also because offerings are transitory by intention: once given, they cannot be used again.

Therefore, the rice used in offerings has an important decorative function, not least because boiled rice and rice dough are rather easy materials to shape and dye. *Jaja,* especially, are often very attractive. Some offerings consist mainly of decorative *jaja* with

23.25   Reflecting the creativity
and humor of the offering
maker, a fantastic head of a
mythic animal decorates the
base of one of the *dangsil*
erected at the decennial festival of Sibetan. The head
consists of a wooden mask
covered with rice dough.
His hair is fashioned from
sheaves of rice. Photograph
by David J. Stuart-Fox,
Sibetan, 1990.

geometric patterns (fig. 23.19). They are fashioned from long rolls of rice dough, made into coils, curves, and spirals, or pushed into a more rectangular shape (*jaja bekayu*). Drawing or painting on very thin sheets of *jaja* to create *kiping* is another technique of using rice in an artistic way in offerings (as in figs. 23.18, 23.20). Furthermore, the artistic arrangement of the *jaja*, together with fruits and flowers, can be very decorative. In principle every woman has the artistic freedom to make offerings as beautiful as she likes.

The *jaja* not only act as decoration for offerings but can also be made into beautiful panels that function as temple decorations (fig. 23.20). In this way, the surroundings of the offerings are transformed into one huge offering itself, as it were. Some spectacular offerings with many *jaja*, especially the *dangsil* or the *sarad* or *pulagembal*, which are made only for special occasions, themselves act also as decoration of the place where the ritual is held. Before they are used in the ritual, these major offerings are first consecrated as living objects, just as always happens with permanent buildings and also with transitory architecture, like the big cremation towers and other objects (fig. 23.24).

Especially in the *sarad*, several of the same iconographic elements of permanent and non-permanent architecture are used, referring to well-known mythological themes. For example Boma (see fig. 23.14), the son of Ibu Pretiwi, the goddess of the earth, is always found above temple gateways, and the turtle and serpent refer to the myth about the churning of the milky ocean. In the process of obtaining the elixir of life by means of rotating the mountain Mandara, the turtle was used as a base and the serpents as rope. More and more often, innovations appear in the design of these offerings. For example, the national symbol of Indonesia, the bird of heaven Garuda, nowadays forms a central element in many *sarad*.

The larger figures in these spectacular offerings often have a wooden core, carved by an artist. Usually they represent just the head, sometimes with hands or claws. These masks are covered either with a kind of colored rice paste and left to dry in the sun, or with rice dough, the material for *cacalan*, and afterward deep-fried in coconut oil (fig. 23.25). Increasingly, artists have become involved in the making of these offerings, which have such an obvious decorative function besides their religious significance. To name just one such artist, Wayan Pugeg, sculptor and architect from the village of Singapadu, often acts as the designer of *sarad* contributed by his village to a major ritual. Wayan Pugeg and other artists and designers work together with the *tukang banten*, but their products show their own individual styles. Non-professionals, too, can be very talented in the art of the offering. For example Sagung Putu Alit from the village of Kerambitan is very creative and she loves to make new designs, especially with rice dough and palm leaf. She invented a new form for the *bantal lantang* (special *jaja* used for wedding ceremonies) that has become very popular in her village.

Clearly influenced by the art of the offering, sculptures made of rice dough are sometimes used as art objects in secular events like art festivals and exhibitions, where they are carried around in processions, and used in contests between villages or districts, or as folkloristic objects to attract tourists. The art of rice in Balinese offerings has many aspects. Rice is not only the main ingredient in offerings but also the carrier of important motifs of life, represented in the offerings, and the offerings themselves are often of striking beauty. In Bali, rice is food, rice means life, rice is turned into art.

•

24.1   Geisha drinking sake.
By Toyohara Kunichika
(1835–1900). Japan.
Woodblock print. H: 37 cm.
Museum of Fine Arts,
Houston 75.356.45; Gift of
Peter Knudtzon.

24

Sake in Japanese Art and Culture

Mariko Fujita

Sake is a traditional Japanese alcoholic beverage brewed from fermented rice and water. The earliest evidence of sake brewing, dated 4800 B.C.E., was found in the Yangtze Valley in what is now China. Around 300 C.E. the technique was brought to Japan, where it was refined. At first sake was used primarily as an offering to the Shinto gods, but gradually it was enjoyed by the general population. In the Sengoku period (1467–1568), modern brewing fundamentals were established and breweries were built throughout the populated areas of Japan. In the Edo period (1603–1868), sake brewing and the enjoyment of sake became very sophisticated. During the Meiji period (1868–1912), sake brewing grew more scientific and the mass production of high-quality sake became possible. Today, although Fushimi (Kyoto Prefecture), Nada (Hyōgo Prefecture), and Saijō (Hiroshima Prefecture) are generally considered the centers of sake manufacturing, there are about twenty-six hundred sake breweries producing about four thousand brands throughout Japan. Sake brewing has expanded around the world as well, with sake breweries in the United States, Australia, Southeast Asia, China, and South America.

This essay examines the traditions of sake brewing and drinking in Japan and sake brewing practices that are specific to Saijō. I first discuss the relationship between sake and deities in Japanese culture. Then I examine the methods and labor practices associated with sake brewing. Finally, I explore the role that sake plays in building and strengthening social relationships. A wide range of art objects associated with sake are presented at the end of the essay (figs. 24.7–24.23).

Sake and Shinto

Sake is more than an alcoholic beverage for Japanese people—as it is closely associated with many Shinto gods. Shinto is not a religion that seeks to control the behavior of believers according to doctrines or commandments but a faith that lets people have direct contact with the deities (*kami*) through worship. Accordingly, worship and ritual ceremonies are considered to be of great importance since they bring the blessings of the *kami* and good luck to people.

Sake brewers respect and worship their guardian deity, the god of sake. Two Shinto shrines in Japan, Matuo Shrine in Kyoto and Ōmiwa Shrine in Nara, are dedicated to the god of sake and are especially important for sake brewers. Matuo Shrine, located in the Arashiyama District of Kyoto, was built during the Nara period (710–794)

Sake brewery sign *(sugidama* or *sakabayashi)*. Japan. Cedar branchlets. W: 32 cm. J. D. Roorda Collection, Los Angeles.

At the beginning of the brewing season, fresh *sugidama* made of green cedar branchlets are hung over the main entrances of sake breweries. When the *sugidama* dries and becomes brown, it is a sign that the sake is ready for drinking.

and is one of the oldest shrines in the city. The water from the well at Matuo Shrine is believed to prevent sake from being spoiled if it is added during the brewing process.

The *sugi* tree has religious significance in Shinto, particularly in connection with Ōmiwa Shrine in Nara Prefecture, which houses a deity of sake brewing. *Sugidama*, also known as *sakabayashi*, are large globes of tightly bound branchlets of Japanese cedar, or more accurately, *Cryptomeria japonica* (*sugi*) about fifty centimeters in diameter (fig. 24.2). They are suspended by a cord in front of sake breweries to signify the arrival of the season's first sake. According to legend, fronds of the *sugi* growing on the grounds of Ōmiwa Shrine were used to make all of the *sugidama* for sake brewers throughout Japan, as they were considered the best for protecting the sake from spoiling. Until about sixty years ago, tanks for sake brewing were made of *sugi* wood, and so were the small boxes called *masu* traditionally used for drinking sake.

The entrance of a sake brewery (*sakagura*) is considered a sacred place and is decorated with a *shimenawa* (fig. 24.3). *Shimenawa*, rice-straw ropes adorned with folded paper streamers (*shime*), are used to mark sacred spaces, such as at Shinto shrines. They keep impurities out and purify the space within.

24.3   A *shimenawa* protects a sake brewery in Saijō, Japan. A *sugidama* hangs above. Photograph by Mariko Fujita, 1999.

The sake deity is considered to protect the success of the brewing process. The sake-brewing season in Saijō runs from October to February. It starts with a festival called Kigansai, for which the heads of the breweries and their brewing masters (*tōji*) gather at a local Matuo shrine in Saijō (fig. 24.4; a branch of Matuo Shrine in Kyoto).

Sake is also an indispensable offering for deities. Large sake barrels (*sakadaru*) are stacked at altars in Shinto shrines throughout the year (fig. 24.6). Sake is an important offering for deities at the household level too. Many traditional Japanese households have a small shrine or altar. The five most customary offerings are rice, sake, water, salt, and evergreen branches. They are arranged in small, symbolic quantities, presented in white pottery containers. Depending upon the region, season, and festival, local produce such as fish or seaweed might be offered as well. It is considered appropriate to present any delicacy to the deities before humans partake of it.

A great deal of sake is, of course, consumed at festivals. The Saijō Sake Matsuri, however, is a festival dedicated to the celebration of sake itself. It is a very popular festival in Japan, attracting more than one hundred thousand people each year. For two days in October, hundreds of different kinds of sake from all over Japan are displayed, and participants sample them for a small fee. Many Saijō breweries open their facilities to the public so that people can observe how sake is made. They also sponsor events such as small art exhibitions. At the city center, concerts and performances are held, and many small vendors sell local delicacies.

Even a modern event like the Sake Matsuri begins with visiting the local Matuo shrine. Sake brewery owners and employees gather at the shrine and undergo a purification ceremony. Then they walk in a procession to the city center, carrying a sacred palanquin with a *sugidama* on it (fig. 24.5).

Sake Brewing and the Artisan Spirit

Good rice, water, and *kōji* are indispensable for brewing sake. Most of the traditional sake brewing locations were selected in part because of their abundant supply of good water, often from on-site wells. Brewery owners often brag that their wells produce the region's renowned water (*meisui*), insisting that it is one of the secrets of their fine brew.

The rice used for brewing sake is quite different from the rice used for cooking. There are nine basic types of rice used to make Japanese sake, each of which yields a specific flavor profile. Yamada Nishiki rice, from Hyōgo, Okayama, and Fukuoka Prefectures, is considered the best and is called the "King of Sake Rice." The main differences between sake rice and table rice are that sake rice kernels are larger, have a special starchy core called *shinpaku*, and are more flexible so that they can be milled to less than 70 percent of their original kernel size. The rice for premium sake is milled to less than 65 percent of the original size.

Kōji, or steamed rice that has been inoculated with spores of a special mold (*Aspergillus oryzae*, or *kōji-kin*), is another crucial ingredient for sake brewing. The mold creates several different enzymes as it multiplies. These break down the starches in the rice to form glucose and are also responsible for some of the flavors and aromas in fine sake. When yeast is mixed in, it "eats" the glucose and produces alcohol. Sake is unique among fermented beverages in that the two fermentation agents work together to naturally ferment sake to 20 percent alcohol. Most sake contains water that has been added to dilute the alcohol content to 15 percent.

24.4   Matuo Shrine in Saijō, Japan, decorated for the Kigansai festival at the start of the brewing season. A large *shimenawa* hangs over the entrance and donated sake barrels (*sakadaru*) are stacked to the sides. Photograph by Mariko Fujita, 1999.

The process of making sake begins with washing the rice and steaming it until it is cooked. The rice is then mixed with yeast and *kōji*. The resulting mixture is allowed to ferment, with more rice, *kōji,* and water added in three batches over four days. This fermentation, which occurs in a large tank, is called *shikomi.* The quality of the rice, the degree to which the *kōji* mold has propagated, and temperature fluctuations are some of the variables that affect *shikomi.* The mash is allowed to sit from eighteen to thirty-two days, after which it is pressed, filtered, and blended.

Fine water, rice, and *kōji* are essential for brewing good sake, but the most crucial element is the team of artisans who make it. Sake is produced by the close collaboration among the brewery owner (*kuramoto*), the head sake brewer (*tōji*), and the brewery workers (*kurabito*). The *tōji* is responsible for the actual brewing and the hiring and management of the *kurabito.* Since sake is brewed only in the winter, the *tōji* and *kurabito* are essentially contract workers. During the spring, summer, and fall, they grow rice or work on fishing boats in their home regions. When the fall harvest is over, or the fishing season ends, they head off to work at sake breweries, often far from home. In Japanese, this traveling for seasonal employment is called *dekasegi.* The pay and status of sake laborers have always been high compared to other forms of seasonal labor. Since the

24.5   A *sugidama* is carried on a
sacred palanquin from the
local Matuo shrine to the
center of the city during the
sake festival in Saijō, Japan.
Photograph by Mariko
Fujita, 1999.

competition for jobs in the sake industry has thus been more intense than in other industries employing *dekasegi* laborers, it is of utmost importance for the brewery owner to
secure an experienced *tōji*.

Even in modern Japanese society, there are many sayings that ascribe the artisan (*shokunin*) spirit to sake brewers, although artisans themselves have been declining in
type and number since the rapid development of mechanized factory production following the Meiji Restoration in 1868. Although many sake manufacturers today incorporate
computers in their operation, the skill, inspiration, knowledge, and experience of the *tōji*
are still considered indispensable. The *tōji* is someone who can tell just how a batch of
sake is turning out by its taste and smell on any given day of the monthlong sake-brewing
process and who can evaluate the development of the *kōji*. These critical skills are, of
course, not acquired overnight; long years of practical experience, beginning with the
most menial tasks, are required to learn the art of brewing.

Traditional sake making requires the total involvement of all the *kurabito*, who
labor for long hours. In the past, *kurabito* lived in the brewery (*kura*) for the entire brewing
season and were not allowed to go out at all once the mixing of the final mash had begun.
The *kurabito* were organized in a strict hierarchical order under the leadership of the *tōji*

24.6   Stacks of *sakadaru* at
Ōmiwa Shrine in Nara,
Japan. Photograph by Roy
W. Hamilton, 1999.

and were assigned tasks according to their ranking. The youngest *kurabito* at the bottom were expected to get up at two o'clock in the morning to fix breakfast for the others.

In the end, responsibility for the quality of the final product rests on the shoulders of the *tōji*. In fact, the job of the *tōji* calls for so much skill and is so demanding that it's hard to find people who can do it. The number of sake breweries in Japan has been declining, and so has the number of *tōji* owing to their advanced age, the lack of successors, and the adoption of mass-production techniques. Some *kuramoto* are working to remedy this situation by, for example, ceasing to hire seasonal "contract" labor and instead offering laborers permanent employment. Others are attempting to automate certain operations, like bottle transport, that do not require a skilled artisan. Some are introducing computers and other new technologies.

It should be noted that the number of female *tōji* is increasing. Until recently, women were not allowed to work in the *kura*. The reason given for the restriction of access was that women were believed to be impure and therefore would pollute the purity of the sake if they entered the *kura*. Women were also thought to distract men from their delicate brewing tasks, which required undivided attention. Necessity and common sense, however, have joined forces to change the situation. Women are now almost common-place in the *kura*, and their products have earned many medals and awards in national tasting competitions.

Sake and Social Relationships

In the Muromachi period (1392–1573), sake drinking achieved the status of a ceremonial art in Japan, nearly comparable to that of the tea ceremony or flower arranging. Strict rules governed the serving and sipping of sake, as well as the procedures for refilling companions' cups and offering one's own to be filled in turn. Many serving utensils were developed as objects of fine art.

Although the art of sake drinking fell into decline after the Muromachi period and the rules became more lenient, drinking sake still has a vital social function, especially when a new relationship is formed. For example, at a wedding, the bride and groom sip sake in a ritualized manner called *san-san-kudo no sakazuki* to formalize their union. This sharing of sake is the crux of the wedding ceremony, akin to the taking of vows in a Christian ceremony. The silence of the bride and groom as they sip the sake implies their mutual promise of lifelong devotion. For this ritual, three sake cups are stacked atop each other. Usually the groom leads off by taking three sips from each cup. Then the bride does the same, sipping from each cup after the groom. Then the couple offers the cups to the family members, customarily in the following order: the groom's father, then his mother, the bride's father, and then her mother. If matchmakers were involved, they drink next, and finally everyone drinks together to signify that the families are joined.

Another occasion when sake plays a ceremonial role is at a *kagami-biraki*,[1] or cask-breaking ceremony. This ritual is performed when a congratulatory message needs to be conveyed, such as at a wedding reception or a dedication ceremony. The lid of a *sakadaru* is broken open in a ceremonial manner, and the people who have gathered partake of the cask's contents. The symbolic importance of the ritual lies in their drinking together and sharing the joy of the occasion. For such a ceremony, barrels large enough to hold eighteen liters (*itto*), thirty-six liters (*nito*), or seventy-two liters (*yonto*) are used.

As these examples show, sake is more than just an alcoholic beverage. It is entwined in Japanese religious life and in the building of social relationships. It embodies the traditional artisanship of its makers, yet takes many modern forms too. Sake has been important in both ceremonial and daily life in the past and present, and it will surely be so in future as well.

•

The Art of Sake

Roy W. Hamilton

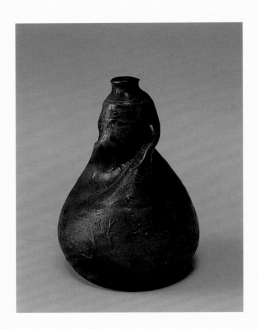

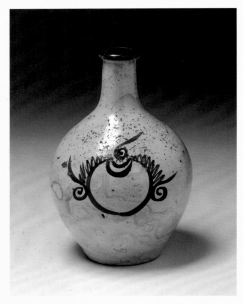

24.7   Sake bottle with Chinese-
        style landscape design.
        Satsuma, Kyushu, Japan.
        Late eighteenth century. W:
        14 cm. The Nelson-Atkins
        Museum of Art, Kansas City,
        Missouri (Purchase: Nelson
        Trust) 32-59/12. Photograph
        by Robert Newcombe.

24.8   Sake bottle with Buddhist
        sacred jewel motif. Japan.
        Eighteenth century. Tamba
        ware. H: 25 cm. The Nelson-
        Atkins Museum of Art,
        Kansas City, Missouri
        (Purchase: Nelson Trust)
        32-58/8.

24.9   Sake bottle. Japan. Collected
        1880s. H: 17 cm. Peabody
        Essex Museum, Salem,
        Massachusetts, E30,338.
        Photograph by Jeffrey Dykes.

24.10   Sake bottle in the form of a
        fox, assistant to the Japanese
        rice deity Inari. Yamashiro,
        Japan. First half of nine-
        teenth century. Hozan ware;
        glazed stoneware. H: 17 cm.
        Museum of Fine Arts,
        Boston 92.5918; Morse
        Collection; Gift by
        Contribution.

24.11   Sake ewer (*chōshi*). Kiyomizu,
         Kyoto, Japan. Early eigh-
         teenth century. Diam: 19 cm.
         Peabody Essex Museum,
         Salem, Massachusetts,
         E20279. Photograph by
         Jeffrey Dykes.

24.12   Sake bottle in the shape
         of Okame, the goddess
         of mirth, also known as
         Otafuku (lit.,"much good
         fortune"). Japan. Nineteenth
         century. Peabody Essex
         Museum, Salem,
         Massachusetts, E79368.
         Photograph by Jeffrey Dykes.

24.13   Sake bottle in double-gourd
         shape with design of pine,
         bamboo, and plum blossoms.
         Japan. Nineteenth century.
         Kiyomizu ware; stoneware
         with enamel overglaze
         touched with gold coloring.
         H: 25 cm. Museum of Fine
         Arts Boston 92.6608;
         Morse Collection; Gift by
         Contribution.

24.14   Sake bottle. Japan.
         Mid-nineteenth century.
         Kiyomizu ware; stoneware
         with green overglaze.
         H: 22 cm. Museum of Fine
         Arts, Boston 92.6015;
         Morse Collection; Gift by
         Contribution.

24.15   Pair of sake casks (*sakadaru*).
         Japan. Edo period, nine-
         teenth century. Lacquer on
         wood. H: 40 cm. Brooklyn
         Museum of Art 86.296.2 and
         86.269.3.

         Sakadaru are typically used
         for weddings or other cere-
         monial occasions.

24.16   Sake ewer (*chōshi*). Japan.
         Edo period, nineteenth
         century. Cast iron, lacquer
         on wood. L: 15 cm.
         Brooklyn Museum of Art
         TL1984.167.1a,b; Lent by
         Robert Anderson.

         Cast iron sake ewers
         could be warmed directly
         over fires.

24.17   Sake shop sign. Japan.
1880s. Wood, bamboo.
H: 60 cm. Peabody Essex
Museum, Salem,
Massachusetts, E15917.
Photograph by Jeffrey Dykes.

Shop signs (*kamban*)
in Japan are often made in
forms suggestive of the
product sold within the shop.
This sake shop sign takes the
form of an actual ceremonial
sake cask (*sakadaru*).

24.18   Votive plaque (*ema*). Ikoma,
Nara Prefecture, Japan. 1929.
H: 21 cm. FMCH X89.867;
Gift of Daniel C. Holtom.

This type of *ema* is offered
by supplicants at the
Buddhist temple Hōzan-ji
in Ikoma, in order to procure the assistance of the
deity Shoten. Due to his
harsh temper, this deity is
considered ideal for helping
the supplicant keep a vow
to abstain from some
unwanted behavior. In the
handwritten message on this
ema, Yamaguchi Genkichi,
a forty-one-year-old man,
pledges to abstain from
drinking sake for three years,
beginning February 10,
1929. The painted image is a
lock over a heart, with a
sake barrel at the top.

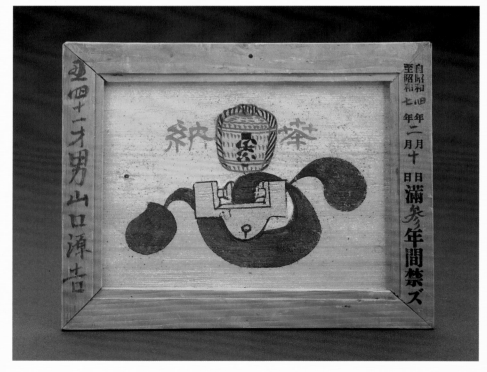

24.19   *A Man Performing the* Shōjō
Dance on a Wine Keg. By
Nishimura Shigenaga
(1697–1756). Japan. 1744.
Two-color print (*beni-e*).
H: 29 cm. The Nelson-
Atkins Museum of Art,
Kansas City, Missouri
(Purchase: Nelson Trust)
32-143/23.

Shōjō, or sake imps, are
felicitous characters from
Japanese No theater that
dance and sing the praises of
sake. Their most distinctive
feature, beyond their love of
drink, is their long, flaming
red hair, which has given
them an otherworldly
appearance to Japanese
audiences since at least the
sixteenth century. In one old
text a *shōjō* defines himself:
"I am not a Buddha, nor am
I a sentient being...I could
be man or beast, or some-
thing from the six realms
of existence. Thus you could
call me human, or an
I-don't-know-what. I just
live in the white waves of
the sea and am called a
shōjō" (Bethe 2002, 219).

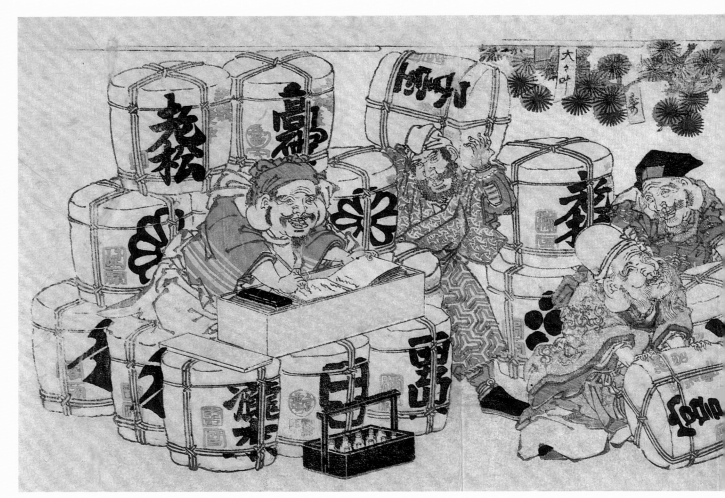

24.20   *The Seven Lucky Gods
         Amusing Themselves with a
         Group of Wine Kegs.* By
         Keisai Eisen (1790–1848).
         Japan. 1815–1820. Wood-
         block print. W: 54 cm.
         Fine Arts Museums of
         San Francisco, Museum
         Collection, 1964.141.963.

         As harbingers of good
         times, the seven lucky gods
         (Shichifukujin) are often
         shown with an abundance
         of riches. Here it is a surfeit
         of sake, itself associated with
         merriment and good fortune.

24.21   *Cat and Mouse* (*Neko to nezumi*). By Shibata Zeshin (1807–1891). Japan. 1873–1891. Hanging scroll; ink and color on paper. W: 46 cm. Honolulu Academy of Arts, The James Edward and Mary Louise O'Brien Collection, 1978, 4667.1.

Zeshin has taken as his subject a common theme from the folk painting tradition of Ōtsu, in which a cat is depicted offering a cup of sake to a mouse, and then waiting patiently for it to take effect.

24.22   Night street in the Yoshiwara brothel district. By Utagawa Kuniyoshi (1797–1861). Japan. 1844–1861. Woodblock print. H: 23 cm. Collection of the Grunwald Center for the Graphic Arts, UCLA; Purchased from the Frank Lloyd Wright Collection. Photograph by Robert Wedemeyer.

Yoshiwara was the so-called pleasure quarter in Edo (now Tokyo) where men sought entertainment. Here a group of drunken revelers carouse, escorted by a female employee of one of the houses of entertainment. A pair of vendors takes the scene in stride, perhaps as an everyday occurrence.

24.23 Portraits of Ichikawa
Danjūrō the Seventh,
Ichikawa Danjūrō the
Eighth, and Kawarazaki
Kanjūrō. By Utagawa
Kunisada, Toyokuni III
(1786–1864). Japan.
Woodblock print.
H: 36 cm. Los Angeles
County Museum of Art
M.79.152.286b.

This is a particularly witty
take on one of the most
common of woodblock
print subjects, portraits
of famous actors. Here the
actors are shown as they
might see themselves
reflected in the smooth
surfaces of their red lacquer
sake cups.

Part Six

Straw Matters

25.1   Sacred rice-straw rope (*shimenawa*) for a home altar. Hirose Rice Straw Craft Association (Hirose Wara Zaiku no Kai). Haguro-chō, Yamagata Prefecture, Japan. 2002. Rice straw, paper. W: 53 cm. FMCH X2002.7.1.

The Hirose Rice Straw Craft Association was formed by a group of elders concerned that their once flourishing local rice-straw crafts were in danger of being lost. The group consists mostly of senior citizens from a rural community in the Shōnai Plain, one of Japan's most productive rice-producing regions (see chapter 17). *Shimenawa* are made in many distinctive regional styles, and the group has fashioned this one in the style characteristic of their local region.

25

Straw Matters

Roy W. Hamilton

Votive plaque (*ema*). Tokyo, Japan. Before 1962. Painted wood. W: 21 cm. FMCH X89.803; Gift of Dr. Daniel C. Holtom.

The sacred tree depicted on this *ema*, marked by a braided chain of rice straw with white paper streamers (*shimenawa*), is known as *enkiri enoki*. This name means, roughly, *enoki* tree (*Celtis sinensis var. japonica*) that severs connections between people. Women prayed to this tree, which stood near Tokyo's Itabashi Station, for divorce, or even for the severing of their husbands' affairs with other women. The original tree was destroyed in the Tokyo fire of 1884, but subsequent generations of trees grew in the same location. The handwritten message on the *ema* is an appeal by a twenty-seven-year-old woman for a divorce from her thirty-three-year-old husband.

I never fully appreciated the importance of rice straw until I walked through a village in Vietnam's Red River Delta one November day after the harvest. Where each junction of village pathways created some open space, a small mountain of rice straw nearly blocked the way. It had recently rained and straw was spread out to dry along the tops of walls and anywhere else where it might catch a bit of sun or breeze. The heady aroma of slowly fermenting wet straw hung thick in the air. Rice straw seemed to be everywhere, engulfing the houses and obscuring the activities of the people who lived in them.

How much rice straw? Rice plants on average produce roughly equal weights of grain and straw, so a world harvest of 400 million metric tons of rice means 400 million metric tons of straw! The straw enriches the soil if it is burned in the fields or plowed under to rot, but in most Asian communities it is far too valuable a resource to be expended in this manner. Probably the largest quantity becomes animal fodder. Mixed with mud, it makes a construction material, or it can be used as a readily available and fast-burning cooking fuel, saved for packing material, or even set out as a medium for

25.3   New Year's decoration
(*shimekazari*). By Saito
Shigeya. Haguro-chō,
Yamagata Prefecture, Japan.
2002. Rice straw, paper. L:
59 cm. FMCH X2002.7.2.

This New Year's decoration
includes a crane and a turtle,
which are good luck sym-
bols. Their tails are made
from rice panicles with the
grains still attached.

25.4   A wayside figure of Jizōbosatsu, a Buddhist protective deity, is surrounded by a wrapper of rice straw. Photograph by Roy W. Hamilton, Tōge, Yamagata Prefecture, Japan, 2000.

growing mushrooms. Potters in India cover their open-air firings with a layer of compressed rice straw to contain the dung-fueled fire (Perryman 2000, 38). In northern Japan, where winters are severe, rice straw is transformed into wrappings that protect plants and even man-made objects from damage by the heavy snow (fig. 25.4). Throughout the rice-growing regions of Asia countless useful articles are made from the vast supply of straw.

Objects made of straw are sometimes invested with sacred meaning, a further reflection of the divine nature of the rice plant. Thai Khorat granary figures (see figs. 3.16, 3.17) are one example. This idea finds its greatest expression in Japan, where rice straw is made into sacred ropes called *shimenawa* that mark and protect holy places. Their protective powers are based on the rice straw itself, as demonstrated in a Japanese myth where a *shimenawa* fails to protect because it turns out to have been made of millet straw (Ohnuki-Tierney 1993, 53). The simplest *shimenawa* are thin twisted strands of rice straw, often with strips of folded white paper attached, that are frequently seen wrapped around large trees on the grounds of Shinto shrines. This custom is based on the ancient use of trees where spirits dwell as natural shrines (fig. 25.2). Ultimately, much more elaborate styles of *shimenawa* developed, including small ones for home altars as well as the huge, famous specimens that hang over the entrances to major Shinto shrines. *Shimenawa* are typically renewed at New Year, and a related use of rice straw is the making of intricate New Year decorations. Ritual objects are sometimes constructed of rice straw in human or animal form, and some Japanese festivals also feature large constructions of rice straw in fantastic shapes (see fig. 17.4). All of these items exploit the stark simplicity and beauty of plain rice straw for their visual impact.

•

25.5   Snow boots. Japan. H: 48 cm.
FMCH X88.1244b,c.

25.7   Snowshoes. Mount Haguro,
Yamagata Prefecture, Japan.
2000. L: 30 cm. FMCH
X2001.7.1a,b

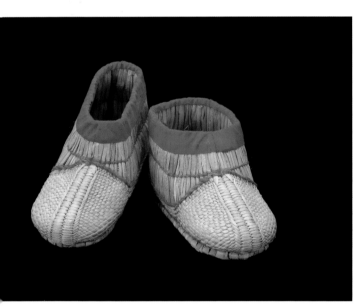

25.6   Child's snowshoes. By Koike
Kazuku. Yamagata Prefec-
ture, Japan. 1970s. L: 20 cm.
FMCH X86.2947a,b; Gift
of Los Angeles County
Museum of Art.

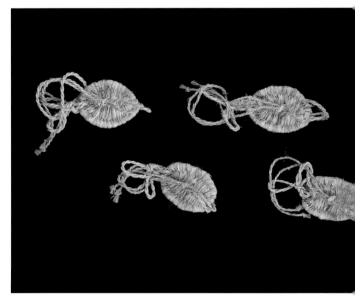

25.8   Set of snowshoes for a cow.
Hirose Rice Straw Craft
Association (Hirose Wara
Zaiku no Kai). Haguro-chō,
Yamagata Prefecture, Japan.
2002. L (each shoe): 15 cm.
FMCH X2002.7.4a–d.

The Hirose Rice Straw Craft
Association makes snowshoes
for horses and cows, once a
functional necessity in snowy
Yamagata Prefecture, but
now an endangered rural
craft tradition. The cow
shoes are distinguished by
the cord that passes through
the cloven hoof.

25.9   Rain cape (*mino*). Japan.
            L: 132 cm. FMCH X99.28.1.

            Entire panicles, stripped of
            their grain, are worked into
            this cape. Their minute
            branches, oriented to point
            downward, aid in the shed-
            ding of water.

25.10   Basket for a baby (*ijime*).
Yamagata Prefecture, Japan.
Mid-twentieth century. W:
68 cm. FMCH X2002.23.1.

Infants in rural Japan were
placed in baskets made of
rice straw. The basket served
as a small playpen, allowing
the baby's caretakers to go
about their tasks.

25.11   Back pad (*bandori*). By
Kozuku Koike. Yamagata
Prefecture, Japan. 1970s. H:
58 cm. FMCH X86.2945;
Gift of Los Angeles County
Museum of Art.

Pads like this are placed on
the back to make carrying a
heavy load more comfort-
able.

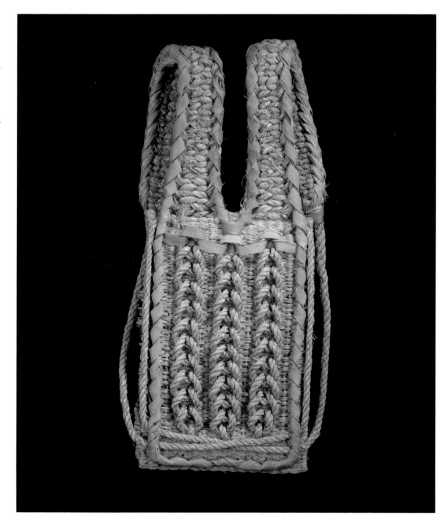

25.12 Rice-straw form (*benkei*). Yamagata Prefecture, Japan. H: 147 cm. FMCH X2001.7.2.

This beautifully worked rice-straw form was purchased from an elderly woman in Tsuruoka, who explained that it is used for drying fish on skewers. The skewers are inserted into the rice straw, like a giant pincushion, and hung over a hearth.

25.13 Loom. Japan. W: 172 cm. FMCH TR2000.4.1a–r; Gift of John McMullen.

This type of loom, now rarely found in Japan, was used for weaving rice-straw mats. The warp and weft both consist of braided strands of rice straw.

Part Seven

Rice, Self, and State

26

Rice, Self, and State

Roy W. Hamilton

Are we what we eat? In the West, this common adage is often taken as a joke, and its true significance is rarely contemplated. Millions in Asia, who feel rice is a sacred food that nourishes them in a way no other food can, may not ponder this question in an overtly philosophical manner, yet they intuitively acknowledge a deep connection between rice and the self.[1] On the most basic level, the eating of a single food as the main component of every meal promotes an awareness that one's body is literally built from rice. More importantly, rice serves symbolically as a key element in the construction of personal identity. The rice plant is held to have a spirit akin to the human spirit, and the life cycle of rice is metaphorically equated with the life cycle of humans. Rice features prominently in human life cycle rituals, and the growing and eating of rice is seen as defining what it means to be human. Since human life depends on rice, the fertility of the rice crop and the fertility of humans are one and the same. Not surprisingly, the pivotal role of rice in the formulation of individual identity expanded to encompass the community and the state. Just as humans are defined by eating rice, many Asian polities have been defined on a practical level by their management of rice production and even on a symbolic level by an imagined relationship, nurtured for political purposes, between the leader or ruler and the spirits or deities of the sacred grain.

This chapter examines the relationship between rice and identity and discusses works of art and objects of material culture that draw their meaning from that relationship. It begins with a look at the use of rice, both materially and symbolically, in human life stages from birth to puberty, marriage, reproduction, aging, and death. This is followed by a case study of Dewi Sri, the Javanese goddess of rice, and her consort (and brother) Jaka Sedana, exploring how this divine couple has been put to use in Java in ever widening circles, from personal life rites, to community economic and social life, and finally in the political formulations of the states of Central Java. Finally, the story of Sri and Sedana leads to a broader consideration of the role of rice in the construction of the state in Asia.

From Birth to Death

Rice is so ubiquitous in Asian rites of passage that it is impossible to give more than a few examples here, drawn from various cultures, but these will suffice to suggest why Asians might justifiably feel they are accompanied by rice throughout their lives, not just as a

26.1   Pair of granary figures (*bulul*). Ifugao peoples. Kababuyan, Luzon, Philippines. Carved wood. H: 46.2 cm. Collection of Thomas Murray.

Bulul normally consist of a male and female pair, but this set is unusual in that it includes an infant. The figures are used in harvest rituals and then placed in the granary. This pair therefore makes explicit the connection between the rice crop and human fertility.

food but also as a potent symbolic agent. Chapter 27 provides a more detailed description of the symbolic uses of rice in the life cycle rituals of a single culture, that of the Tamil people of southern India.

In the traditional central Thai rituals for childbirth, rice is scattered around the room when a women is in the final stages of labor (Anuman Rajadhon 1961, 121–22). Thus the baby is born into a space already marked with the sacred grain that will become a lifelong companion. In northern Thailand, a Tai Yong mother will give a taste of rice to her newborn infant before she breastfeeds it for the first time, acquainting it first with the food that will provide sustenance for a lifetime. By the age of one month the baby will be eating rice regularly (Trankell 1995, 133).

In Kerala, the celebration of a young woman's first menses involves a display of a large quantity of unhusked rice as a fertility symbol invoking her future reproductive power. Rituals in Kerala often include sprinkling the earth with a liquid colored red with turmeric and lime. This "blood" represents the menstrual flow of the goddess Bhagavati, the preeminent goddess of Kerala (and regional variant of the pan-Indian goddess Kali). When soaked with the goddess's blood, the earth is said to give birth to rice plants in the same way that a mother gives birth to a child after experiencing menstruation (Caldwell 1996, 210–15).

Rice in various forms appears repeatedly in engagement and wedding celebrations (fig. 26.2). For example, rice wine decorates a magnificent bridal robe from Japan (fig. 26.3). The scene of the sake nymphs (*shōjō*) is primarily intended to invoke felicity and joy, but these emotions in a wedding context surely also carry a suggestion of fertility. Chapter 28 provides a further exploration of one of the most intimate associations ever made between rice and women's bodies, the use of rice-related patterns evocative of agricultural fecundity on women's garments in Edo period (1603–1868) Japan. The pregnancy rituals that are conducted for the Thai Rice Goddess, Mae Phosop, in imitation of the equivalent rituals for pregnant women, have already been mentioned (see fig. 1.1). The connection between rice fertility and the fertility of the human couple is equally powerfully evoked in an unusual pair of granary guardian figures (*bulul*) from the Ifugao region of the Philippines (fig. 26.1).

Likewise in aging and death a strong relationship between rice and humans may be expressed. In Japan, a sumo wrestler's sake bottle turned into a reading lamp becomes a metaphor for aging (fig. 26.4). After death, in many diverse Asian religious traditions rice is provided to the corpse to provision the journey to the afterlife. In the Tai Yong cremation ceremony, for example, the corpse is supplied with both cooked and unhusked rice, while the attendees throw puffed rice. Afterward, the leftover rice offerings that have been symbolically tasted by the deceased are considered to be impure and must be consumed by the family in order to end the liminal period of several days during which the deceased waits to be reborn (Trankell 1995, 161).

In some cultures, most notably those influenced by Confucian practices honoring ancestors, rice offerings may continue to be made to the deceased as they gradually attain ancestor status. In the memorial practices of the Baba (Malaysian) Chinese community, for example, bowls of rice and chopsticks are placed daily on the altar for family ancestors. In the first year after death only one bowl is prepared, but thereafter more are added to accommodate the friends that the ancestor will have made in the other world (Anderson 2001, 29). In Song dynasty (960–1279) China, beautiful ceramic urns were sometimes filled with rice to represent the soul of a deceased ancestor (fig. 26.5).

If rice and human identity are so intimately entwined, then conversely it might be expected that the absence of rice would suggest something less than human.

26.2   The traditional Vietnamese wedding process consists of a sequence of ceremonies typically spread out over a number of months. The ceremony called *lễ ăn hỏi*, takes place after a formal proposal has been made but before the actual wedding ceremony. It is considered the most important event in the entire process because its purpose is to solidify the relationship between the two family lineages. The groom's kin assemble a number of different gifts including wedding "cakes" made of glutinous rice with a bean filling, shown here wrapped in green paper in front of the groom (left) and the bride's younger brother. These gifts are delivered by a delegation to the bride's home, where they are placed on the altar for her lineage ancestors. After the ceremony some of the gifts are returned to the groom's family and others are shared among the bride's kin. The wedding cakes thus symbolically bind the ties between the two groups, including not only the living members but also their ancestors. Photograph courtesy of Trần Thị Thu Thủy, Hanoi, Vietnam, 2000.

An example comes from the Tai Yong, who normally consume only the rice they have grown within the confines of the family, which is also the rice production unit. The sole exception, other than giving rice to Buddhist monks or the temple in order to earn religious merit, is to give rice directly to poor people who have no rice of their own. Such individuals, belonging to no family group that produces rice, lack all standing by which an individual is recognized as a member of society (Trankell 1995, 133–34). Although such individuals are in a sense discounted as standing outside society, this practice nevertheless assures that they are provisioned with rice to eat.

Sri and Sedana

The traditional Central Javanese harvest ritual (Methik) conducted in the field prior to the cutting of the grain took the form of a wedding ceremony for the Rice Goddess Dewi Sri and her brother/consort Jaka Sedana. The first-cut stalks, representing the incestuous divine couple, were brought together in a "rice stalk wedding" (Pemberton 1994, 205; Jay 1969, 210). The stalks were then clothed in a batik cloth and carried in procession under a ceremonial umbrella to the field owner's house, where the "newlyweds" were installed in a "bridal" chamber at the rear center of the house. This sacred space essentially constituted a shrine for Sri. Offerings were made to her on a regular basis until the seeds contained in the straw image, mixed with other seed rice, were planted in the following crop cycle (Pemberton 1994, 208). These rituals, which equate the fertility of the divine couple with the fecundity and continuity of the rice crop, could be readily observed as late as the 1980s, but they are now rare.[2] In chapter 32, Rens Heringa details a version that can still be observed in a village in rural East Java, where the seed rice is stored in the same position in the house in two clay pots representing Sri and Sedana. The pots are linked together with a coil of cotton thread, signifying their marital union (see fig. 32.2). The entire assemblage is known by the name *pedaringan*, a word usually glossed as "granary."

The fertile divine couple, often referred to in Java by the hyphenated name Sri-Sedana, is in many ways an imagined model for the ideal human wedded couple. The preparations made for a Javanese bride by the ritual specialist (*dhukun*) on the eve of her wedding connect the bride to Sri (Pemberton 1994, 210). The bride spends the night meditating to enhance her radiance (see chapter 18 regarding Sri's origin as divine energy or power [*sakti*] for her husband, Vishnu), thereby effecting her symbolic transformation into Sri. The following day the bride and groom sit together in state. Older sources indicate that this took place on mats laid in front of the sacred chamber at the rear of the house called *senthong tengah* or *kobongan*, a space "designed to enable the inhabitants to receive the goddess Sri" (Rassers 1959, 247). This is of course the same space where Sri was worshiped and the seed rice was placed after the harvest. The chamber itself sometimes contained a couch or framed bed (fig. 26.6), and more recent descriptions tend to call this structure itself the *kerobongan*, a variant of *kobongan* (Achjadi 1989, 152; Jessup 1990, 263). These differences presumably reflect local variation, differences in status and wealth, and the evolution of wedding practices over time. Alternative names for the *kerobongan* include *petanen* ("place of agricultural activity") or *pedaringan,* the same name used in East Java for the seed storage pots (Achjadi 1989, 152).

In short, the bride and groom evoke Sri and Sedana, representing hope for the fertility of humans in the same way that the seed rice stored in the form of the wedded Sri and Sedana embodies the fertility of the rice crop. The Javanese bride and groom are also often said to represent a king and queen for a day, and they are dressed as lavishly as circumstances allow in order to express this ideal while they sit in state on the *kerobongan.*

26.3   Bridal robe (*uchikake*).
Japan. First quarter of the
twentieth century. L: 170
cm. Silk; *yūzen*-dyed,
painted, stenciled,
embroidered with silk and
gold-leaf-wrapped thread.
Neutrogena Collection,
Museum of International
Folk Art 1995.93.482.
Photograph by Pat Pollard.

The *shōjō* sip sake from
enormous sake cups filled
from the giant vat. See figure 24.19 for more about the
redheaded sake imps.

This too brings to mind the original Hindu ideology of Sri as the wife/queen of her husband/king, Vishnu, a further indication that the Javanese bride in her finery is an evocation of Sri.

Rather standardized images of the ideal wedded couple are widespread in Javanese art, especially in the form of pairs of carved wooden figures and as paintings on glass (fig. 26.7). The glass paintings were common items of Javanese popular art in the early twentieth century (Fischer 1994, 36) and still can still be seen hanging in some Javanese homes. The wooden figures too have become common items of popular and tourist arts. Technically the figures are known collectively as *loro blonyo*, sometimes translated as "anointed pair" (Rassers 1959, 253) or "inseparable pair," but they are also explicit representations of Sri and Sedana (Jessup 1990, 59, 263). Even the title "inseparable pair" recalls the unsuccessful efforts of the supreme deity to keep the incestuous couple apart in some versions of their story (as described in chapter 18).

To follow these ideas further, a short digression into Javanese domestic and court architecture is necessary. The traditional rural Central Javanese house has an enclosed, private central room (*dalem*). At the rear of this room are three smaller store-rooms (*senthong*), also called the *jromah*. The middle one of these, the *senthong tengah*, comprises the innermost sacred space in the house and is regarded as the domain of Sri (Waterson 1990, 186). The scale and grandeur of this space varies according to the means of the household. No matter how humble the house, there will be some sort of partition (*gebyog*) separating the *dalem* from the *jromah* (Keeler 1983, 2). As circumstances allow, the partition may be more or less finely carved, painted, or gilded. If the *senthong tengah* is small, the storeroom itself comprises Sri's framed nuptial bed. If the room is larger, it may be separated from the *dalem* with an elaborate partition and have a freestanding *kerobongan* within it, also beautifully framed. No matter how grand or small, it is this space in which the straw figure representing Sri was placed after the harvest and where the bride meditates on the eve of her wedding, to be joined by the groom the following day. Thus Sri-Sedana, the bride and groom, and the storage place for the rice seed are all bound together in the complex of ritual practices involving the sacred heart of the rural farmhouse.

As Javanese architecture evolved to encompass other forms of dwellings, the *senthong tengah* and *kerobongan* and their meanings evolved as well. From the fourteenth century onward, the trading centers that developed on the north coast of Java began to rival the importance of the inland agricultural areas. The wealth generated by this trade allowed leading citizens to build splendid houses famed for the elaborateness of their carved and painted woodwork (fig. 26.10). The rather humble *senthong tengah* of the farmer's house grew into an ostentatious inner alcove in the north coast merchant's house. The space that the farmer used to store his most precious commodity, his seed rice, the merchant used as a storeroom for his valuables.

In the courts of Central Java, the palaces (*keraton*) that belonged to the various sultans also each had a private inner sanctum, analogous to those of the farmers and merchants. The royal *kerobongan*, which resemble large, fully enclosed four-poster beds, were situated at the back of these chambers, in a position analogous to the *senthong tengah*. There the sultan was supposed to retire one night of the year to consort with his super-natural spouse. We could be forgiven for leaping to the conclusion that this must have been Sri, the bride of kings. Actually, the sultan in theory mated with Nyai Lara Kidul, the otherworldly Queen of the Southern Ocean. Both Sri and Lara Kidul, however, could lay claim to being the symbolic progenitrix of the Javanese state, and the two of them are inextricably entangled in a web of mythology and ritual that defies clear separations.[3] Some

26.4

Old Age. By Shibata Zeshin
(1807–1891). Japan. Late
1870s. Ink and light color
on paper mounted as a
hanging scroll. L: 123 cm.
Honolulu Academy of Arts;
The James Edward and
Mary Louise O'Brien
Collection, 1977 (4601.1).

Zeshin made this painting
when he was himself
approaching old age. He
chose as his subject two
poignant symbols of age: a
pair of reading glasses and a
sumo wrestler's sake bottle,
retired from use and con-
verted into a reading lamp
(Link 1979, 87).

writers have in fact opined that Sri and Lara Kidul are transformations of one another (Tjahjono, in Waterson 1990, 186). While Lara Kidul may have presided in the royal bridal bed, the sultans also kept beautiful *loro blonyo* sets that served as regalia of the state (fig. 26.7).

Published reports (Poeroebaja 1939, 319) and photographs indicate that the *loro blonyo* were displayed in front of the *kerobongan* in the palaces, although according to other sources they were at least in some cases kept in the chief minister's residence (Jessup 1990, 264). Beyond the court setting, the *loro blonyo* and the *kerobongan* appear to have been closely associated. According to colonial era reports (G. A. J. Hazen, in Rassers 1959, 253), and also in the memory of elderly Javanese, the *loro blonyo* figures were placed in front of the *kerobongan* and formed a part of the shrine to Sri. The figures would not have been found in every house, but primarily the homes of aristocratic families.[4]

In Javanese studies, rural cultural practices are often characterized as poor-cousin imitations of the more refined practices prevailing at court—witness the "king and queen for a day" ideology. In the case of the Rice Goddess Dewi Sri, however, it may have been the other way around.[5] Sri was probably first and foremost a rural agricultural "Rice Mother" who became deified as a Hindu *dewi* and came eventually to serve the state as regalia in the form of the sultan's *loro blonyo*. In his brilliant study of modern Java, John Pemberton (1994, 197–235) shows how Sri-Sedana and the imagery of the Javanese bridal couple was used by the ruling elite during the time of the "New Order" Suharto period (1965–1998) to enhance their legitimacy with a mantle of "tradition." In the process, the *kerobongan* became a stage set that can be installed in large halls, totally removed from its former context as a sacred space in the house (fig. 26.9). As middle-class families have rushed to adopt this prevailing version of the "traditional" wedding, rent-a-*kerobongan* have become a feature of modern life. Seed storage, the promise of a wedded couple's fertility, the emblems of state, and society weddings—all have become bound together in the imagery of Sri and her bed.

Rice and the State

In 1957 the German economist Karl Wittfogel published his theory of "Oriental Despotism," in which he claimed that the construction and maintenance of major irrigation systems for rice agriculture encouraged the centralization of power in premodern Asia and led to despotic states. Wittfogel's theory, though influential in his time, has by now been debunked several times over. One of the key cases that he cited was the Khmer kingdom of Angkor (ninth through fifteenth century), whose vast and famous temple complex was possible only with the production of enough rice to feed a huge corps of corvée laborers. Among the many impressive features of the site are the artificial lakes, the largest of which, the eleventh-century West Baray, is about 8 kilometers long and 2.2 kilometers wide. Wittfogel assumed that the main purpose of the lakes, like the "tanks" of ancient Sri Lanka, was for irrigation of the rice crop. More recently this has come into question. Chinese emissaries who visited Angkor in 1296–1297 and wrote extensive descriptions made no mention of the lake water being used for irrigation (Higham 1989, 341). Archaeologists have pointed out that no mechanism for delivering the water to the fields has been identified, that the waterworks may have served religious or transport functions rather than as a means of irrigation, and that adequate rice may have been grown in rainfed bunded fields without any irrigation at all (as it is in much of the area today). This debate remains unresolved.

26.5   Funerary jar. Jiangxi
Province, China. 1250–1320.
Glazed porcelain; Qingbai
ware. H: 63.5 cm. Victoria
and Albert Museum
c.225-1912.

Jars with phoenix figures on
their lids were placed in
pairs in the tombs of the
wealthy and, particularly in
southern China, were sometimes filled with rice offerings for the souls of the
deceased (Kerr 1991, 64).

26.6   A smaller scale replica of a
kerobongan, or wedding bed,
is used in a wedding
exchange in Tengger Wetan,
Tuban Regency, East Java.
Newly harvested rice, a gift
from the groom's kin to the
bride's, is piled into the
elaborately framed bed-like
structure and carried in procession to the bride's house
about a week before the
wedding. Once inside the
bride's house, the rice-filled
bed is taken to a space in
front of the kitchen that
serves as the essential
"female" space. Photograph
by Rens Heringa, 1996.

Another example that Wittfogel cited was Bali, where in precolonial times (up to 1906) several small kingdoms competed to dominate the rich irrigated rice lands of the southern slope of the island. Where he erred here is that irrigation in Bali is organized through a complex system of irrigation societies called *subak*, which still exist today and have never been a function of the state (Lansing 1991, 7). Finally, the strongest refutation of Wittfogel's theory comes from a place he did not consider, the highlands of Luzon. There, successive generations of farmers living in tiny independent hamlets, with no state organization of any kind, built the most impressive system of terraced and irrigated rice fields anywhere in the world.

While it therefore cannot be claimed that any particular form of state organi-zation developed in response to the needs of rice agriculture, there are nevertheless many ways in which states and rice were, and continue to be, linked in Asia. Chapter 29 dis-cusses the role of agriculture in imperial China, whose foundations were first laid in the millet-growing regions of the north. In later times, however, while the capital and its huge bureaucracy remained in the north, the state became critically dependent on rice grown in the south: "The whole story of the Grand Canal, which took definitive form first in the Sui [581–618], was essentially the building of a main artery to bring tax grain from the economic to the political centre of gravity of the country" (Needham 1971, 227).

The Grand Canal (fig. 26.11), which stretched a total of 1,770 kilometers, is one of the greatest engineering feats of the premodern world. It rises to a height of 42 meters above sea level using a system of locks to traverse a mountain barrier dividing northern and southern China. When the most complex part of the canal, located where it crosses over this barrier, was reconstructed in 1411, the work was completed in two hundred days by a force of 165,000 laborers (Needham 1971, 315). The imperial governments main-tained systems of granaries all along the canal's route, where grain could be stored in the event that high or low water levels rendered transport temporarily impossible.

Although the term "green revolution" has been coined to characterize the changes that have occurred in modern rice and other grain farming in the past thirty-five years based on the development of fast-maturing high-yield varieties, the changes that took place in Song dynasty China were no less sweeping and have been likened to an earlier green revolution (Bray 1984, 597–600). Due to repeated nomadic invasions of the north during the later part of the Tang dynasty (618–905 C.E.), the population of China had been gradually shifting to the south. By the time of the Song, an outright majority resided in the south, in prime rice-growing regions. In 1012 new varieties of faster-ripening rice plants were imported from the kingdom of Champa, located in what is today Vietnam. The new varieties allowed double cropping in the fertile Yangtze Delta and elsewhere in southern China. These changes finalized the transfer of the economic center of China from the north to the south, where it has remained ever since. Without the Grand Canal, and without the massive importation of rice from the south to the north, the Chinese state could not have survived into modern times.

Throughout China's history, the clearing of land for agriculture defined the borders of the imperial state. The influx of Han Chinese settlers in newly cleared areas brought stability, while the opening of new land provided a safety valve in Chinese society against the dangers posed by refugees and expanding landless rural populations (Bray 1984, 95). At the far borders of the state, especially in the south, minority groups such as the Yao and Mien were repeatedly pushed further into marginal mountainous lands. Various Yao groups possess scrolls granted by the Chinese emperor that give them the right to move about, clear mountain land for swidden agriculture, and not be subject to

26.7   Figures representing the
ideal wedded couple (*loro
blonyo*). Java, Indonesia.
Carved and painted wood.
H (male figure): 52.5 cm;
H (female figure): 51.5 cm.
National Museum of
Scotland, A.1991.65a,b.

This *loro blonyo* set, of a
quality that suggests it may
have originated in one of the
royal palaces of Java, was
reportedly present in
England by the nineteenth
century, if not earlier. How it
arrived there is not known.

26.8   Painting of the ideal wedded
        couple (*loro blonyo*). Java,
        Indonesia. 1930s. Glass,
        paint, wood. W: 47 cm.
        FMCH LX74.217.

        These paintings are some-
        times known as "reverse"
        glass paintings because
        the fine details must be
        painted on the back surface
        of the glass first, and the
        larger background elements
        added later.

26.9   A Javanese bride and groom
        sit on a settee, surrounded
        by a gilded frame, at their
        wedding reception in Jakarta.
        This formal setting represents
        a modern transformation of
        the traditional practice of
        seating the bride and groom
        in front of Sri's framed sacred
        space at the rear center of the
        house. Photograph by Judi
        Achjadi, 1984.

26.10   Interior partition (*gebyog*). East Java, Indonesia. Ninteenth or early twentieth century. Carved wood, paint, gilt, glass. W: 11.76 m. FMCH X2001.25.1; Gift of Carolyn Kimball Holmquist.

This unusually long partition, reportedly from Gresik, East Java, must have come from an exceptionally grand house. The work is not as intricate as the famous carvings of Kudus, on the north coast of Central Java, but the bold design and execution are characteristic of East Java. The three-part division, featuring a central alcove flanked by spaces on either side accessed through the arched doorways, is reminiscent of the division of the sacred rear portion of the traditional Javanese farmhouse into three *senthong*. Indeed, when originally marketed by an art dealer, *loro blonyo* figures were placed in the alcove, suggesting that this central access was intended to function as a *senthong tengah*. The alcove is quite shallow, however, and has some other unusual features, including sliding doors at the rear with glass windows in them. Aristocratic houses had various floor plans, and it is likely that this *gebyog* may have been located closer to the front of the house, dividing public areas from private. It is as if the entire form of the house has been informed by the spatial conventions of the sacred storeroom dedicated to Sri.

26.11   Detail of *Map of the Grand Canal in Shandong (Shandong Yunhe quantu)*. China. Nineteenth century, before 1854. Ink and color on paper. L: 326 cm. Library of Congress, Map and Geography Division G7822.G7.

This map shows technical details including dikes, locks, and floodgates. The color shading of the canal itself, from blue to brown, indicates the amount of silt in the water, while the inscriptions describe waterways connected to the canal and the behavior of floodwaters (Yee 1996, 69). The map may have been used by those who administered the canal, which was essential for transporting tax paid in rice from southern China to the imperial capital in the north.

taxation wherever they went (Jonsson 2000, 68). It was in part their practice of swidden rather than irrigated rice agriculture that defined these groups as being outside the Han Chinese state (although they differed in language and cultural practices as well). As the Han population expanded, many minority groups moved through the mountainous terrain into Southeast Asia. One Yao group, for example, settled in the 1880s in the territory of the small kingdom of Nan, located in northern Thailand (Jonsson 2000, 70). Whereas the Tai people of Nan farmed wet rice in the valley bottom that ran through the center of the kingdom, the Yao occupied the surrounding hills. As foreigners and swidden rice farmers living on the outskirts of the Tai State, the Yao were considered nonsubjects. Thus the state was in this case essentially defined by the limits of wet-rice agriculture.

Poetic evidence for the centrality of agriculture to Chinese conceptions of human society can be found in a shaman's summons for the return of the wandering soul of an ailing king, composed in the third century B.C.E. In the summons, after the soul has journeyed in one direction, the shaman calls it back, saying "The five grains do not grow there; dry stalks are the only food" (Hawkes 1959, 104). The implication is that a desirable way of life cannot be had without agriculture. After the shaman calls the soul back from each of the cardinal directions, the remainder of the summons consists of a recital of the joys of homecoming: "All your household have come to do you honor; all kinds of good food are ready; / Rice, broom-corn, early wheat, mixed all with yellow millet..." (Hawkes 1959, 107).

All over the rice-growing regions of Asia, the practical aspects of rice agriculture impacted the formulations of the state, and vice versa. For example, from the time of the kingdom of Pegan (ninth through thirteenth centuries C.E.) onward, the Burmese state was always centered in northern Burma because a well-developed irrigation system there made rice agriculture feasible even though the region is quite dry. Only after British colonial efforts in the late nineteenth century drained swamps in the south and opened the area for settlement did the political focus shift southward (Aung-Thwin 1990, 5, 63). The settling of low-lying swamplands in Nepal's Terai region by Tharu minority groups was in some ways similar. As Tharu village communities opened the region to rice cultivation, the Nepalese kings gave them official documents certifying their status according to several different types of land ownership

26.12   *Amaterasu Ōmikami.* Japan. Probably 1890s. L: 136 cm. Hanging scroll; ink and color on paper. Peabody Essex Museum, Salem, Massachusetts, 11,675. Photograph by Jeffrey Dykes.

This scroll depicts the Shinto deities enshrined at Ise Grand Shrine, the most important shrine in Japan, and offers many keys to the symbolic importance of rice in Japan. At the top is Amaterasu Ōmikami, the goddess of the sun and deity of the imperial family. At the right is Toyouke-no-ōkami, the goddess of grains, holding a sheaf of freshly harvested rice. At the left is Sarutahiko-no-mikoto. The actual characters over the head of this deity read Saru-Ta-Hiko-Kami. The character *Ta*, meaning rice field, is pictographic, a rectangular set of four rice fields divided by dikes. At the bottom, a rice-straw rope (*shimenawa*) joins the sacred male and female rocks of Futami, off the coast at Ise, a famous site with symbolic meanings related to fertility.

26.13   *Moon Viewing.* Japan. Circa 1870. Album leaf, woodblock print. W: 48 cm. San Diego Museum of Art 64:137c.

The name of the artist and the full significance of this print are unknown, but it appears to be a Meiji-period political satire showing workers toiling to turn a giant rice mill, while wealthy citizens, perhaps the prototypes of Japanese industrialists, greedily pat out the rice balls at the bottom of the mill.

and labor management (Krauskopff 2000, 182–84). No matter which system characterized a particular community, the documents also regulated the tax that was owed to the king, payable in rice.

Japan also had a very extensive system of rice tax collection. During the medieval period (1185–1603), rice seed was distributed in the spring by the government and also by religious authorities. In the fall it was paid back, with interest, in the form of offerings of the new crop to the Buddha and to the Shinto deities (Ohnuki-Tierney 1993, 67). In the Edo period, the feudal system consisted of an urban warrior class ruling over village farmers, based on the collection of a 30 to 40 percent tax on the rice harvest paid directly in grain (Okada 1989, 36).

The relationship between the state and rice in Japan not only involved the practicalities of production and taxation but was also highly symbolic. Japan's imperial system originated with agrarian leaders whose political authority resided in their perceived ability to intercede with supernatural powers in order to achieve a bountiful rice harvest: "For this reason many scholars consider the emperor first and foremost as the officiant in rituals for the rice soul (*inadama no shusaisha*) who ensures the blessings of the deities for the new rice crop on behalf of the people" (Ohnuki-Tierney 1993, 45).

In October a representative of the emperor makes offerings of the newly harvested crop at Ise Grand Shrine, the most important Shinto shrine in Japan. The main deity enshrined at Ise is the Sun Goddess, Amaterasu Ōmikami, who is regarded as the

26.14   *Those Who Eat Well Will Be Victorious (Ăn no đánh thắng)*. By Phạm Thanh Liêm. Hanoi, Vietnam. 1974. Silk screen on paper. H: 76 cm. Collection of Tom Patchett, Los Angeles.

The inscription reads, "According to Ho Chi Minh's teachings: Those who eat well will be victorious" (Theo lời bác dạy: Thực túc binh cường). This classic propaganda image, copied many times over, depicts the wartime valor of women in Vietnam's communist state, who gather the rice harvest while keeping their weapons ready.

26.15   *Because Oppression Is Down to the Bone, Hear the Weeping in the Fields (Pagkat kaapihan ay sagad sa buto, panangis sa bukid ay ulinigin mo)*. Philippine Peasant Institute. Philippines. Early 1990s. Offset print. H: 53 cm. Center for the Study of Political Graphics.

Political protest poster from the land reform movement in the Philippines. The banner in the photograph reads, "Enforce true land reform" (Ipatupad, tunay na reporma sa lupa). Translation from Tagalog courtesy of Tita Pambid.

ancestress of the imperial family and the Japanese state. Among the other important deities enshrined at Ise is Toyouke-no-ōkami or Toyoukehime-no-kami, the goddess of grains, who is often depicted with a sheaf of rice (fig. 26.12).

Likewise the installation of a new emperor in Japan is closely bound up with the rice agricultural season. In the spring two rice fields are selected by divination, one on the northwest side of Kyoto, the old imperial capital, and the other on the southeast side. These two fields symbolically represent the nation. In the fall, the new emperor offers the rice harvested from these fields to the deities, and then partakes of a private meal with the deities to consume this sacred food together (Ohnuki-Tierney 1993, 48, 50). Afterward, the public symbolically shares in the special relationship between the emperor and the deities through public feasting.

These practices indicate that the identity of the Japanese state is closely tied to rice agriculture. Similar ideologies are found throughout the rice-growing regions of Asia, reiterated through the rituals of state. In the Indian kingdom of Puri (Orissa), a bride marrying the king entered the kitchen in the royal household for the first time on the fourth day of her marriage and prepared rice offerings for the ancestors of the king. Thus her first official act was to offer the life-sustaining food to the royal lineage. In the Islamic sultanates of Central Java the most important annual holiday was Garebeg Mulud, the celebration of the Prophet's birthday. On this occasion, the sultan sent a procession bearing towering cones of rice known as *gunungan* through the streets of the city. These offerings represented the ruler's contribution to the fertility of the realm and were apparently ancient in origin, as an inscription dating to 878 C.E. describes similar rice cones called "food lingam" (*anna linga*) being offered to the Hindu deity Brahma. The name *gunungan*, meaning mountain-like, is itself derived from pre-Islamic concepts of the Hindu sacred mountain Mahameru, another symbol of divine rule. At the end of the procession, the offerings were torn apart in a scramble by the ordinary citizenry and carried off, spreading the blessing throughout the land.

Many Asian kingdoms had royal plowing ceremonies in which the monarch plowed the first furrow to mark the opening of the agricultural season. These ceremonies are quite ancient and probably originated in India, as they are mentioned in the classical Indian epic *Ramayana*. In the Buddhist tradition they are mentioned in the Jataka stories, the tales of the life of Prince Siddhārtha before he achieved enlightenment as the Buddha. In one of the stories, the young prince shows supernatural talents at the annual plowing ceremony conducted by his father, the king (Gerson 1996, 21). The Thai royal plowing ceremony, Raeg Naa (lit., "first plowing"), continues to this day in Bangkok.

Because rice has played so prevalent a role in conceptions of the state, rice imagery has been an ideal tool for expressing political messages in the modern era. In Meiji period (1868–1912) Japan, rice imagery appears in woodblock prints that satirized both cultural conventions and modern developments (fig. 26.13). In India the Bengali artist Abanindranath Tagore (1871–1951) produced his famous painting *Mother India (Bharat Mata)* in response to nationalist agitation following the British partition of Bengal in 1905. Mother India is shown in the form of a four-armed Bengali goddess clutching the symbols of Indian self-sufficiency, including a sheaf of rice. In the second half of the twentieth century, as the use of political posters became widespread, it was inevitable that political views of all stripes would be expressed using the idiom of rice (figs. 26.14, 26.15).

•

27.1   A young novitiate into
Brahminism receives gifts of
rice from senior matrons in
the Aiyengar community in
the *upanayanam,* or "thread
ceremony." Photograph by
K. Suresh, Madras, 2001.

27

Rice in the Human Life Cycle: Traditions from Tamil Nadu, India

Nanditha Krishna

Rice has been cultivated in India for nearly six thousand years. Pottery with carbonized remains of rice grains and husks from the fourth millennium B.C.E. is among the earliest indications of its presence in the region, and there is ample evidence of rice agriculture by the third millennium B.C.E. The domestication of rice in the thickly forested Gangetic Plain of that time was probably of the slash-and-burn (*jhum*) variety still followed by many tribal communities in India.

In 1977, India became a self-sufficient rice producer as a result of the green revolution. Today rice is the staple food of 65 percent of India's population and its cultivation is a major source of employment. It constitutes about 52 percent of the nation's total food grain production and 55 percent of the total cereal production. India has the largest area devoted to rice cultivation in the world, of which 33 percent is rainfed lowland, 45 percent irrigated lowland, 15 percent rainfed upland, and 7 percent is prone to flood. Since nearly half of the land used to grow rice is rainfed, production can be erratic. Lack of rain often leads to drought, while flash floods in eastern India can damage the crops.

The Indian reverence for rice is embedded in both language and sacred texts. The Sanskrit word for raw rice is *vrihi* and for boiled rice is *annam*. The goddess of rice is Annapurna, whose name means "full of rice" and who is pictured holding a bowl in one hand and a long spoon or ladle in the other. The importance of rice is evident in the fact that the word for food is also *annam*. Ritual or charitable feeding is called *annadana*—the gifting of rice. There is no reference to rice in the *Rig Veda*, but the *Atharva* describes the sanctity of the grain, describing it as a healing balm, "the sons of heaven who never die" (*Atharva Veda* 6.140, 2; 8.7, 20; 9.6, 14). According to the *Taittiriya Brahmana*, Annadevata—the god of rice—is the first progenitor of immortality (*amrita*) and of sacrifice, an essential ritual in the Vedic period involving offerings to the sacred fire.[1] The *Taittiriya Upanishad* (2.1) says that the Supreme Being, Purusha, is formed of rice (*anna*) and the vital essences (*rasa*). The *Upanishad* goes on to celebrate the greatness of rice, saying that all that is born is born of rice and whatever exists on earth is born of rice, lives on rice, and merges into rice. Rice is the firstborn among all beings; that is why it is called the medicine that relieves the body of all discomforts. Those who venerate rice as Brahman[2] are said to attain rice, or achieve a mystical union with Brahman, thereby equating rice with Brahman (*Taittiriya Upanishad* 2.2). The *Upanishad* includes several strong injunctions related to rice—do not look down on rice, do not neglect rice, multiply rice

27.2   A child in the Aiyar community is fed rice for the first time during the *annaprasanam* ceremony. Photograph by M. S. Subramaniyam, Madras, 2000.

many times over, obtain a great abundance of rice—and states that he who knows the discipline and glory of rice attains that which he seeks (3.7–10). In all of these references, rice is regarded as equivalent to food and even Brahman himself. The *Mahabharata* says that there is no gift (*dana*) greater than that of rice, because all beings are born of rice and obtain sustenance from rice; the one who gives rice gives life itself (Bajaj and Srinivas 1996), while even the *Bhagavat Gita* affirms that all beings are formed of *anna; anna* rises from the rains and the rains rise from the sacrifice (3.10–16).

Rice particularly symbolizes auspiciousness, wealth, and prosperity. In Tamil Nadu and other South Indian states, the Rice Goddess is called Annalakshmi (*anna / annam*=rice, and Lakshmi=prosperity). As in Sanskrit, the Tamil word for boiled rice (*soru*) also means food, while the word for raw rice is *arisi*. (The word *arisi* entered the Greek and Latin languages as *oryza*, whence the English word *rice*.) Annalakshmi holds a small sheaf of rice in one hand, and Dhanyalakshmi (goddess of grains and cereals, including rice) holds several sheaves in her hand. There are many varieties of rice, one of which is celebrated as the goddess Ponni (Ponniyamman; see chapter 19) in parts of Tamil Nadu. In the *Mahabharata,* Krishna, the eighth incarnation of Vishnu, the preserver, gives the princess Draupadi an *akshaya patram,* or bowl that is always full of rice, a concept that is repeated in the Tamil epic *Manimekalai* where Aputran, a sage, owns a similar bowl that he disposes of when he is stranded on an island and unable to share all the rice. The bowl later comes into the hands of Manimekalai, the heroine of the epic.

Finally, the importance of rice cultivation in Tamil culture can be gauged by the fact that long ago the performance of functions such as weddings, entering a new home, and so on were discouraged in the lunar month of Adi (mid-July through mid-August), when the seed was sown, and Margali (mid-December to mid-January), when the crop was harvested. In time, people forgot the reason for the ban, and the months have now come to be thought of as inauspicious.

I. RICE IN RITUAL AND IN THE HOME

Many Hindu rituals involve the use of rice. Generally the rice used in rituals is paddy, or unhusked rice, but since paddy isn't available in cities, husked rice is considered an acceptable substitute. Only for death ceremonies is cooked rice used. Paddy and uncooked rice are seen as pure, while cooked rice is regarded as polluted. Here are brief descriptions of how rice is used for blessings, offerings, and invocations, followed by explanations of some particular forms in which rice is used for ritual purposes.

A. AKSHADAI

The most frequent ritual use of rice is for a blessing, when rice grains mixed with turmeric powder (*akshadai*) are sprinkled on the head of a young person by an elder. (The Sanskrit meaning of *akshata* is unbroken or unhusked barley corns, the auspicious grain in the Indus Valley, where the *Rig Veda* was written.) So revered is *akshadai* that it can even substitute for flowers, clothing, and jewelry when making an offering to the deity.

B. NAIVEDHYAM

This is the ritual offering of food, consisting of rice, salted red lentils (*tuvaram paruppu*), and *ney* (ghee, or clarified butter), along with a sweet pudding (*payasam*) made of rice, milk, and sugar or jaggery. It is offered to the deity at every home and temple. After the ritual, the three are mixed and consumed, with the first portion being fed to crows, who are believed to represent the souls of the ancestors. The goddess Lalithambika, in particular, is supposed to be very fond of sweet rice pudding.

C. KUMBHAM OR KALASAM

With the aid of two types of sacred pots (*kumbham* or *kalasam*), a priest invokes the gods and ancestors. The pot may be made of mud, brass, bronze, silver, or gold, depending on the wealth of the *yajaman*, the performer of the sacrifice. It is filled with water, representing Varuna (the god of water), and mango leaves, signifying prosperity, and placed on rice—preferably paddy—and surrounded by eight other grains, the whole making up the *navadhanyam,* or nine sacred grains that represent the nine heavenly bodies (seven planets plus Rahu and Ketu, the monster snakes that cause eclipses of the sun and moon).

During this ritual, the gods (*devas*) are invoked and brought into the sacred pot of water. The gods—represented by the *kalasam*—may not be placed on the plain ground; only rice is a fitting seat (*asanam*). A banana leaf is spread on the floor, paddy is placed on it, a layer of wheat grain above, another layer of rice above that, black gram above the rice and black sesame seeds (to keep away evil demons) on top, just below the *kalasam.*

D. NAVADHANYAM

Rice is one of the nine sacred grains variously associated with heavenly bodies, days of the week, and deities. The *navadhanyam* are used to invoke the planets and their presiding deities whenever the *kalasam* are set up.

Grain	Heavenly Body	Day	Presiding Deity
wheat (*godumai*)	sun (Surya)	Sunday	Shiva
rice (*arisi*)	moon (Chandra)	Monday	Parvati
yellow gram (*tuvarai*)	Mars (Angaraka)	Tuesday	Kartikeya
green gram (*pasi*)	Mercury (Budha)	Wednesday	Vishnu
soybeans (*kadalai*)	Jupiter (Guru)	Thursday	Dakshinamurti
white beans (*mochai*)	Venus (Shukra)	Friday	Lakshmi
sesame seeds (*ellu*)	Saturn (Shani)	Saturday	Yama
horse gram (*kollu*)	snake or eclipse (Rahu)		Durga
black gram (*ulundu*)	snake or eclipse (Ketu)		Ganesha or Indra

27.3

27.3   A young girl in the Aiyengar community is blessed with rice by her parents in the *ritu kala* ceremony, which celebrates her coming of age. Photograph by K. Suresh, Madras, 1995.

E. HAVIS

This is boiled rice offered to the sacrificial fire (*homam*), which is essential to all Hindu rituals. It is served to Agni, the god of fire, in spoonfuls, along with sacred twigs (*samit*) and ghee (Sanskrit, *ajyam*). Agni is considered to be the bearer of human offerings to the gods.

F. KOLAM

The ritual designs (*kolam*) created on the floor in front of each home are made of rice flour. The rice flour keeps away the evil eye and demons, forming a protective wall before the house. Birds and insects partake of the rice flour, which pleases the souls of ancestors.

G. NELKADIR

This is a large sheaf of paddy hung up outside the house to keep away the evil eye and demons.

II. RICE IN SAMSKARAS

Samskaras are the Hindu life rites and ceremonies that must be performed by or for every individual. Differing from caste to caste and region to region, they can be divided into *samskaras* associated with pregnancy, childhood, education, marriage, and death. The ideal life is built around the desire to be good, fulfill one's duties, enjoy material gain, think good thoughts, and, finally, retire in expectation of repeating this cycle in the next life. The *samskaras* support the individual at every stage. Since rice is sacred, it is used for all the religious ceremonies connected with the various *samskaras*. The following descriptions focus on the role played by rice in the principal *samskaras*.[3]

A. PRENATAL SAMSKARAS

Prenatal *samskaras* are performed during a woman's first pregnancy. They both protect the unborn child and ensure the birth of a son.

1. Valaikappu

Valaikappu is held in the fifth or seventh month after conception. During the course of this ritual, the pregnant woman's arms are filled with bangles. Each married woman (*sumangali*) present is given a pair of bangles and also adds a pair to the pregnant woman's arms.

A small bowl of rice grains mixed with *ney* is kept in front of the pregnant woman, along with little plates of sandal paste and red *kumkum* powder. Each married woman visitor applies the *kumkum* powder to her forehead and then dabs the sandal paste onto the arms of the pregnant woman, sprinkles rice grains on her head, and wishes her prosperity, an easy pregnancy, and the birth of sons. The rice grains used by the women are called *shobhana akshadai*. Five types of rice preparations (*masakai sadam*) and *kapparisi*, a mixture of rice, jaggery, sesame, and millet, are made in advance as a feast for the soon-to-be mother.

2. Simantham

The *simantham* (derived from "hair parting," or *simantonnayana*) ceremony may also be performed in the fifth or seventh month of pregnancy. It is a long and important ritual meant to ensure a safe delivery and the birth of a son.

During this ceremony, the parents of the pregnant woman distribute different kinds of rice (curd rice, tamarind rice, etc.) and *kapparisi*. A *kalasam*, containing water with turmeric powder is placed on a pile of paddy, the gods are invoked, and the ancestors are invited to the ceremony. This is followed by several rituals that use various grains to symbolize the woman's fertility, leading up to the parting of the hair, a rite emphasizing the formation of the mind of the baby, which is believed to take place in the fifth month of pregnancy. Thereafter, the husband tells his wife to gaze at a mixture of rice, sesame, and ghee, seeing therein children, cattle, prosperity, longevity, and good fortune (*mangalyam*).

B. SAMSKARAS OF CHILDHOOD

1. Namakarmam

The *namakarmam* is a naming ceremony. Rice plays an important role in this rite, performed on the tenth or twelfth day after the birth of a child. The child's name may be determined by the position of the planets, a grandparent's name, or other factors. During the *namakarmam* the baby is placed on the lap of the father's sister, who whispers the chosen name into the baby's ear, while the father or the priest traces it with his finger on a heap of paddy placed on a banana leaf.

2. Annaprasanam

Annaprasanam (fig. 27.2) is the ceremonial feeding of solid food to a baby for the first time. It takes place at home in the sixth month after birth. A rice mixture is first placed on a sacred fire for purification and then offered to Vac, the goddess of speech, before it is fed to the child. The combination of ghee and rice is intended to ensure brilliance, while a blend of curd and rice can assure good sense.

3. Karnavedham

A child's ears are first pierced by a goldsmith while seated on the lap of his or her maternal uncle for the *karnavedham* ritual. Afterward, the goldsmith is given a bag of rice and yellow lentils.

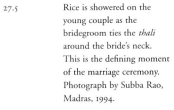

27.6   The *thali* is the key symbol in the Vaniya Chettiyar wedding ceremony. Photograph by Subba Rao, Madras, 1994.

27.7   The sacred coconut with the *thali* on a tray of rice sits before the couple. The priest will place the coconut in the hands of the groom and tie a sacred turmeric-dyed thread around his wrist. The groom then places the coconut in the hands of the bride and ties a similar thread around her wrist. Photograph by Subba Rao, Madras, 1994.

27.4   The painted pots of sanctified water are placed on a base of rice during a wedding in the Vaniya Chettiyar community. Photograph by Subba Rao, Madras, 1994.

27.5   Rice is showered on the young couple as the bridegroom ties the *thali* around the bride's neck. This is the defining moment of the marriage ceremony. Photograph by Subba Rao, Madras, 1994.

4. Chudakaranam

This ceremony marks a child's first haircut (*chaula*) at the age of three. It is initiated by the father and completed by the barber. For his efforts, the barber is given a bag of rice, yellow lentils, and the cloth used to wet the child's hair.

5. Puberty *Samskara*

Ritu kala is the celebration of a young girl's coming of age (fig. 27.3). A steamed rice cake (*puttu*) is given to the girl throughout her first menses, beginning on the second day. She is also given jewelry and her first sari, in recognition that she has become a young adult. After the ceremony, family elders and her parents shower rice on her head as a symbol of blessing.

C. EDUCATION SAMSKARAS

1. Akshararambham, also known as *Vidyarambham*

This is a very important ceremony signifying the beginning of a child's education. In former times it took place in a child's fifth year. Now it is done before he or she enters school. The ritual takes place with the child seated on the father's lap before a heap of raw paddy placed on a banana leaf. Holding the child's first finger, the father first traces the word *Om* in the rice. Then he writes "Salutations to Ganesha" or "Salutations to Narayana," depending on the family's religious sect or preferences. At the conclusion the priest is gifted with rice and lentils.

2. Upanayanam

Upanayanam is the "thread ceremony," which initiates a boy into his caste. In the past it signified the departure of a child to the *gurukula*, the home of his teacher, for his education. Today it has lost much of its original meaning. The word *upanayanam* means "taking the pupil to the teacher" and marks the beginning of the stage of *brahmacharya* in the life of the young boy, a period when he commences his formal education, observes celibacy, and acquires knowledge. In ancient times, this stage lasted from the age of seven until marriage and was spent mostly in a hermitage under the supervision of a guru.

Various ceremonies during the *upanayanam* involve rice:

- *Kalasa avahanam* is the placing of the sacred pot on a bed of paddy surrounded by eight other grains, the whole making up the *navadhanyam.* The gods are invoked by the *avahanam* prayers, and the ancestors by the *nandi* ceremony. The gods include the twenty-eight constellations (*nakshatras*) and the ten *dikpalas,* or protectors of the directions: Indra, Varuna, Durga, Ganapati, Kshetrapalaka, Abhayankara, Vastu, Triyambaka Rudra, Vani Hiranyagarbha (Brahman), and Lakshmi Narayana.

- Along with other boys, the young novitiate is honored with a *kumarabhojanam*, a brunch consisting of varieties of rice and sweets and no salt. This prepares him for the long and difficult ceremonies to follow.

- Next the boy must have his head tonsured. The barber who carries this out is given rice, yellow lentils, jaggery, and the wet towel used during the haircut.

- The highlight of the *upanayanam* is the *brahmopadesam*, when the boy is taught the sacred Gayatri mantra invoking the sun. All the persons present shower *mantrakshadai*—rice grains mixed with turmeric powder and blessed by the priest—on the novitiate.

- The boy has now been initiated into the phase of studentship. He must leave home with nothing except the clothes he wears; henceforth he lives by begging for his food, to learn the quality of humility. (Nowadays, the rite is mainly symbolic, as boys rarely live with their guru. Those who are sent to a Veda Pathashala to be trained as priests continue to live on public donations.) At the ceremony's conclusion the boy goes to the women present and asks, three times, "Bhavati bhiksham dehi,"meaning, "Lady, please give me food." The women who are attending the *upanayanam* bring along some rice, which they pour into his bowl three times (fig. 27.1). (Often a large vessel of rice is made available from which the women give him the rice as he begs for it, and into which he pours it back.) This rice, known as the *bhiksharisi*, must be cooked and fed to the novitiate.

D. MARRIAGE RITES

1. Kalyanam

The marriage ceremony (*kalyanam*) is celebrated with great pomp and style. There are several variations in each caste; I have tried to document as many customs prevailing in Tamil Nadu as could be found that use rice. The first few are common to nearly all the communities in the region.

A few days before the marriage, nine kinds of grains must be raised in nine mud pots. This is called *paligai valarthal* (Tamil). It serves to both affirm the agricultural process and invoke fertility. The *paligai,* or potted grains, is later worshiped at the wedding. After the wedding, the *paligai* is generally immersed in water.

Nearly all communities commence the wedding with the *kalasam* placed on a base consisting of layers of sesame, rice, black gram, rice again, wheat grains, and paddy on a banana leaf. Rice, coconut, flowers, and turmeric are given by the bridegroom's sister to the bride, who ties it in the *pallu* of her sari, the portion that hangs over her shoulder. This signifies fertility and prosperity. The ceremony of the seven sacred steps (*saptapadi*) involves the pouring of cooked rice (*havis*) into the fire, followed by the *nel pori* ritual, when the bride's brother pours fried puffed rice (also called *nel pori*) over the bride's and groom's hands, after which it is poured into the fire.

The following rituals are common to nearly all communities:

- A fistful of rice is placed in a pot and cooked over the sacred fire by the couple to signify the beginning of marriage.

- Sanctified and painted pots of water are placed on a base of rice during the marriage ceremony as a mute witness to the events (fig. 27.4).

- The *thali*, or marriage pendant, is generally kept on or tied around a coconut and placed on a tray of rice symbolizing fertility, before it is taken off and tied around the neck of the bride. The equivalent of the wedding ring in European and American culture, the *thali* can never be removed. Rice is showered on the bride and bridegroom as the latter ties the *thali* around his bride's neck (figs. 27.5–27.7).

Various Tamil Nadu communities have developed their own unique wedding rituals involving rice:

- Brahmins perform a ceremony called *pidisuttal*. Rice balls are made of cooked rice mixed with turmeric or *kumkum* powder. The senior women in the bride's and groom's families bless the couple with the rice balls. They pick up each ball and rotate their arms in a clockwise direction with the ball in their right hands and then throw them to the four quarters.

- The Vanniyars and Wodeyars spread paddy on the floor and place a mat on top. The bride and groom sit on the mat during the ceremony.

- In the Pillai community, rice is spread on the floor and a banana leaf is set over it, on which the bride stands (Santhi 1994, 283).

- The Balija Naidus pour raw rice on the couple. This is known as the pouring of leftovers (*sesham*). Later, the rice is given away to the barber (fig. 27.8).

- The Nayakkars pile rice on a banana leaf. After the tying of the *thali*, the elders take the rice in their hands and pour it thrice on each partner.

- Among the Vannan (washermen) community, the bride's arrival in the bridegroom's house (where the wedding is to take place) is the start of the *Mariammanukku pataittal* ceremony. Paddy is cooked in a *kappu*—a new pot tied with a yellow string—over a new stone stove built for the ceremony. It is then milled and cooked with mutton and vegetables and offered to the goddess Mariamman. Paddy is spread on the marriage platform and covered with a sheet. The bride sits on it and is anointed with oil, *sikkakai* powder from acacia tree pods, *kumkum* powder, turmeric, and sandal paste in a ceremony called the *nalangu*. The same rites are performed for the bridegroom. Thereafter, the main ceremonies begin (Santhi 1994, 306)

- Immediately after the wedding, the bride and groom enter the groom's house in a ceremony known as the *grihapravesham*. In many communities in the central and southern districts of Tamil Nadu, a Lakshmi *deepam* (a brass handheld oil lamp sculpted with the image of goddess Lakshmi) is placed on a wooden rice measure (*marakal*) filled with rice, facing east. The girl lights the lamp, signifying that she brings the goddess of prosperity with her into the house. In the northern districts she enters the house carrying the Lakshmi *deepam* on a *marakal* filled with rice, signifying the entrance into the house of prosperity.

- In many communities in the northern districts of Tamil Nadu, paddy or raw rice is kept in a large rice measure (*padi alakku*) on the doorstep. The bride kicks it as she enters—the distance traversed by the spilt rice determines the extent of prosperity she will bring to the family.

27.8   A bride and groom of the
        Balija Naidu community
        pour rice on each other as
        a symbol of prosperity and
        fertility. Photograph by
        G. Nagaraj, Madras, 2002.

2. *Grihapravesham*

A couple moving to a new house performs the *grihapravesham*, a ceremony similar to the one of the same name carried out after a wedding. It starts with carrying the *kalasam*, and the sacred fire is fed by the *havis*. To signify prosperity, rice is spread out on the floor, and then the owner and his wife enter the house. The priest invokes the deity of construction (*vastu purusha*) by putting paddy in the *kalasam* and chanting *slokas,* hymns in praise of the gods.

3. *Sashtiabtapurti* and *Satabhishekham*

These rites are performed in the sixtieth and eightieth years of a man's life, respectively. They are simpler versions of the wedding ceremony, during which rice is used in same way as in a wedding ceremony.

E. DEATH CEREMONIES

The ceremonies surrounding a person's death ensure a smooth passage for the soul. They generally take place over the course of thirteen to sixteen days. Monthly rites are performed thereafter for the entire year of mourning, concluding with a ceremony at the end of one year. As with the ceremonies marking other life events, death ceremonies in Tamil Nadu may vary according to caste and local tradition. (Of all the castes, Brahmins have the most elaborate customs.) At several points in the ritual observances rice plays a key role.

1. Day 1: *Dahanam*

Members of all castes observe the first day of death ceremonies by pouring half-cooked, saltless rice (*vakkarisi*) on the mouth of the corpse. By this act, the family gives its last offering. The rice is half-cooked and saltless to discourage the soul from lingering.

At this point, Brahmins proceed to complete the cremation as fast as possible. Non-Brahmin castes, however, place a *marakal* filled with paddy and a Lakshmi *vilakku* (Sanskrit, *deepam*) beside the dead body. Rice is spread on the ground and pounded with a pestle by women in a tradition called *vakkarisi kutthal.* The deceased's son and his wife, followed by those present, circumambulate the body thrice, holding the paddy-filled *marakal* and the lamp, and dropping th*e vakkarisi* onto the mouth of the dead person. After the cremation or burial, rice is pressed into cow dung that had been placed on the body before it was taken from the home; this represents the departed soul. It is applied to a wall inside the house, where it is worshiped for ten days.

2. Day 2: *Sanchayanam*

On the second day the bones are collected, bathed in milk, and immersed in the sea or river or at least a pond. Then, among the Brahmins, a donation (*danam*) of rice, green gram, and raw banana is given to four Brahmins. More Brahmins are added every day for the *danam*—five on the third day, six on the fourth day, and so on until fifteen Brahmins are given *danam* on the twelfth day. Non-Brahmin castes give *danam* to a single Brahmin only on the sixteenth day.

From the first to the tenth day, a *pindam*—a large ball of boiled rice, sesame seeds, and water—is thrown into the river or sea or tank to feed the fish and, thereby, satisfy the soul. The rice assuages hunger and the sesame seed water quenches thirst, as the soul (*sukshma sharira*) wanders around, waiting to move to another world.

3. Day 10: *Dashas or Pattu*

On the tenth day the wandering soul takes on a full body and is very hungry. About five and one-half pounds (one *padi*) of cooked rice is mixed with raw banana and either drumstick (the fruit of the horseradish tree) or a type of greens called *avithi keerai* and decorated with a circle of sweets, uncooked Bengal gram, and other savories. The whole is, again, thrown into the water.

A *shanti homam* ritual is performed with the sacred fire to bring peace to the family and an *ananda homam* to bring back happiness. Then a *punyavachanam*, a ritual cleaning to take away the pollution is performed. Finally, a *kalasam* signifying prosperity and good fortune is set up again on a bed of rice and other foods.

4. Day 11: *Rudra Puja*

The day starts with the *kalasa puja* to Shiva or Rudra, one of the manifestations of Shiva, in which two pots are decorated and filled with pure water and the gods are invoked to enter the *kalasa*. A male calf, representing Nandi, the sacred bull and vehicle of Rudra, is fed with rice, while thirty-two handfuls of rice are offered to the fire to invoke the soul of the deceased. It is believed that the soul spends the eleventh day in Rudra *loka*, the heaven of Rudra. One Brahmin—the *ottan*, representing the deceased's spirit (*pretam*)—is fed from rice cooked separately for him alone.

5. Day 12: *Punyavachanam*

On this day, fifteen Brahmins are given a gift of rice to enable the soul to pass through the sixteen stages (*mandapas*) on the path to the *pitrulokam*, the home of the ancestors. Two separate *homams*, one to the ancestors and another to the spirit of the dead person, are lit and two separate vessels of *havis* are prepared.

Seven rice balls (*pindam*) are prepared. One is large and shaped like a banana; it represents the dead person. Of the other six, three represent the immediate male ancestors (father, grandfather, and great-grandfather) and the other three the immediate female ancestors (mother, grandmother, and great-grandmother). In an elaborate ritual, all the *pindam* are joined together. The person performing the ceremonies (generally the eldest son or male heir) maneuvers each *pindam* on the floor with the entire right arm moving toward the left arm. The rice balls are mashed together into one mass and then thrown into a water source to feed the fish. By this action, the soul is sped on its journey to the *pitrulokam*. The soul of either the great-grandfather or great-grandmother (depending on the sex of the dead person) is now released to be reborn.

6. Day 13: *Devataradhana*

For Brahmins, the rites end on this day when the soul finds rest in *pitrulokam*. The *kalasam* is set up on rice and worship is offered to the nine sacred grains. The *danam* or gift is presented to one or more Brahmins. The gift may be of money, clothes, silver and brass vessels, fruits, coconuts, and rice, depending on the wealth and desire of the donor, but rice must always be included.

7. Day 16

The rites for the non-Brahmin castes continue till the sixteenth day. Fried puffed rice (*aval*) is offered to the soul on all or certain days, depending on the caste. *Pindam* are offered on the sixteenth day.

8. Monthly Rites: *Masyam*

For the next year, two days are set aside every month to observe the rites for the dead. On the first day (*shoda kumbham*) one *pindam* is made, and on the second day (*una masyam*) seven *pindam* are prepared. They are placed on a bed of cut *darbha* grass and put into water.

9. Annual Rite: *Devasam*

Seven *pindam* are made for this ceremony; six are put into water, while one large *pindam* is kept out to feed the crows. A *homam* is lit and fed with *havis*, rice, banana, and a rice sweet called *soji appam*. The year of mourning is over.

III. RICE IN FESTIVALS

Many festivals that are celebrated regularly also involve the use of rice. They all share three common elements. The food that is offered to the deity, sanctified, and then eaten (*naivedhyam*) always consists of rice, the *kalasam* is always placed on paddy, and rice dishes and rice-based sweets are always served. Several of these festivals are briefly described below, with notes on additional ways in which they use rice.

A. PATTINETTAM PERUKKU

The eighteenth day of the month of Adi is celebrated as the day of planting the new rice seed. Four kinds of savory rice dishes, in which rice is mixed with spices or lime or tamarind, etc., and one sweet rice dish are prepared for this day.

B. VARALAKSHMI VRITAM

This festival is probably of Telugu origin, but it is very popular among women of the northern districts of Tamil Nadu. It is celebrated on a Friday after the new moon in the month of Adi.

The *kalasam* is filled with rice, coins, a box of *kumkum* powder, turmeric, a small mirror, black beads, green glass bangles, betel leaves, areca nuts, and turmeric, and five mango leaves and a coconut are placed on the mouth of the *kalasam*. Lakshmi's face may be drawn on the *kalasam* or a nearby wall or attached separately. The *kalasam* is placed on a heap of rice spread over a plantain leaf. The *puja* is performed with *akshadai* and flowers.

The rice dishes for the *naivedhyam* consist of *payasam*, cooked rice, and steamed dumplings (*idlis*) made with raw rice and black gram that have been soaked in water separately and then ground with water and left to ferment. The day after the festival, the rice inside the *kalasam* is made into *payasam* and eaten by the family.

C. AYUDHA (SARASWATI) PUJA

Ayudha means "weapons" or "the tools of one's trade." Saraswati is the goddess of learning, who is revered on Ayudha Puja day. This ritual is performed on the ninth day of the Navaratri festival, which honors the goddesses Durga, Lakshmi, and Saraswati. Puffed rice flakes, jaggery, and Bengal gram are offered to the goddess and then shared by family and friends.

27.9   While worshipping the god Ayyappa, ghee (clarified butter) is poured into a coconut shell placed on rice. Then, the coconut shell and rice are packed in a cloth bundle and taken to Sabarimalai (in Kerala) and offered to the deity. This Ayyappa *puja* is conducted by Brahmins. Photograph by S. Ganesan, Madras, 1990.

D. KARTIKAI DEEPAM

The same foods offered during Ayudha Puja are offered at this time.

E. PONGAL

This is the most important rice festival in the year. It is the harvest festival of Tamil Nadu, when the fresh rice crop is gathered. A bronze or brass vessel is decorated with sandal paste and dots of wet *kumkum* powder. Then a bunch of ginger and another of turmeric, with the leaves, is tied around the mouth of the vessel. Milk, water, green gram, jaggery, and the freshly harvested rice are boiled in the vessel, and the mixture is allowed to overflow. As it runs over the sides of the pot, everyone shouts "*pongal-o-pongal*," meaning "it is boiling." The overflow signifies the prosperity brought about by the new rice harvest. Freshly boiled new rice is also offered as *naivedhyam.*

F. MATTU PONGAL, ALSO KNOWN AS KANU PANDIGAI

On the day after Pongal, the rice cooked for the *naivedhyam* is mixed with salt, curd, and ginger and made into little balls, some colored red with *kumkum* powder, some colored yellow with turmeric powder, and some left plain. Along with sweet *pongal,* or rice boiled with milk, and pieces of sugarcane, the rice balls are offered to birds, cows, and calves.

G. URANI PONGAL

This celebration honors a village's own guardian deities. The first rice of the new harvest is collected from every family in the village and ritually offered to the Sun God. Then it is collectively cooked in the temple as *pongal,* symbolizing the increase in and sharing of prosperity. A portion of the *pongal* is offered to the gods and goddesses and distributed as *prasadam* to those who contributed to the festival. After the ritual, uncooked rice is given to the officiating priest or fed to cattle.

The Seven Virgin Sisters (Sapta Kannimaars), who share a temple in the village and are equivalent to the Seven Divine Mothers (Sapta Matrika) of the Central Indian tradition, are each offered a different rice mixture:

Narayani (Kamini)	sweet rice (*payasa annam*)
Maheswari	white *pongal* (*ven pongal*)
Kaumari	tamarind rice (*puliyodarai*)
Vaishnavi	curd rice (*thayir sadam*)
Varahi	sweet *pongal* (*sarkarai pongal*)
Indrani	coconut rice (*thengai sadam*)
Chamundi	rice and Bengal gram (*akkara vadisal*)

H. AYYAPPA PUJA

Pongal is also associated with Makara Sankranti, when the sun enters the northern hemisphere. Men may take a forty-day vow of abstinence before visiting Sabarimalai in Kerala on January 14, Makara Sankranti day, to observe a mysterious light that appears on the hill (women may go only after menopause). The pilgrims, numbering in the millions, carry a cloth bundle containing rice sanctified by prayers and a coconut shell as their offering to Ayyappa, Lord of Sabarimalai (fig. 27.9).

Rice plays an important role in the human life cycle within the Hindu culture. It is an offering to the gods, and it also represents the gods and even the souls of the dead. Whether it appears in a domestic ritual, an important ceremony, or a festival, rice is given a position of honor, for it is the abundance of rice that sustains life and ensures happiness and prosperity in the home.

•

番五

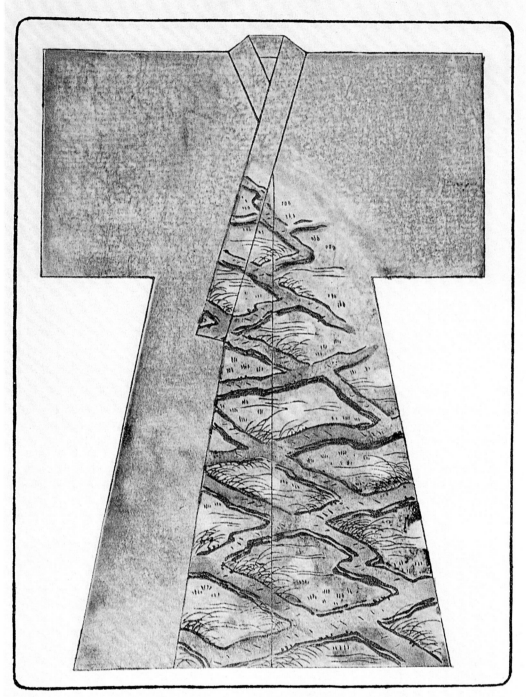

28

Wrapping the Body with Images of Rice: Kimono Patterns from the Edo Period

Toshiyuki Sano

Rice was first introduced to the Japanese islands approximately three thousand years ago, and it has been widely cultivated ever since. In fact, the Japanese often refer to their country as "The Land of Vigorous Rice Plants" (Mizuho-no-Kuni), an expression honoring the fertility of the land that provides rice as food and wealth for its people. This epithet stands in contrast to another, "The Land of the Rising Sun" (Hi-Izuru-Kuni), which suggests the country's power and sacred character. Rice is so vital to the Japanese nation that the golden five-yen coin depicts a sheaf of harvested rice. Thus, in addition to being the country's most important food, rice is a key symbol in cultural discourse, ritual, and practice (Ohnuki-Tierney 1993).

It is not surprising therefore that this leading cultural symbol and its appurtenances appear as motifs in the decoration of cloth and clothing, one of the most important of Japanese art forms. In the course of analyzing patterns from thirty-six sample design books (*hiinakata-bon*) that were in commercial use between 1667 and 1800 (all of which have been recently compiled and reprinted),[1] I have identified twenty-seven rice-related patterns for kimono design. The kimono in question are mostly of the type known as *kosode,* the principal outer garment worn throughout the Edo period (1603–1868). Additional examples can be found in sample books containing family crests or shop and shrine emblems. These depict motifs for use on kimono as well as various craft items, including *noren* (curtains hung in the entrances of shops or rooms).

Use of Rice-Related Patterns

The *hiinakata-bon* referred to in this essay were illustrated with woodblock prints, and according to Louise Cort, they "focus almost exclusively on women's *kosode*, possibly because the composition of the painted, dyed, and embroidered designs encompassed the entire garment" (1992, 186). Such books were distributed to kimono traders, merchants, dealers, and wealthy customers in major towns. Most of the *hiinakata-bon* were subject to hard use. They wore out and were replaced by new books. As a result, many of the original volumes have been lost. Some, however, were saved with family treasures or perhaps forgotten and preserved in storage. A number of the surviving books appear to have been well cared for, having been mended and in some cases combined with other design books. This suggests that they were used primarily in placing orders for patterns or for reference in customizing designs.

28.1   Placing images of rice paddies—each of which reflects a full moon in a clear sky—on the left front panel of a kimono is quite an exceptional design. This scene, however, is well known from its frequent use in haiku. This pattern was found in one of the earliest polychrome sample design books to be printed. One of fifty patterns from *Saishiki hiinakata kokonoe nisiki* (*Colored Sample Designs, Nine-Time Print*), 1784. Reproduced from Yamanobe (1974d). Courtesy of Gakusha-Kennkyusya.

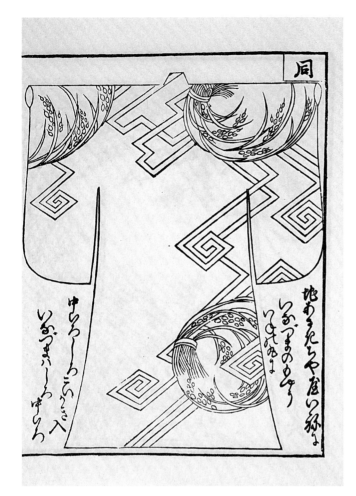

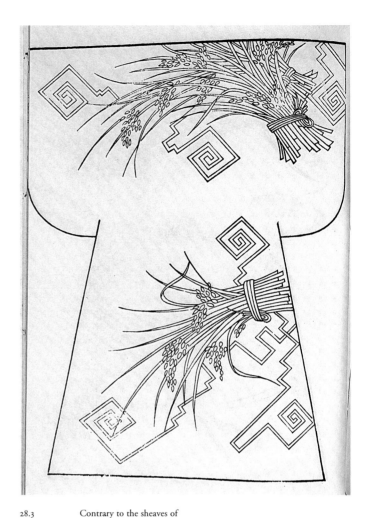

28.2   The Japanese term for lightning, *inazuma*, contains within it the word *ina*, or rice plant. Many centuries ago farmers believed that the reproductive power of lightning helped rice plants to thrive. The pattern seen here embodies this belief by showing the healthy bundles of rice in conjunction with strong, geometric flashes of lightning. One of 192 patterns from *Shokoku on-hiinakata* (*Country Sample Designs*), 1686. Reproduced from Yamanobe (1974a). Courtesy of Gakusha-Kennkyusya.

28.3   Contrary to the sheaves of rice seen in the designs on figure 28.2, those in this pattern are depicted with longer, sharper leaves. The angular, zigzag flashes of lightning make the image extremely dynamic and vivid. One of eighty patterns from *Tōryū moyō hiinakata matunotsuki jyochū-tachi mon-zukusi* (*Contemporary Style, Sample Designs, Moon with Pine Trees, Service Women, Every Pattern*), 1697. Reproduced from Yamanobe (1974b). Courtesy of Gakusha-Kennkyusya.

Each sample design book contained approximately a hundred patterns. The choice of these patterns was presumably determined by the kimono merchant and by *e-shi*, a class of professional artists who emerged in the early Edo period. Initially *e-shi* resided only in Edo and Kyoto; however, starting in the middle of the Edo period, they began to appear in Osaka and later in other major towns. Tachibana Morikuni (1679–1748), one such *e-shi*, was known for his drawings of agricultural subjects based on his close observation of rural Osaka. This choice of inspiration was unlike that favored by the majority of Kano school artists, who preferred to copy agricultural scenes from Chinese drawings (Reizei et al. 1996, 101). The everyday urban and rural scenes published by Tachibana Morikuni and some of his contemporaries in turn served as models for the patterns of other *e-shi*. Typically, *e-shi* discussed the type of patterns to be included in a new book with the merchants who hired them. In some cases, *ukiyo-e-shi* (woodblock print artists) were contracted instead, because of their artistic sensibility and their reputed ability to spot trends. The *e-shi*'s work might entail writing short explanations to accompany patterns as well as detailing their colors and themes.

The patterns depicted within the borders of an outspread kimono were typically printed on one page of the *hiinakata-bon*. The kimono was shown from the back with the sleeves extended so that their patterns would be visible as well. Exceptions to this format existed, however, and some patterns were presented on two consecutive pages, one devoted to the front view and another to the back. Such an arrangement must have been helpful for the kimono designers and their clients in picturing how the patterns would appear. It should be remembered, however, that the illustrations in the book were samples or suggestions and that the finished patterns on the dyed—or in some cases painted or embroidered—kimono would not be identical to them.

Numerous sample pattern books were published from 1684 to 1704 (the Jōkyō and Genroku eras), which coincided with the technological innovation of using rice paste for resist dyeing, as in the *yuzen* dyeing of Kyoto (Ueno 1974a, 15). Of the thirty-six reprinted *hiinakata-bon* that I have consulted, twenty feature at least one rice-related pattern: one book has four, three others have two, and the remaining books have one each. Fortunately for the purposes of this study, publication of these books is spread throughout the time period under consideration. It should be noted, however, that design books were published with greater frequency during the early eighteenth century, a time of heightened artistic and cultural activity in Japan. More than 50 percent of the sample books analyzed contain rice-related patterns, suggesting that rice imagery, although it varied over time, was long considered indispensable to kimono design.

Year	1662–80	1681–1700	1701–20	1721–40	1741–60	1761–80	1781–1800
Number of Sample Design Books Containing Rice-Related Patterns							
	1	3	10	1	1	1	3
Total Number of Sample Design Books Available							
	3	8	11	6	3	2	3

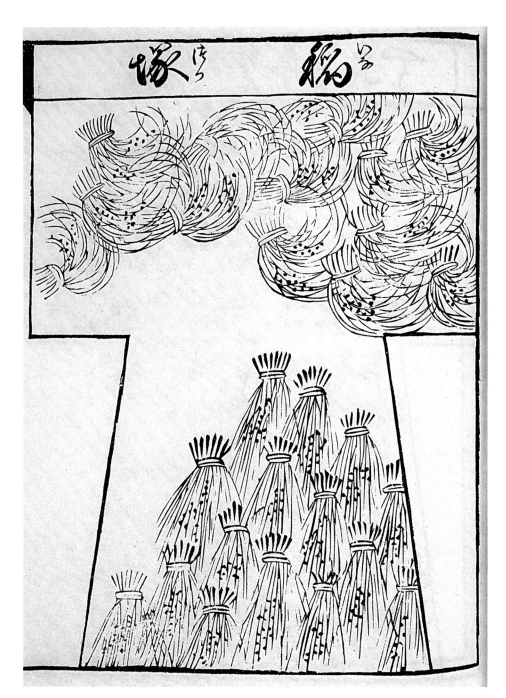

28.4   In this simple yet dynamic
        kimono pattern, sheaves of
        rice representing the sky
        appear to fly through the
        air, while bundles represent-
        ing the ground remain sta-
        tionary and neatly stacked.
        One of 120 patterns from
        Tōfū bijyo hiinakata
        (*Contemporary Style,
        Beautiful Women, Sample
        Designs*), 1715. Reproduced
        from Yamanobe (1974c).
        Courtesy of Gakusha-
        Kennkyusya.

Characteristics of Rice-Related Patterns in Sample Design Books

Principal elements of rice-related patterns seen in the *hiinakata-bon* considered here are: (a) bundles of harvested rice (*inataba*; figs. 28.2–28.6), (b) straw rice bags (*kome-tawara*; figs. 28.7–28.9), (c) rice planting (*taue*; fig. 28.10), (d) irrigation devices for rice paddies (fig. 28.11), and (e) watered rice fields (fig. 28.1). They occur with the frequency indicated in the chart below:

Year	1662–80	1681–1700	1701–20	1721–40	1741–60	1761–80	1781–1800
(a)		3	9	1		2	2
(b)	1		1				2
(c)			1		1		
(d)			2				
(e)			1				1
Total	1	3	14	1	1	2	5

The most frequently employed theme—the harvest season—is represented symbolically by drying bundles of rice stacked on the ground or hung from bamboo poles. Rice-related patterns are often associated with specific times of the year, such as spring or fall. Rice paddies in summer and winter, however, offered few activities that would lend themselves to treatment as kimono designs. Pattern elements such as straw rice bags could occur in any season, although they were most often seen following the harvest when they were used for collecting or paying taxes with rice.

Other symbolic elements often associated with those quantified in the chart above include:

1. sparrows (*suzume*; a symbol closely tied with the harvest; see fig. 28.6);
2. treasure boats loaded with sacks of rice (*takarabune*; a symbol of wealth; see figs. 28.8, 28.9);
3. lightning (*inazuma*; thought to be the coupling of heaven and earth, it symbolized the reproductive power that would produce an abundant rice harvest; see figs. 28.2, 28.3);
4. pine or bamboo trees (*matsu* or *take*; considered auspicious plants; see fig. 28.8);
5. lotus flowers (*hasu-no-hana*; a sacred symbol of the planting season);
6. cranes or turtles (*tsuru* or *kame*; auspicious animals);
7. various rural elements (rustic hamlets, a child riding on an ox, and so on); and
8. numbers (lucky numbers—three [*san*] or five [*go*] are common; seven [*nana*] and nine [*kyu*] also occur).

Changing Fashion in Rice-Related Designs of the Edo Period

Various rice-related designs also occur among the tremendous number of family crests (*kamon*) used throughout the Edo period (fig. 28.12). Family crests are stylized and contained within a relatively small geometric shape, such as a circle, square, or diamond. Their emblems consist of a single stylized item, multiples of the same item, or a combination of items. These emblems may be shared by different family lines.

Family crests, which are passed along from generation to generation, are often depicted on both front panels and the center back of the outer kimono. They are also found on other objects, including *noren* (cloths used as a space dividers), *furoshiki* (wrapping cloths), and *fukusa* (small cloths used in tea ceremonies or for wrapping gifts). Items such as tombstones, which are tended by the deceased's family throughout the generations, are also adorned with *kamon*. The origin of each family's emblem is often difficult

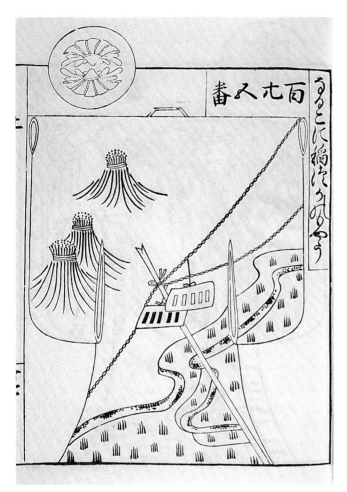

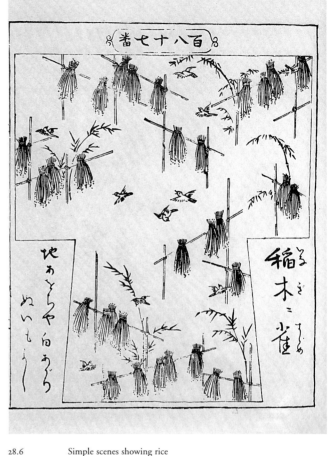

28.5   This pattern depicts the period following the harvest when wooden noisemakers (*naruko*) are used to keep sparrows away from the rice. One of one hundred patterns from *Tōse-moyō isai hiinakata* (*Contemporary Style, Detailed Sample Designs*), 1705. Reproduced from Yamanobe (1974b). Courtesy of Gakusha-Kennkyusya.

28.6   Simple scenes showing rice bundles hanging from bamboo poles are positioned around a center space, which represents the sky. Three sparrows are shown flying in this space in search of food. One of ninety-five patterns from *Shin hiinakata kyō-kosode* (*New Sample Designs, Kyoto-Style Kimono*), 1770. Reproduced from Yamanobe (1974d). Courtesy of Gakusha-Kennkyusya.

to determine. Shinto shrines such as Inari and Inafuri, however, are associated with myths concerning the origin of rice. As these shrines established branches throughout Japan, they took with them their crests incorporating rice motifs.

Of the rice-related kimono patterns illustrated in this essay, only figure 28.2 has a design resembling a family crest. Each rice bundle depicted is enclosed in a circle and the lightning is rendered in geometric patterns. Given that this kimono design was created in 1686 and that no later patterns in the sampling at hand share these characteristics, it appears that kimono pattern makers began to avoid such stylized symbols around 1700. After that time, patterns seem to have depicted more realistic subject matter, at least for a time.

In the decades preceding 1700, rice-related patterns were generally simple and the combination of rice images and other objects was based on folklore or drawn from the artist's imagination. Patterns in the first two decades of the eighteenth century, however, depicted realistic landscapes that highlighted auspicious plants, such as pine trees and bamboo. As drying rice appears to have been a major theme during this period, sparrows—regularly observed in the rice fields, especially at harvesttime—became a standard element in such patterns. While realism characterized designs created throughout the eighteenth century, a new pattern appeared in its last four decades. This was associated with celebratory occasions—a treasure boat and a phoenix or auspicious animals such as the crane or the turtle. The appearance of the stylized pattern associated with joyous occasions suggests that picturesque kimono patterns had come into use by the wealthy. This may be the reason why *hiinakata-bon* were not published after 1820, a time when ordinary people came to favor plaid patterns on their kimono (Ueno 1974B, 74).

The mythical image of rice was initially used in kimono patterns; then, the seasonal changes in rice production were given attention and carefully observed, especially the rice drying after harvest, which became a major theme of these patterns. Following the long, difficult six-month period of planting, tending, watering, and harvesting the rice came a peaceful, quiet time, a moment to appreciate an abundance of food, wealth, and happiness, as well as for festivities. Over time these realistic depictions of rice production lost favor with kimono pattern makers—and doubtless their clients as well—and they began to favor patterns suggestive of good fortune and employing images of trees considered to be auspicious. Later more direct, even decorative, expressions of celebration and good fortune were added to the repertoire. Some old styles were retained and new styles were continuously introduced.

Wrapping the Body with Rice-Related Images

Before trying to understand what it means to wrap the body with kimono patterns bearing rice-related imagery, it is useful to examine what types of rice-related images were not found attractive—and therefore not used by kimono pattern makers or worn by consumers. Rice as a food does not appear to have been used as a theme for kimono decoration. *Mochi* cakes, sake, or cooked rice, for example, seem to be absent from the repertoire of the *e-shi*, as do objects related to rice as a food, such as chopsticks, rice bowls, rice-cooking pots, and rice scoops. This fact, combined with the kinds of rice-related images used in kimono patterns, suggests that happiness, good luck, and wealth were most frequently intended to be communicated through the use of such patterns. Rice storage bags are symbolically equated with gold coins; therefore, the treasure boats carry the bags in place of money. An abundance of rice bundles is symbolic of good fortune; a big harvest was considered to be good luck bestowed upon the farmers, and consequently the townspeople, by nature. When the harvest was large, farmers were able to feed their families, pay taxes with the bags of rice, and make rice cakes and sake for offerings to gods and for their own festivities.

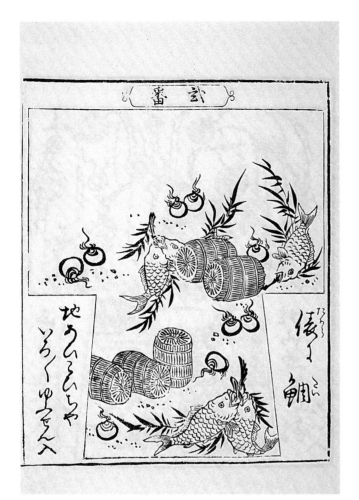

28.7   Five straw rice bags, five fish, and nine Buddhist jewel motifs appear in this pattern. Carp,[2] believed to swim up steep waterfalls, symbolize strength of body and mind. Large carp streamers are raised to celebrate Boy's Day, a traditional seasonal festival that is still observed on May 5th. The fish and rice seem ready to be cooked and are thus suggestive of a large happy gathering. One of ninety patterns from *Shin hinakata akebono zakura* (*New Sample Designs, Morning Cherry Blossoms*), 1781. Reproduced from Yamanobe (1974d). Courtesy of Gakusha-Kennkyusya.

28.8   This pattern was likely used for celebratory occasions as it contains many auspicious symbols: a phoenix, a pine tree, various decorative motifs in the middle, and a treasure boat filled with five straw rice bags near the bottom of the back of the kimono. The entire pattern seems designed to invite good luck. One of ninety-five patterns from *Shin hiinakata kyō-kosode* (*New Sample Designs, Kyoto-Style Kimono*), 1770. Reproduced from Yamanobe (1974d). Courtesy of Gakusha-Kennkyusya.

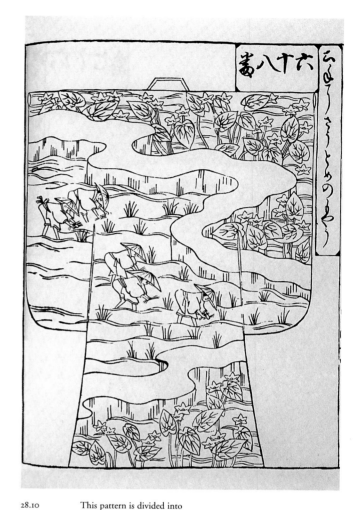

28.9   The kimono design shown above is replete with items associated with celebratory occasions, as is the pattern in figure 28.8. Unlike that pattern, however, the central theme here is a treasure boat (*takarabune*) full of straw rice bags and accompanied by two auspicious animals—a crane and a turtle. One of eighty-six patterns from *Shin hiinakata chitose sode* (*New Sample Designs, Millennium Sleeves*), 1800. Reproduced from Yamanobe (1974d). Courtesy of Gakusha-Kennkyusya.

28.10   This pattern is divided into two segments: in the central portion of the design, young women plant rice; on the top and bottom of the kimono, however, wild lotus flowers flourish. The human world is portrayed as filled with cultivated, staple plants, while in the natural world sacred plants thrive. A man-made path meanders between and separates the two realms. One of 114 patterns from *Tanzen hiinakata* (*Padded Kimono Sample Designs*), 1704. Reproduced from Yamanobe (1974b). Courtesy of Gakusha-Kennkyusya.

28.11   Seven wild geese fly over the
fields while two men swing
a bucket of water to irrigate
the growing rice. Pine trees
appear along the path at the
horizon line. One of 110
patterns from *Shin-zu kayō
hiinakata kōmoku* (*New
Edition, Flower and Sun,
Sample Designs Details*),
1708. Reproduced from
Yamanobe (1974b). Courtesy
of Gakusha-Kennkyusya.

28.12   A family crest (*mon*) with a rice pattern. Drawing after Kotani 1942.

It seems natural, therefore, that women wore *kosode* with rice-related images on those occasions when the participants wanted to invite the wealth, good luck, or happiness associated with a large harvest. It is curious, however, that urbanites would envelope themselves in scenes of rural life. It may have been that townspeople during the Edo period held some affection for rural existence, a feeling that might have been reinforced by a hierarchical system in which the peasant class was positioned above that of merchants. Following interpretations by Reizei et al. (1996, 130–31) that explain why drawings of rice cultivation on sliding doors (*fusuma*) and folding screens (*byobu*) were favored during the Edo period, it is likely that this feeling was not completely romantic in nature but rather reflective, part of a rethinking of the original foundation of the nation and a belief that governance should be based on rice cultivation. Increasingly wealthy and independent merchants and craftsmen bolstered their growing occupational pride with images of rice cultivation. These images permitted them to experience the presence of deities who resided far from the towns where they made their livings. It is important to note here that the scenes depicted on kimono patterns portrayed aspects of rural existence, but not peasants themselves, who were rarely represented. Having said this, it should be added that urban dwellers had long used romanticized images of the scenic rural countryside in other forms of artistic expression, such as poems, haiku, and *tanka* (a poetic form longer than a haiku).

As rice-related patterns were found almost exclusively on women's garments, it is pertinent to examine their relationship to the female body image. It is also interesting to consider whether any other tradition exists of symbolically connecting female fertility with the rice cycle. Stone artifacts in the form of human genitalia that are placed beside paths (though not necessarily paths to rice fields) are believed to symbolize agricultural success as well as human prosperity. These, however, take the form of male as well as of female genitalia. It seems, therefore, that few visible associations between rice production and female fertility exist.

Women customarily wore *kosode* in spring and autumn. Autumn, when the land proved especially fertile, was the season most frequently depicted in rice-related kimono patterns. This suggests that women wearing these *kosode* were supposed to, expected to, or hoped to be fertile. As is well known, female infertility has long been an issue of grave concern in Japanese society because a woman who was incapable of reproducing could not provide a male heir for her husband's family.

Such associations between production and reproduction and between the land and the female body seem to have been overtaken, however, by the more dominant symbolism of wealth and happiness embodied in the image of the treasure boat in the late eighteenth century. Perhaps these notions of accumulation and felicity were conflated with that of female fecundity. Rice-related patterns began to disappear from kimono patterns by the time the last surviving sample design book was published in 1820. Until then, however, they had been consistently employed over a span of 140 years. Furthermore, the theme and content of these patterns changed over time implying that rice had multiple symbolic meanings that varied in response to the outlook of the middle-class urbanites who wore them. Within the context of Edo period Japan, rice was such a source of symbolic power that it could provide a continuous source of artistic inpiration to the artists of the *hiinakata-bon*.

•

㑊介醋不快故欄諺而愈其疾也

又云孫光憲家婢抱小兒不覺落炭火上便以醋泥傅之無痕

子母秘錄治姙娠月未足胎死不出醋煑大豆服三升死兒立便分解如未下再服又云醋二升格口灌之

丹房鏡源米醋煑四黃化諸藥丹砂膽礬味蜀本酢酒有數種此米醉也

衍義曰醋酒糟為之乞鄰者是此物然有米醋麥醋棗醋米醋最釅入藥多用穀氣全也故勝糟醋產婦房中常得醋氣則為佳酸益血也磨雄黃塗蜂蠆亦取其收而不散也今人食酸則齒軟謂其水生木水氣弱木氣盛故如是造靴皮須得此而紋纖故知其性收斂不負酸以之說

稻米

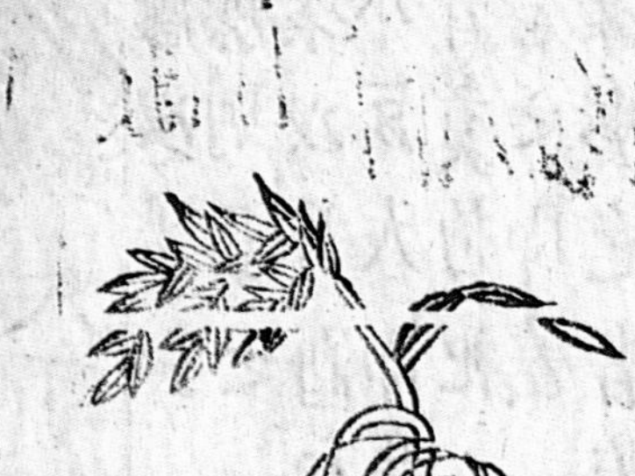

29

Images of Rice in Imperial Chinese Culture

Francesca Bray

Contrary to popular Western belief, rice is not the traditional staple food of all China. The heartlands of China span an area roughly half the size of the continental United States, divided into two broad climatic zones. The northern provinces have a continental climate, with fiercely cold, dry winters and hot summers marked by heavy storms. From the Huai River Valley south through the Yangtze region and down to Guangzhou, the climate is subtropical to tropical: winters are mild, summers hot and humid, and rains abundant and spread throughout the year. The domestication of crops took place during the same period in both the north and the south: by 5000 B.C.E. farming was well established in both regions. But while the Neolithic farmers of the south were growing rice in marshy depressions, northern farmers were growing millets, small-grained cereals suited to the dry climate.

The two zones maintained distinct agricultural and dietary traditions through the imperial era. By medieval times the population of rice eaters was beginning to outstrip that of the eaters of millet, wheat, and sorghum in the north, and by the mid-Song dynasty (960–1279) the rice produced in the southern provinces had become the lifeblood of the Chinese imperial economy (fig. 29.1). However the early dynasties that laid down the framework for China's cultural tradition had been based in the north, and despite its indisputable material importance, rice plays a minor role in Chinese symbolic imagery. While in other regions of Asia rice and its associated images figure prominently in folklore, in religion, in poetry, and in the symbolism of fertility and well-being, in China popular rice imagery, verbal and visual, is conspicuous by its near-total absence. There is no goddess of rice, no shrine, no spirit of rice residing in the seed grain. Nor is there even a material repertory of tools, containers, pots, or culinary techniques specifically associated with rice.[1]

29.1 A rice plant illustrated along with an accompanying text enumerating the medical properties of rice appeared in the *Zhenglei bencao* (26/3A), compiled by Tang Shenwei in 1108. Reproduced from Tang Shenwei (1600), courtesy of UCLA Richard C. Rudolph East Asian Library.

Why No Rice Gods?

Early China was a cultural and political mosaic.[2] People in the south spoke different languages from the northern tongue that was to become the official language of the Chinese empire; their political organizations differed from the northern states, and they worshiped different deities (Rawson 1996; Bagley 2001). The rice regions of the south and of what is now Sichuan generated rich cultures and powerful states. After the collapse of the northern dynasty of Zhou (founded circa 1050 B.C.E.), during the period known as the Warring States (475–221 B.C.E.), the southern states of Wu (the Yangtze Delta), Shu (Sichuan), and Chu (Hunan and the middle Yangtze) were among a dozen or so contenders for sovereignty

籍田圖

天子親耕

三公

甸師

卷七

四

29.2   In this depiction of the New Year imperial plowing ceremony, the emperor is shown holding the strut of a plow at the left. This illustration originally appeared in the *Nongshu* (7/4A-B) by Wang Zhen, published in 1313. Reproduced from Wang Zhen (1969) with permission of Yee Wen Publishing Co., Ltd., courtesy of UCLA Richard C. Rudolph East Asian Library.

over the Central States, as China was then known. By about 250 B.C.E. Qin in the north and Chu in the south had devoured all the surrounding states and were locked in a final battle for supremacy. In 225 B.C.E. the armies of Qin vanquished the forces of Chu, and in 221 B.C.E. the first unified Chinese empire was established.

The first Qin emperor not only conquered the southern states, but standardized government, laws, weights, measures, language, and script throughout his dominions. The Qin rulers were extravagant and cruel and were overthrown in 206 B.C.E. by the founder of the Han dynasty, which lasted until 220 C.E. The Han rulers took imperial standardization and cultural unification to a still higher level. They chose Confucianism as the state creed and sponsored the compilation, critical editing, and distribution of official histories and of the ancient Confucian classics. These works formed the core curriculum for the scholars who flocked from around the realm to the imperial university in the capital, hoping that their education would gain them a place in the new bureaucratic administration.[3]

History is written by the victors. Official histories traced the lineage of the Han dynasty, and of Chinese civilization as it was then conceived, back to an ancient past firmly located in the north, where millets were the staff of life. As the Chinese state pushed its boundaries southward, first in the Han and then in the Tang dynasty (618–905), the encounter between north and south was one of conquest and subjugation and was most frequently represented as advanced northerners bringing civilization to savages (Schafer 1967; Diamond 1988). Such images have died hard. In the 1970s many Chinese archaeologists still believed that farming had been introduced to southern China, and rice domesticated there, by migrants from the north with more advanced cultures (Ho 1975).

Following the unification of the empire in 221 B.C.E., popular culture in China was subjected to a high level of official control. In religious terms, the doctrine of Confucianism was rooted in the cult of ancestral spirits within the family, and in the cult of heaven and of the gods of soil and grain at the official level. The emperor was the Son of Heaven and performed religious rituals on behalf of the whole nation; magistrates performed analogous ceremonies in the local Confucian temples on behalf of the people in their charge. Taoism and Buddhism were officially approved, though carefully controlled. The beliefs and practices of the "Three Schools" of Confucianism, Taoism, and Buddhism were seen as complementary, not exclusive. The gentleman-scholar, for example, was commonly said to be a Confucian in his office, a Taoist in his garden, and a Buddhist when contemplating death.

Beyond the Three Schools, the state as well as the educated elite who dominated local society regularly conducted campaigns to eliminate "superstitious" folk beliefs and "unorthodox" practices. As the Chinese state consolidated its hold on the rice-growing regions beyond the northern heartlands, it faced the problem of incorporating not only non-Chinese groups like the Miao or Yao but also local Chinese populations whose regional culture differed in significant respects from Confucian orthodoxy. The usual approach was a combination of education and accommodation. Village schools would be set up, and magistrates would give public readings from Confucian moral tracts. Gentry and officials would officiate at weddings and funerals and take the opportunity to introduce the local population to orthodox liturgy. Local gods or spirits would be declared avatars of some Confucian deity, so that the local cult could be assimilated into a national network of the acceptable forms of religion. These processes were to some extent dialectical: there are certainly instances in which the assimilation of new cultural traditions modulated the nature of orthodox "Chineseness." But the overwhelming weight was on the side of officially construed orthodoxy, and that orthodoxy was rooted in northern tradition (Johnson et al. 1985).

29.3   A polder surrounds a block of rice fields; drainage channels run through the middle; and houses nestle among the willow trees planted along the high surrounding dykes. This illustration from E'ertai's *Shoushi tongkao* of 1742 (14/5 B) was based on one in Wang Zhen's *Nongshu,* first published in 1313. Reproduced from Wang Heyan (1744), courtesy of the Louise M. Darling Biomedical Library.

In a society in which even peasant farmers value antiquity and learning, it is not surprising to find that the main elements in the cultural repertory were usually expressed in archaic terms. For example, two key sources from which the Chinese traditional lexicon of poetic and ritual imagery was drawn, *The Book of Songs* and *The Book of Rites,* are both products of northern culture. These two works were supposedly written in the early Zhou dynasty (circa 1050–circa 770 B.C.E.) and edited by Confucius in the sixth century B.C.E.; in the Han dynasty (206 B.C.E.–220 C.E.) they were annotated by leading scholars and incorporated into the canon of texts that all candidates for office needed to study. When these authoritative and widely known texts mention bountiful harvests or the food prepared for ritual offerings, they speak of millets and not of rice. Through the imperial era, in order to secure a good harvest, on the first day of spring the emperor plowed a symbolic furrow at the altar of the gods of soil and grain outside the capital (fig. 29.2), and magistrates throughout the land made sacrifices at the local altar to the same deities. But these altars were a tradition dating back to the Zhou dynasty, whose founding ancestor was Lord Millet. So although in south China these fertility rituals were conducted for farmers about to sow their rice, the name of the god of grain was the same throughout China, and his name was not Lord Rice but Lord Millet.

The Rice Landscape

Even two thousand years ago there was a clear popular image of what we might call a distinctive southern rice landscape, an image associated with abundance and wealth. Although the terms *gu* (grain in the husk), *mi* (milled grain), and *fan* (cooked grain) are ambivalent, referring to different grains in the mouths of northerners and southerners, Han northerners referred to the south as "the land of rice and fish," and by the Song dynasty people regularly contrasted the dry-crop landscapes of the north (*di*) with the irrigated rice-growing landscapes of the south (*tian*). Associated in people's minds with these simple terms for field types were a range of demographic, social, economic, and political characteristics that contributed to a progressive divergence between backward north and dynamic south.

In recent decades archaeology has started to rewrite almost every aspect of the early history of China. Excavations have copiously documented the achievements of early southern cultures and their important contributions to—or differences from— "Chineseness." The theory that plant domestication and farming had been introduced to the southern territories by more civilized northerners had to be reconsidered when, in 1976, extensive remains of domesticated rice were found at a site called Hemudu in the Yangtze Delta. These dated back to 5000 B.C.E., just as early as the millet remains from the first farming villages of the north. There is now no doubt that rice has been the staple grain in China's southern regions for thousands of years. Recent archaeological finds suggest that domesticated rice may have been grown as long ago as 8000 B.C.E. along the middle reaches of the Yangtze, and numerous early sites containing rice remains have been found along the Yangtze from the coast up into Sichuan, along the Huai River, and down the southeastern coast—in other words, throughout the region that is still growing rice today (Wang and Sun 1996; An 1999). Remains of domesticated water buffalo, cattle, dogs, chickens, and pigs have also been found in these sites. Unlike northern villagers, who built the foundations of their houses in shallow pits in the soil or on low platforms of rammed earth, the southern populations built their houses on high stilts safe from floods, like modern Southeast Asians, and continued to do so until well into the historical period.

The rice-growing regions made a number of contributions to the early formation of a mainstream Chinese cultural tradition. By the late Neolithic, in about 2500 B.C.E., the Liangzhu culture of the lower Yangtze had perfected the art of carving elegant jade disks and cylinders (Ebrey 1996, 19). These jade forms were adopted as symbols of rulership by the kings of the northern dynasties of Shang (circa 1550–circa 1050 B.C.E.) and Zhou. And the technology of iron casting, which played a key role in the foundation of imperial rule, was first developed some time around the fifth century B.C.E. in the "barbarian" southern states of Wu and Chu (Wagner 1996).

The splendors of several southern kingdoms of the Warring States period, and their profound differences from northern culture, have been revealed through recently excavated tombs. Chu, for example, had its own written script, but when the Qin and Han rulers imposed northern language and script throughout the empire, and a unified history of China was composed from a northern perspective, much valuable information about the southern traditions was lost. In the absence of any historical documents, the splendid carvings and paintings of religious figures, for example, remain mysterious, and if rice deities or spirits were a central part of the culture, they have yet to be identified.

An ancient southern poem "Summoning the Soul," written by a poet of Chu, mentions rice dishes among the lavish delicacies offered to the dead:

29.4   A large sluice crosses a river, with irrigation channels leading off into paddy fields. This is the kind of distribution system developed at Dujiangyan in present-day Sichuan in the third century B.C.E. This illustration originally appeared in the *Nongshu* (19/2A-B) by Wang Zhen, published in 1313. Reproduced from Wang Zhen (1969) with permission of Yee Wen Publishing Co., Ltd., courtesy of UCLA Richard C. Rudolph East Asian Library.

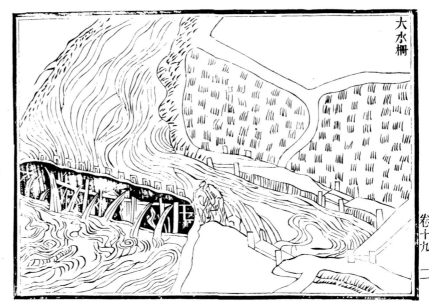

29.5   This chain-pump is operated by two men. The illustration originally appeared in the *Nongshu* (19/7A-B) by Wang Zhen, published 1313. Reproduced from Wang Zhen (1969) with permission of Yee Wen Publishing Co., Ltd., courtesy of UCLA Richard C. Rudolph East Asian Library.

水轉飜車

卷十九 十

29.6 Waterwheels could be used
to drive chain-pumps for
irrigation. They were also
harnessed to millstones, the
bellows for metal foundries,
and multiple trip-hammers
for fulling cloth or mixing
clay for potteries. This illustration originally appeared
in the *Nongshu* (19/11A-B) by
Wang Zhen, published 1313.
Reproduced from Wang
Zhen (1969) with permission
of Yee Wen Publishing Co.,
Ltd., courtesy of UCLA
Richard C. Rudolph East
Asian Library.

O soul, come back! Why should you go far away?
All your household have come to do you honour, all kinds of good food are
ready:
Rice, broom-corn, early wheat...
Geese cooked in sour sauce, casseroled duck, dried flesh of the great crane;
Braised chicken, seethed tortoise, high-seasoned, but not to spoil the taste,
Fried honey-cakes of rice-flour and malt-sugar sweetmeats...
[Hawkes 1959, 107]

One famous southern rice dish dates back to equally ancient times. The legendary death of the faithful minister Qu Yuan (circa 340–circa 278 B.C.E.), who threw himself into the Yangtze when he was unfairly dismissed by his prince, is still commemorated
today by dragon-boat races and by eating *zongzi*, delicious packages of rice mixed with
meat and lotus seeds, wrapped in lotus-leaf parcels, and steamed. Folklorists have argued
that, like the water-spraying festivals of modern Tai minorities in the southwest, this ceremony is in fact a fertility ritual to ensure a plentiful rice harvest (Eberhard 1968, 92ff.).

The material practices of rice cultivation in the late Warring States and early
empire are quite well documented.[4] Clay tomb models from Guangzhou and Guizhou,
dating to about 100 C.E., show that farmers dug small irrigation tanks adjoining their rice
fields in which they grew lotuses and water chestnuts and raised fish and turtles. The
models often show the rice plants in neat rows, suggesting that transplanting was already
practiced. By digging tanks, farmers were able to extend the rice-growing area beyond natural marshes to fields built on slopes or in the hills (Bray 1984, 110).

29.7   Four men operate a chain-pump, as pictured in a late thirteenth-century version of the *Gengzhi tu* by the early Yuan artist Cheng Qi. On the right, a man can be seen raising water with a bucket attached to a weighted pole. Freer Gallery of Art, Smithsonian Institution, Washington, D.C.; Purchase, F1954.21.

Natural marshes, on the other hand, were prone to flooding. As early as the Warring States period, large tracts of floodlands in river deltas were being reclaimed for rice cultivation in the kingdoms of Wu and Yue, in the Yangtze Delta and along the southeast coast. First a high dike (polder) was built encircling the land, and then bunds were used to subdivide the land inside into small fields; the channels (fig. 29.3) that ran between the fields and were linked to the river by sluices in the dike served for irrigation or for drainage, depending on the need (Bray 1984, 115–19) .

Meanwhile, two thousand kilometers inland in Sichuan, between 250 and 230 B.C.E., the provincial governer Li Bing and his son Li Ergang constructed a magnificent irrigation scheme based on contour canals at Dujiangyan, where the feeder streams of the upper Yangtze tumble down from the Tibetan foothills into the fertile Chengdu Plain (fig. 29.4). The system, which irrigates an area of about sixty-five by eighty kilometers of rice land is still functioning today (Needham et al. 1971, 288).

These different types of fields and water control allowed the spread of rice farming in subsequent centuries into terrains that were not naturally suitable for rice. As well as storing water in tanks, farmers created gravity-fed irrigation systems by diverting water from streams or springs. Sometimes it simply flowed along ditches; sometimes hollowed bamboos, laid end to end, were used to carry water across valleys or slopes to fields a considerable distance away. Water-raising equipment was necessary to make full use of poldered fields or fields that lay above the natural water source. The famous chain-pump, often called the "dragon's backbone," was probably first used in Han times, and by the Song period it was ubiquitous in the south (figs. 29.5, 29.7). These pumps were easy to dismount and move; they could be pedaled by one or more people, or driven by an ox. Waterwheels (fig. 29.6) were much more expensive than foot-powered chain-pumps and could not be moved, but they raised greater quantities of water and were driven by the flow of the stream itself (Needham and Wang 1965, 330–62) .

If asked to describe a typical Chinese landscape, most Westerners would probably mention terraced fields, yet terraces seem to have been a rather recent development in China proper, though they may date back millennia in parts of Southeast Asia. Terraces are first mentioned in Chinese documents by generals or officials sent to Yunnan, Sichuan, and Jiangxi in the eighth and ninth centuries c.e. (Bray 1984, 125–26). However, as land became scarce in plains and river valleys, settlers opened up terraced fields all over the south. In his *Nongshu* (Agricultural treatise) published in 1313, the agronomist Wang Zhen writes:

> There are mountains where the slope is extremely steep, without even a foothold, at the very limits of cultivation where men creep upwards bent close to the ground. There they pile up the soil like ants, prepare the ground for sowing with hoes [because the fields are too narrow to plow], and step carefully while they weed for fear of the chasm at their side. Such fields mount like the rungs of a ladder. If there is a source of water above the field, then all types of rice may be grown… [Wang Zhen 1969, 7/20B]

Wet-rice fields require considerable labor to construct, and several years of cultivation, before they reach the peak of fertility. However, once they are established, they can be kept at high levels of productivity with relatively small inputs of manure or other fertilizer (fig. 29.8), and soil erosion is less of a problem than with dry farming (Geertz 1963, 32; Bray 1986, 28ff.). If rice is first sown in a seedbed and then transplanted into the main field, yields are raised and the amount of time the crop occupies the field is reduced. This opens the opportunity for growing a second or even a third crop in the same field. From late Song times a common combination in the lower Yangtze was summer rice and winter wheat. In the uplands along the Guangdong-Vietnam border in the seventeenth century, farmers grew two crops of rice (exporting the surplus downriver to Guangzhou) and one crop of oilseed, turmeric, or sweet potatoes. When the rice was transplanted, it was common to put fish eggs in the field. The young fish ate insect larvae that would damage the rice plants, and as they grew they provided tasty dishes in their own right. Ducks were also given the run of the rice fields, eating pests and some of the fish and providing eggs in return. In the lower Yangtze the bunds between the rice fields were planted with mulberry trees, whose leaves were fed to silkworms, and the silkworm moltings and droppings provided rich fertilizer for the fields (Bray 1986, 131–34).

In other words, rice farming in southern China sustained densely populated and productive landscapes (Bray 2000, 24–41). The common saying "In the north, much land and few people; in the south, many people and little land" signified not only that the population was denser in the south but also that an acre of farmland in the south could feed many more people than an acre in the north. Northerners envied the fertility and wealth of the "lands of rice and fish" and migrated south in large numbers over the centuries. For a thousand years the governments of the Song, Yuan, Ming, and Qing dynasties were able to maintain their capitals and defend their borders in the north thanks to generous rice surpluses from the rich harvests of the south, brought north along the Grand Canal (see fig. 26.11). By 1403 about half a million tons of tax rice were being shipped annually along the Grand Canal to Beijing in special tax barges, built with watertight compartments and a capacity of 25 metric tons; by about 1600 the quantities had so increased that new barges were built to hold 45 tons (Brook 1998, 53). In 1607 the future grand secretary, Xu Guangqi, compared the Chinese empire to a father with one industrious son (the rice-growing south) and one idle son (the north); the father, complained Xu, simply relied on the industrious son to support them all and thus encouraged the wastrel son in his idleness (Übelhör 1969, 44–47).

雲　畦　心　滋　芋　力　坊　勤　沃　浥
　　稼　期　地　漑　分　由　重　水　水
　　如　千　利　庶　薙　用　轕　菅　土

浥陰

穢草聞吳兒灑灰傳自祖
田〻皆沃壤活〻流膏乳
塍頭烏喙泥谷口鳩喚雨
敢望稼如雲工夫盖如許

29.8   A man applies fertilizer to the rice seedlings in the nursery bed before transplanting. This illustration from a Qing dynasty edition of the *Gengzhi tu* was based on an album dating to the reign of the Yongzheng emperor (r. 1723–1735). The Seattle Art Museum, Eugene Fuller Memorial Collection.

29.9   This copy of Lou Chou's painting—from a late thirteenth-century version of the *Gengzhi tu* by the early Yuan artist Cheng Qi—shows women beating the rice panicles with sticks to thresh out the grain, while the men winnow away the chaff and carry the grain off to the granary in big baskets. Freer Gallery of Art, Smithsonian Institution, Washington, D.C.; Purchase, F1954.21.

Rice as a Political Symbol

The southern provinces began to rival the millet- and wheat-growing north in their tax contributions in around 800 C.E., yet the harvests of the north continued to be spoken of as the principal source of the nation's wealth. The Song dynasty's loss of the northern provinces to the Tatars in 1126 precipitated a crisis that permanently transformed perceptions of the relative importance of the two economies. The government, the propertied elite, and huge numbers of northern peasants fled south, and suddenly the southern territories were no longer remote provinces—they had become the heartlands of the Chinese state. Not only were these southern rice landscapes coaxed into accommodating the newcomers and yielding food for an extra two million mouths, their abundance helped build the most magnificent capital city that China had ever seen (in what is now Hangzhou, known to Marco Polo as Kinsay), and their productivity powered several centuries of economic growth (Bray 2000).

The loss of the north was hardly unforeseen, and well before 1126 Song administrators had been encouraging the development and dissemination of improved rice technology in the southern provinces. One of the most famous measures was the introduction to the Yangtze Delta in 1012 of new varieties of quick-ripening rices from Champa in Vietnam (Bray 1984, 492–95). This transformed production patterns, allowing two crops of grain to be grown in a year. Seeds of the new varieties were distributed to farmers through the district *yamen* and written instructions on their cultivation methods were circulated to magistrates and to literate "master farmers," chosen for their skill and experience to fill a minor official post that included the duty of improving farming techniques in their village. They were to instruct their peers not only in new techniques such as improved sowing and fertilizing methods or crop choices but also in the organization of mutual aid and so on. Agricultural treatises were printed by the Song government and circulated, and the state also provided infrastructural support such as new irrigation networks, as well as low-interest loans, lower taxation, and tax rebates for farmers who opened up new rice land. The effects of these policies were so dramatic that the historian Mark Elvin referred to them as the "Song Green Revolution" (Elvin 1973).

29.10   This illustration shows the plowing of the rice field. The Qianlong emperor (r. 1736–1796) had this version of the *Gengzhi tu* inscribed on stone tablets in the gardens of the Summer Palace outside Beijing, probably in about 1753. The pictures include new poems in the calligraphy of the emperor himself. Photograph © Ph. Sebert, Paris. Collection of Alan Kennedy and Kokoro Oriental Art.

It was in the Southern Song period (1127–1279), after the loss of the northern provinces, that the virtues and skills of rice farming became prominent icons of a quintessentially Chinese social order. The Song government took the local landscape to its bosom and gave rice a prominent symbolic role in official discourse. The new political cult of the rice economy was crystallized in a set of forty-five scroll paintings of rice farming and silk production, entitled *Gengzhi tu* (Agriculture and sericulture illustrated; figs. 29.7, 29.8–29.10 and see also fig. 2.40), painted by an official named Lou Chou (1090–1162), when he was the magistrate of Yuqian County in the lower Yangtze region. An imperial envoy learned about Lou's paintings when he was on a tour of inspection and informed the emperor. In 1144 Lou was summoned to court to present the pictures to the emperor, Gaozong (r. 1127–1163), who was so delighted with them that he ordered them to be copied onto a screen in the inner court of the palace (Wang Chaosheng 1995, 33–36).

Earlier Song emperors had commissioned frescoes of sericulture for the palace in the capital at Kaifeng (now lost to the Tatars), and Gaozong himself had previously declared that in his opinion "rearing silkworms should be done in the imperial palace, so that all will know the hardships of farm work" (cited in Wang Chaosheng 1995, 33, 35). Since the early empire, the taxation system had required peasant families to pay their dues in grain and in cloth—men and women both contributed through their labor to the state economy. As the common saying went, "Men plow, women weave" (*nangeng nüzhi*): every family needs food and clothes to survive, and in cosmological terms this fruitful complementarity of male and female labor mirrored the fecundity of yin-yang interactions. In terms of political reproduction, these products fed and clothed the state. Tax grain was redistributed as official wages and army provisions. Tax cloth (which could be coarse hemp or simple silk, depending on the locality) was used for uniforms and official robes and also served as a form of currency; silk thread was levied for the imperial manufactures that made sumptuous gauzes and brocades for imperial wear and for presentation as gifts to foreign powers (Bray 1997, 183ff.).

When the Song government fled south, the key silk-producing region of Shandong was lost, and because so much silk was needed to establish and defend the new state, and to buy off the enemies threatening the northern borders, great efforts went into building up silk production in the south (Kuhn 1987, 351ff.). It turned out that the region around the new capital Hangzhou and along the lower Yangtze was ideally suited for silk-worms as well as rice. Mulberry trees were planted densely along the edges of the rice fields and on sloping land, and this lower Yangtze landscape, so richly productive of grain and cloth, sustained rapid economic growth while maintaining the gendered division of labor that was the foundation stone of the Chinese political and cosmological order.

Lou Chou's elegant pictures and poems not only provided accurate technical illustrations of all the processes of growing rice and making silk, which Lou and his official colleagues felt would be useful for disseminating the knowledge of advanced practices to regions where techniques were less developed. They also movingly portrayed the sweat, dedication, and sacrifice that went into producing the rice and silk with which peasants paid their taxes. The paintings depicted the village roots of imperial wealth and power; the poems reminded the ruling class that these goods were produced at a cost and should not be abused. In other words, the scrolls depicted an ideal social contract in which the peasantry toiled to produce the rice and silk upon which the imperial government depended, and the emperor and his officials—duly grateful for the sacrifices entailed—devoted themselves to promoting the welfare of the people.

29.11   The harvest feast, illustrated by a 1593 woodblock print in the *Bianmin tuzuan* (Collection of pictures for the convenience of the people), a household encyclopedia that begins with its own, simplified versions of the *Gengzhi tu* pictures and poems. Reproduced from Guang Fan (1959, 1/8 A), courtesy of the UCLA Richard C. Rudolph East Asian Library.

29.12   This illustration of a boy on a water buffalo originally appeared in the *Nongshu* (13/7A) by Wang Zhen, published 1313. Reproduced from Wang Zhen (1969) with permission of Yee Wen Publishing Co., Ltd., courtesy of UCLA Richard C. Rudolph East Asian Library.

The *Gengzhi tu* immediately acquired iconic status. In 1210 Lou's grandsons had the original paintings copied onto stone so they could be passed down to posterity. Even so, both the originals and the Song copies were lost long ago. Yet today we still have quite clear knowledge of the original *Gengzhi tu*, for it was reproduced in many forms and versions over the centuries (figs. 29.10, 29.11), ranging from elegantly colored paintings on silk, to imperial woodcut editions with new poems added by the commissioning emperor, to rough and freely rendered woodblock versions in popular encyclopedias (Wang Chaosheng 1995; Kuhn 1976). Its circulation was not restricted to regions where rice was grown or sericulture practiced, for it served (whatever historians of technology may say) as an exemplar rather than a technical handbook. Indeed we might note that its representations of technical processes and actors hardly changed, though in reality the techniques of agriculture and sericulture continued to develop over the centuries.

The imagery of the *Gengzhi tu,* and versions of its poems, were so widely disseminated as to become familiar at most levels of Chinese society, like other expressions of neo-Confucian ethics and values. Here we should note that the rice landscape so vividly rendered in the *Gengzhi tu* never entered the repertory of what we refer to as Chinese landscape painting. These paintings by literati commonly took as their themes towering mountains, river gorges, and remote lakes, peopled only by lone fishermen, traveling scholars, and Buddhist hermits; the motif of the small boy perched on the back of a water buffalo and playing a flute as they wander together through the lonely valleys symbolized spiritual tranquillity, and so this rustic scene did occasionally enter the elite aesthetic repertoire as well as the farming manuals (fig. 29.12). However, though the Chinese literati were well aware that their wealth was rooted in the farmland that they owned, they produced no equivalent of the European genre that depicts a landscape of fertile farmlands (with or without the owner and his wife proudly seated under an oak; Clunas 1996). When Lou Chou painted his scrolls, it was in his working role as magistrate, not in his leisured role as aesthete. Furthermore what he portrayed were the technical and social features of rice farming, not its folklore (though that must surely have existed). The *Gengzhi tu* is in many respects the richest single source we have on rice in imperial Chinese culture, yet its significance is not as a repository of peasant traditions or popular beliefs, but rather as a model of an ideal society, as a core symbolic element in the formation of a political tradition that evolved and endured over almost a thousand years.

•

Part Eight

The Future of Rice

30.1   A freezer vault at the
International Rice
Genebank, Los Baños,
Philippines. Photograph
by Ariel Javellana,
courtesy of IRRI.

30

The Future of Rice

Roy W. Hamilton

Anyone who grew up in America in the third quarter of the twentieth century will recall the frequent reminders from parents or teachers not to waste food because people were starving…in China…in India…in Bangladesh…somewhere, usually, in Asia. While the causes of famine are complex and typically involve bad leadership and inadequate infrastructure, the root cause of the problem in this instance was that rice production was simply failing to keep up with Asia's burgeoning population. In 1966 the International Rice Research Institute (IRRI), headquartered in Los Baños, Philippines, released IR8, the first modern "high-yield" variety of rice for irrigated tropical lowlands. The new plant was shorter than most traditional varieties of rice, making it less susceptible to weather damage. It matured more quickly, producing its grain in four and one-half months rather than the five to seven months required for most traditional varieties. And it bore a heavy load of grain, making up a much greater proportion of the plant's weight than in the traditional varieties.

The high-yield rice, together with similar improvements in wheat and corn, launched what quickly became known as the "green revolution." Using IR8 and other even faster maturing high-yield varieties released by IRRI in the ensuing years, together with applications of fertilizer and pesticides, farmers in all of Asia's rice-producing countries increased their yields. In many cases the increases were dramatic, as the shorter maturation time allowed double cropping in areas that had previously produced only one crop per year, and triple cropping in areas that had previously produced two.

The green revolution spread quickly. In Bali, for example, by 1974, 48 percent of the rice fields were planted with the new varieties, and by 1977 this had risen to 70 percent (Lansing 1991, 112). China, India, and Indonesia, the nations with the largest populations and the toughest demographic problems, reached self-sufficiency and even became net exporters of rice. These remarkable achievements, together with improvements in infrastructure that allowed food to be delivered where it was needed, have for the most part eliminated large-scale famine in Asia's rice-producing countries.[1] Rice yields per acre in Southeast Asia were 83 percent higher in the period from 1991 to 1993 than they had been from 1964 to 1966 (Paterno-Locsin 1999, 120). Today approximately 90 percent of the total rice harvest in Asia comes from modern high-yield varieties.

The green revolution had barely started, however, before it began to attract criticism. Some of the problems were environmental. IR8 proved to be highly susceptible to insect pests. Many areas experienced increased problems with rats. More ominously, the pesticides and fertilizers needed for high-yield crops, if improperly stored or used, damaged rice lands and caused health problems for farmers. Rice researchers have been relatively quick to grasp the scientific problems and work toward solutions. Scientists, both at IRRI and at the national-level rice research facilities that now exist in all major rice-producing countries, regularly experiment to develop-pest resistant varieties, for example, or new field management procedures. But economic and cultural problems, which also soon began to be reported, have proved to be more intractable. Farmers were sometimes treated arrogantly by government agricultural agents or found that they were losing some of their autonomy when making decisions. The new technology was often accompanied by a shift from subsistence production (for a family's own use) to commodity production (for sale in the cash economy), which worked to benefit some people but not others:

> The economic changes following the spread of the green revolution have clearly favored the farmers already in an advantageous position, i.e. those with some capital at their disposal and with sufficient landholdings for credit collateral. [T]he landless and near-landless peasants have been negatively affected via large-scale elimination of agricultural employment opportunities. [Antlöv 1986, 166]

The reorganization of harvest procedures that accompanied commoditization in Java provides an example of how the green revolution in one case worked to the detriment of some elements of society (Antlöv 1986). The traditional Javanese form of harvest, known as *bawon*, was open to all. The rice stalks were carefully cut using a finger knife (*ani-ani*, see figs. 2.27–2.32), a tool women in particular were skilled at using, in order to avoid "hurting" the rice and thereby offending the Rice Goddess. The bundled sheaves of grain were then carried by men from the field to nearby village granaries. By simply showing up at a field where the harvest was taking place, and joining in the work, both men and women could earn a small portion of what they reaped in return for their labor. This system provided a crucial safety net for the landless and the poor. The new varieties of rice, which shatter easily from the stalk, are cut with a sickle, threshed and bagged directly in the field, and taken away by truck to commercial warehouses. The labor is organized by agents under contract with the landowner. The workers are mostly male and are paid a cash wage rather than a share of the crop. This new system has taken hold because it benefits everyone directly involved in it. The landowner's cash labor costs are less than the value of the portion of the crop he formerly had to pay. The labor agent gets a cut. The harvest workers make more in cash than they formerly made as a percentage of the crop. The losers, of course, are those who are no longer involved: the poorest and weakest, many among them women, who have lost their right to claim a share of the crop by showing up to work. The result is a falling living standard for some among the poorest segment of society.

Although changes of this sort have been closely linked to the widespread adoption of modern high-yield varieties of rice, the increasing commoditization of rice agriculture would surely have occurred with or without the new varieties of rice. The real issue is the process of change from subsistence farming to commercial enterprise driven by development and globalization. For this, plant breeders cannot be held responsible. Commoditization is not in fact a new challenge. Japanese rice agriculture gradually shifted from a subsistence and barter basis to a cash and commodity system over the course of Japan's medieval period (1185–1603). Much of the resistance that arose to this process had to do with the symbolic importance of rice for the Japanese people (Ohnuki-Tierney 1993, 69). Similar patterns can be found in developing countries today, where farmers sometimes produce mainly the new high-yield varieties of rice for cash sale but continue to produce a smaller quantity of their traditional indigenous varieties for their own domestic use or for ritual purposes.

Many other problems have also beset rice agriculture in recent decades. Ten cities in Asia's rice countries now have metropolitan populations of over ten million, and most of them sprawl over prime lowlands.[2] China alone has forty-five cities of over one million, and India thirty-eight. As the populations of Asia's cities surge, some of the best rice lands are gobbled up for residential and recreational uses (fig. 30.2), even while the total area planted to rice is at a historic high point due to widespread use of irrigation. Problems of land use and ownership are also not limited to the urban fringes. Conflicts have broken out even in sparsely settled swidden agriculture areas where temporary use rights were important but permanent private ownership of the land was not recognized. Modern administrators often disregard such traditional management practices and in the most abusive cases declare the land unowned, selling it to private interests over the objections of the communities that had managed it for generations. Some communities in Borneo, for example, have lost their communally held lands to timber or oil palm companies in this manner.

As levels of education rise across Asia, many children of farming families aspire to government or professional jobs while the backbreaking labor of rice farming seems less and less appealing. Pride in farming evaporates as production moves from independent subsistence to low-wage labor. Vietnam has risen in recent years to become the second largest exporter of rice (after Thailand), largely because farming is still the occupation of the majority, but this will be difficult to maintain in the future given Vietnam's spiraling population, quickening pace of development, and emphasis on education. In recognition of these sorts of problems, IRRI recently started a program in the Philippines to increase respect for rice farming through school programs that teach children how complex and engaging it can be to produce rice successfully. Still, it will be a major challenge in the future to produce more rice with less labor.

Economic stagnation and aging infrastructure also present problems in some rice-growing countries. Burma, which was the world's leading exporter of rice prior to World War II, dropped to fourteenth place by the late 1990s largely due to the government's failed economic management. In Sri Lanka's dry zone, where the system of thousands of small independent irrigation tanks was developed from ancient times, population growth by the 1980s had reduced the proportion of families that had access to tank-irrigated rice fields from 100 percent to 66 percent (Bryde 1986, 86). The traditional system of cooperative labor for maintaining the tanks broke down, and control over irrigation passed from the local community to appointed government officials whose authority was not recognized within the villages (Bryde 1986, 100–101). To make matters worse, the modern varieties of high-yield rice proved less tolerant of poor growing conditions. Many farmers could not adequately pamper the new varieties of rice plants, yet government policies resulted in the traditional varieties becoming unavailable. Many of the poor resorted to subsisting on millet.

30.2   *Golf Plan.* By Alfredo Esquillo Jr. (1972–). Philippines. 1997. Mixed media. W: 122.5 cm. Collection of Joan T. Simpson. The protective garb of the central figure identifies her as a rice harvest laborer. The artist asks us to consider the wisdom of converting rice fields to golf courses, which has been happening at an increasing rate on the outskirts of Asia's fast-growing cities.

The loss of traditional indigenous varieties has in fact been one of the hottest environmental issues for the past thirty years. Many rice-farming communities that converted completely to modern varieties found it deeply painful to lose their stock of traditional varieties, which they regarded as an inheritance from their ancestors. Additionally, the loss of diversity limits possibilities for future generations of plant breeders. To stem the damage, efforts are underway to collect seed samples of Asia's incredibly diverse rice varieties. Seed collecting became increasingly complex after the 1993 international Convention on Biological Diversity, which recognized that genetic resources are subject to sovereign rights and established procedures for sharing benefits derived from their use. Nevertheless, collecting of traditional varieties of rice has continued under the coordination of national governmental bodies. In Laos alone, more than thirteen thousand varieties were collected from 1995 to 1999, reflecting the great diversity of Lao rice environments, ranging from tropical floodplains to steep mountain slopes (IRRI 2000, 40). The country's dry hillside farms accounted for the most samples, confirming the importance of these highly localized agricultural systems, which have often been considered backward or environmentally damaging, for the genetic future of rice.

At IRRI, the International Rice Genebank now maintains samples of 110,000 varieties of rice from around the world (IRRI 2000, 42). For long-term storage, each sample, consisting of 120 grams of seed, is hermetically sealed into aluminum cans and stored in a vault kept at −20 degrees centigrade. These seeds are expected to remain viable for about one hundred years. Seeds of the same varieties are also kept more readily accessible in a shorter-term "active" storage bank at 4 degrees centigrade (fig. 30.1). Seeds from this bank are provided free of charge to any serious requester who agrees to honor the sovereign rights of the country from which the seed originally came by not seeking patents or other intellectual property rights and by sharing with the country of origin any profits arising from commercial exploitation. The viability of the seed in this bank is tested on a regular schedule, and the seeds are planted out and harvested as needed to keep a viable supply replenished.[3]

Some scientists have expressed a concern that increases in yield gained through traditional plant-breeding methods may be approaching their limits, although not all agree with this. One of the few relatively low-tech methods still promising greater yields is the use of hybrid seed, which must be produced and supplied anew to farmers each year, rather than saved from the previous season by the farmers themselves. Hybrid seeds yield 15 to 20 percent more than the best inbred varieties, and in China they now account for over half of the land planted to rice, but they probably will not be widely adaptable until a system is developed for farmers to produce their own hybrid seed. In the late 1990s, rice scientists began warning of complacency born of the success of the green revolution and raised the specter of famine returning to Asia. The constant pressure of intensification on favorable irrigated lands has been shown in some cases to degrade the health and fertility of rice fields, resulting in declining productivity, while technological breakthroughs for unfavorable rainfed lands have proved elusive (Pingali et al. 1997, 2).

Just at the time when traditional research seemed to be running out of options, however, new developments have overtaken the field of rice research at a stunning pace. In 1999, scientists at the Institute of Plant Sciences in Switzerland isolated two genes from a daffodil and a third from a bacterium and spliced them into a rice plant's genetic material. The result was "golden rice," rich in vitamin A, which offers hope for the two million children under age five who die annually around the world due to vitamin-A deficiency (Pingali et al. 1997, 10). IRRI is now working to breed the new characteristics into widely planted varieties.

In April 2000, the agro-biotech firm Monsanto announced that it had produced an incomplete draft of the entire sequence for the genome of rice. This is of tremendous potential importance because the genome sequence provides the starting point for systematically unlocking the secrets of individual genes, and ultimately for the genetic engineering of new crop varieties as well as other potential agricultural and ecological advances. A publicly funded consortium called the International Rice Genome Sequencing Project (IRGSP) had been working to decode the rice genome, but was not expecting results until 2008. Monsanto's announcement "shocked, then worried, but ultimately delighted" the rice research community (Normile and Pennisi 2002, 34). The concern was that biotech corporations might claim proprietary rights over the genome, which many people around the world see as the rightful heritage of the hundreds of generations of Asian rice farmers who have nurtured and developed the astonishing number of varieties of rice. Moreover, rice was proving to be an ideal research plant, with a relatively small genome (about fifty thousand genes composed of 430 million bases, as opposed to 16 billion bases for wheat), and much of it is held in common in the genome of other cereals, including wheat, barley, and maize. This means that rice genome research will produce key insights for understanding the genetics of other grains. As one scientist put it, "Wheat is rice" (Normile and Pennisi 2002, 32). The international community's fears were ultimately not realized because Monsanto agreed to provide its data to other researchers, and the IRGSP effort to produce a more complete and accurate genome sequence was helped along. In April 2002, the journal *Science* reported that two other efforts to produce more complete draft sequences of the rice genome had succeeded, one by a Chinese research institute and another by a private laboratory. These were still drafts, containing many gaps and errors, but in December 2002 IRGSP announced the completion of a more accurate sequence. Clearly rice research is moving forward at an unprecedented pace and has entered a new era. On another front, researchers announced in November 2002 that they had used a gene from the common bacterium *Escherichia coli* to create rice plants capable of producing trehalose, a sugar that helps plants withstand stress from drought, salinity, and cold (*Los Angeles Times* 2002, A16). Scientists now dream of varieties of rice that can produce well under such extreme conditions, or even perennial rice plants that yield grain year after year without replanting.

No matter how one feels about the negative aspects of the green revolution, the development and spread of high-yield varieties of rice over the past forty years must be credited with banishing the specter of mass starvation from Asia in our time, and that is a monumental accomplishment. Given the recent advances in plant genetics, it seems reasonable to hope that rice will continue to be the food of choice that feeds Asia's vast population in the future. Amid all of the environmental, economic, and scientific debates, however, what is rarely discussed is the cultural cost that has almost universally accompanied the transformation of rice growing from subsistence farming to commodity production and ultimately to agribusiness.

Lisa Klopfer, who conducted research in the Minangkabau region of Sumatra, where pervasive changes in rice agriculture took place in the 1970s, observed that the younger generation of rice farmers had adopted what was considered a "modern" attitude

30.3   *Lucky Monoleum.*
By Kim Chong-hon (1946–).
Korea. 1981. Oil on canvas.
H:90 cm. Seoul City
Museum of Art.

toward rice. In short, they treated it as nothing more than a market commodity, removing it from the web of closely interrelated cultural practices that had encompassed rice production in traditional Minangkabau society (1994, 164–65). Although her research was conducted in an Islamic community, the debate about modernity could have happened anywhere in Asia and is in no way unique to Islam.

> The shift [in attitudes toward rice] is associated with a number of other changes: the pressure to adhere to reformed, modernist Islam; the pressure to behave in a "modern" manner as prescribed by the government and government approved media; the realities of planting new high yield rice varieties that no longer allow for synchronized planting and harvesting; the increase in rice production and consequent lack of storage space and increase in the direct sale of rice right from the field. [Klopfer 1994, 163]

Belief in the rice spirit was seen by the government and religious authorities as "primitive" and associated with poverty. The change from a single synchronized annual crop to year-round multiple cropping destroyed any sense of a seasonal cycle, removing the rationale for annual ceremonies to open irrigation canals, prepare the fields, or cut the first grain. Conservative elderly people were shocked and dismayed by the lack of respect shown to the rice spirits by such modern procedures as the use of diesel-powered rice mills, or the sale of the crop directly from the field without first inviting the rice spirits properly home from the fields to the granaries (Klopfer 1994, 159–162). The beautiful Minangkabau granaries (see fig. 8.2) fell into disuse.

The shift in attitudes between the generations is widespread in many cultures, based on the new realities of commodity production as opposed to subsistence farming. Though there may be a sense of loss on the part of the older generation, the younger generation quickly adapts. Some Javanese farmers, for example, justified the change from finger knives to sickles by saying that the new varieties were foreign rice and therefore the Rice Goddess would not be offended by the use of the sickles (Collier 1973, 42). One farmer in 1983 incisively yet matter-of-factly summed up his experiences, which were recorded by John Pemberton (1994, 214). The traditional practice in his community was to braid the tallest of the newly harvested stalks of rice into a long-necked figure of the Rice Goddess Dewi Sri. The image of the goddess was mounted in the field so that as the harvest progressed, she could look out and survey the productivity of her lands. The farmer ran into difficulty the first time he attempted this ritual procedure with the much shorter stalks of the new high-yield varieties: "When we tried using the new rice stalks to braid Sri, her neck was missing. She came out looking like a dwarf, so we didn't bother with it any more" (Pemberton 1994, 214).

Some of the issues of cultural loss facing rice-farming communities throughout Asia are examined in the remaining chapters of this book. Chapter 31 surveys the ambivalent attitudes with which the Ifugao of the Philippines are facing the incipient loss of their rice rituals. Chapter 32 documents the decline of the Rice Goddess in one Javanese village, while chapter 33 presents the work of a puppet master who is trying to preserve her legacy.

30.4   *Mr. Rhu in Eunhang Dong.*
Lee Jonggu (1954–). Korea.
1991. Acrylic on paper. H:
138 cm. Collection of the
artist.

A look at the most urbanized and industrialized of the rice-producing countries, especially Japan and South Korea, offers some indications of what the role of rice agriculture may be in the future in other Asian nations. The very early shift from subsistence to commodity production in Japan is quite instructive. During the Edo period (1603–1868), the city of Edo (today Tokyo) is thought to have been the most populous in the world. The huge urban middle class that developed relatively quickly soon became removed from the rural routines of rice agriculture. Japanese art of this period shows a nostalgic romanticizing of all things rural and associated with rice, as witnessed by some of the woodblock prints in this book (see figs. 2.13, 2.21) and by the rice patterns that decorated the garments of urban women who were in no way involved in rice agriculture (see chapter 28). The rice landscape endured as an idealized Japanese landscape through the Meiji period (1868–1912) even as industrialization began to spread over the land. Ironically, the Japanese maintain their rice-based agricultural identity today even though the country is now highly industrialized and rice agriculture is only a tiny part of the total economy (Ohnuki-Tierney 1993, 81). While virtually no Japanese person today would seriously express a belief in the mountain deity *yama-no-kami* descending in the spring to become the rice field deity *ta-no-kami*, and few could explain the connections between rice agriculture and Japan's imperial system (see chapter 26), rice remains a potent symbol in everyday life in Japan. At tremendous cost, the domestic rice industry is protected from foreign imports in order to maintain rice agriculture as a viable part of the Japanese landscape and keep Japanese-grown rice in the supermarket. Rice is still considered the ideal food, though in actuality people eat less and less of it as time goes on. Rice cake offerings continue to be ubiquitous at New Year and the streets are full of stalls selling rice-straw decorations.

Industrialization had proceeded even more quickly in South Korea, where in the span of a generation or two the nation has been transformed from a largely agricultural society to an urban one. Chapter 34 discusses the sweeping social changes that have accompanied the modernization of the agricultural sector in Korea and captures the sense of nostalgia and loss that has resulted. Chapter 35 examines the changing meanings of key icons from Korea's agricultural past in contemporary society. The same idealization of the rural agricultural past that appeared in Japanese art in the Edo period can now be found in contemporary Korean painting (fig. 30.3). A painting by the artist Lee Jonggu, which conveys the ethos of an urban society looking back at its rural roots, is suffused with an overwhelming sense of nostalgia, inevitability, and loss (fig. 30.4). The old men left alone working in the rice fields, the luxury car whizzing along the dike.…blink again and they will be gone. Asians will continue to eat rice, but it is doubtful that in the future they will continue to fashion images of the Rice Goddess to survey her fields.

•

31

Let's Hope the Bile Is Good!

Aurora Ammayao with Gene Hettel

> Naam-ami ya naguyud di himbulan ta potlang;
> biliyon da'y pinanal;
> ya napyuk ya kimmali da:
> "Ay mabolnat di page."
>
> After a month and a half,
> they visit and inspect the rice seedlings,
> and the leaves are bent down and they say:
> "*Ay*, we transplant already the seedlings." [Dulawan 2003, ll. 487–90]

It seems like only yesterday that I last heard the chanting of the *Huuwa'n di Nabugbugan di Page*, the Ifugao account of the *Myth of the Origin of Rice* (as it has been translated; see Dulawan 2003). In my mind the voice of my father, Ammayao Dimmangna, rises above that of the other priests (*mumbaki*). This memory is particularly meaningful to me as I recall the life of my father, who passed away in April 2002. Nearly forty years earlier, in late June 1963, my father and his fellow *mumbaki* officiated at the annual Ingngilin, the rite (*baki*) for the rice harvest. They were in the granary of the richest or leading family (*tumoná*) in Tucbuban village of my home town (*barangay*) Amganad, nestled in the Cordillera Mountains of north central Luzon in the Philippines. Our family had no rice fields of its own, but my mother (Indanum Kinadduy), sisters, and I still lent a hand to our wealthier landowning neighbors in the various field activities throughout the Ifugao agricultural calendar. My father, a respected *mumbaki* at the time, officiated during many of the rituals associated with these activities (fig. 31.1).

> Ipiya da'y baya da ya mangin-innum da;
> painglayon da ya tobalon da'y
> nundomang ya nunhalug;
> hi nabugbugan di page.
>
> They pour the wine and they drink;
> after a while, they pray to and invoke
> their ancestors on both sides
> for the rite on the origin of rice. [Dulawan 2003, ll. 494–97]

I remember hearing my father chanting one portion of the myth in which the ancient Ifugao brothers Wigan[1] and Kabigat pour and drink rice wine (*baya*). They are

31.1

Mumbaki Ammayao Dimmangna drinks rice wine during the Ingngilin/Ani harvest ritual. The striped cloth, or *uloh an bayya-ong*, that he wears is part of the traditional garb of the *mumbaki*. Photograph by Mervyn Keeney, Amganad, Banaue, Ifugao, 1975.

celebrating after having transplanted—for the very first time—our esteemed Ifugao rice, which was obtained earlier in the story in trade for fire from the gods of the Skyworld. As a twelve-year-old girl, I disliked the recitation of this ancient folk epic as an excuse for drinking rice wine—which is exactly what my father, the other *mumbaki,* and all the men present were doing at this stage of the harvest rite. I thought they were having all the "fun" above in the shade of the granary—telling stories, arguing, and getting drunk. Below in the fields under an intense midmorning sun, we women and girls, perhaps thirty-five in number and ranging in age from eight to seventy-eight, were actually harvesting the rice—much of which would end up as more wine to drink.

Across the valley I could hear reverberations of the loud squeals of the unfortunate pig that was being ritually sacrificed by one of the officiating *mumbaki.* This meant that it would be another three hours until early afternoon, when my fellow field hands and I could participate—at least in a small way—in the ongoing rite underneath the granary on the hill above us. Then we would be able to have our lunch, consisting of only some of the less-than-choice pieces of pork. The *mumbaki* and their male assistants would be given the best pieces of meat—although as one of the officiating *mumbaki,* my father would also get to take home a few of the finest cuts for the rest of us to eat later.

It was during those days that I asked myself: what kind of culture is this? I saw no merit or purpose in preserving such traditions, not that there was any sign of them disappearing back then. Nearly forty years later, the situation has changed dramatically. Many aspects of Ifugao culture seem to be on the brink of oblivion. My late father had given up the ways of the *mumbaki* for Christianity in the mid-1980s. In contrast, I, with three children of my own, have come full circle and feel that all Ifugao will lose an essential part of themselves if their culture is permitted to slip away.

Rice Rituals—Relegated to the Fading Memories of Retired *Mumbaki*

At one time detailed rituals covered all important facets of Ifugao life. There were rituals for birth (Wa'lin or Bagol); engagement (Moma); marriage (Tanig); prestige (Baya or Bumayah), such as inheriting rice fields; sickness (Honga); an aging father's final blessing over his children (Yabyab); headhunting (Ngayo), in a bygone era; and death (Nate). The various rituals associated with the growing of rice tied to the agricultural calendar, were once considered the most important of all from Lukat to Kahiu (Barton 1946; Dulawan 2003; see box, pp. 453–54). Now it seems that some of the nonagricultural social rites (such as Honga and Tanig) are weathering the encroachment of modern society into our hinterlands better than most of the rice rites. The rice rituals, for the most part, have been relegated to the fading memories of retired *mumbaki* turned Christian, such as my father, or descriptions in musty old books written more than half a century ago.

Ifugao rice rituals evolved over centuries and were unfailingly performed throughout the year to protect the rice plants from pests and envious spirits, to ensure a good harvest, and to bless the rice. Traditional Ifugao deeply believe that these rites were taught to their ancestors by the gods, especially Dinipaan (the god of agriculture). The performance of these rites follows Dinipaan's teachings about the religious nature and significance of the culture of rice and is a means of honoring the covenant that the forebears of the Ifugao forged with the gods. The Ifugao once believed that neglecting to perform any or all of these rites would result in crop failure and/or grave illness in the family. Nowadays, however, few if any of these rites are observed on a regular basis; Kulpe/Kulpi and Ingngilin/Ani appear to be surviving the longest.

The names of these rituals and their content vary by location within Ifugao Province. The particular rites listed in the box on pages 453–54 are a composite of those

Mumbaki Buyuccan Udhuk sacrifices a pig during the Ingngilin/Ani harvest ritual. In the foreground, some of the first harvested rice bundles are also offered to the gods. The latter are represented by the *bulul* figure that is partially visible behind the raised hand of the participant at the left of the photograph. Photograph by Gene Hettel, Lugu, Banaue, Ifugao, 1995.

Representative Rice Rituals of the Ifugao

1. Lukat*
(October)—This rite accompanies the first weeding of the fields. With the sacrifice of a chicken, the *mumbaki* invoke the ancestral spirits and the deities to make sure that the field dikes are strong enough to hold the water so that the crop "abounds in life."

2. Pudung*
(October–November) Just before the seedbed is planted, *runo* grass stalks are stuck upright and set all around the bed within the field. With a chicken sacrifice, the *mumbaki* implore the ancestral spirits and the deities to allow the field to be continually flooded, so that the crop will be plentiful when the seasons change and pests (such as rats, mice, birds, and other crop destroyers) and diseases will be "ashamed" to come to the field.

3. Loka*/Lukya†
(October–November)—The *mumbaki* perform this rite, sacrificing three chickens, in a family's granary just before the start of the first working phase of the rice-growing season. This rite is the first one of the agricultural calendar in some districts and marks the first time that rice may be taken out of the granary for family consumption or for sale (Lukya) and as seed for planting (Loka). The *mumbaki* invokes the gods and ancestral spirits to bless the rice so that even a small amount will be enough to satisfy those who eat it.

4. Ugwid*/Hipngat†
(October–November)—The *mumbaki* perform this rite in the granary either before men begin spading the fields (Ugwid) or after the general field cleaning (Hipngat), when the fields are robust with vegetables planted on the mounds of decayed rice stalks and grasses. The purpose is to invoke the gods to bless the rice in the granary so that it will be safe, last a long time, and give the family strength and prosperity.

5. Panal†
(November)—This ceremony includes the sacrifice of four chickens in the granary. During the Panal (lit., "sowing") the *mumbaki* ask the gods to bless the rice seeds so that they will all sprout and grow into robust seedlings. The day after the ritual performance, a ceremonial period of idleness is observed. After this holy day of rest, the seed bundles called *binong-o* are taken to the seedbed (*panopnakan*). After being separated from the bundles, the panicles are carefully laid by hand on the mud four to six inches apart from one another in a vertical line. The first seed-laying in a given village must be done in the fields of the richest family (*tumoná*); other owners' fields are seeded later.

6. Bolnat*†
(November–December)—In this rite, performed in the granary immediately prior to transplanting the rice seedlings, the *mumbaki* offer three or four chickens to the gods and ancestral spirits, asking them to bless the seedlings so that they will not wilt and die.

7. Kulpe*/Kulpi†
(December–January)—When the transplanting season is over, one or more *mumbaki* perform this rite, moving from house to house in the village and/or adjacent villages. It culminates at the granary of the *tumoná*, where the public feast (*hamul*) takes place. (In some districts stories from Ifugao's oral literature [*liwliwa*] are chanted for entertainment while people wait for food to be cooked.) Villagers contribute chickens for the ritual, which marks the end of labor in the rice fields. Men and women are now free to do other necessary work—the men start clearing land for their swidden farms (*habal*) while the women weave.

8. Hagophop*†
(February–March)—About one month after the Kulpi, the *mumbaki* ritually sacrifice three chickens at the granary to open the weeding season (*ahikagoko*). They petition the gods to make the rice plants *humaping*, or robust with many tillers. After the rite, the women enter the fields to remove weeds and to replace dead or stunted rice plants with seedlings maintained in the *inhuj-un*, or reserve seedling bed.

9. Bodad[†]
(March)—Chickens are offered during this rite, which is performed in the granary during the wall-cleaning season (*ahidalu*). It is at this time that the rice plants are about to develop grains. The *mumbaki* petition the gods to make the plants bear as many grains as possible.

10. Paad*[†]
(April–May)—The *mumbaki* sacrifice a chicken during this rite, which takes place at the granary while the rice grains are maturing in the field. The people promise to abstain from eating fish, shellfish, snails, and other aquatic foods and legumes until after Kahiu/Kahiw. During the Paad, the gods are beseeched to make sure the rice plants yield a plentiful amount of grain.

11. Pokol*/Ngilin[†]
(June–July)—This rite is performed on the eve of the harvest in either the house (Pokol) or a given rice field about to be harvested (Ngilin). The Pokol is a feast that is meant to ensure the general welfare. It lasts all night and includes an *alim*, or ritual chant performed by high-ranking *mumbaki*, and invocations shouted very late at night. The ritual resumes the following morning before dawn. In Ngilin, the *mumbaki* offer a chick to the gods of covetousness and envy, the Umamo. The carcass is skewered and attached to the stem of a *bilau* reed with the leaves intact. The reed is then inserted into the dike of the main paddy where the rice is to be harvested before the female reapers arrive. The main purpose of the Ngilin is to implore this particular set of gods not to "covet" the rice harvest.

12. Ingngilin*/Ani[†]
(June–July)—On harvest day, the ritual celebration is centered at the granary (fig. 31.2). While the women harvest the rice below in the field, the men gather at the granary to drink, talk, and argue in between the various phases of the ceremony performed by the *mumbaki*. They narrate the *Myth of the Origin of Rice* with variations that involve the discovery and planting of the original Ifugao rice by, alternatively, two brothers named Balituk and Kabigat (or Wigan and Kabigat) or only Wigan. If the rice field owner is affluent or is the first planter or leading family (*tumoná*) in the village or region, a pig is offered in addition to chickens. When a pig is sacrificed, a more elaborate version of the rite is performed. The pig is cut up and the meat is boiled until it is heated clear through. The women are called in from the field and each is given a piece of meat, which they carry to a basket of cooked rice. They squat around the basket and eat. Huge servings of the choicest cuts of meat are placed in baskets for the *mumbaki* and their assistants.

13. Upin[†]
(July)—This is a simple rite performed after the harvest season. The gods are invoked to bless the rice, the granaries, and the houses in the village. The *mumbaki* ask the gods to protect the people from sickness, famine, and pests and to help the community to be prosperous, healthy, and peaceful. The following day is a sacred day (*tungo*). No one may go to the rice fields for any reason whatsoever.

14. Tuldag*
(July)—When the village's rice crop is dry and ready to be stacked in the granary, a three-day period of ritual idleness is declared during which no one may leave the village. Each household that has a granary performs the Tuldag. In some districts, the rite includes the making of rice cakes sweetened by sugarcane juice.

15. Pompon *
(July)—Immediately after the Tuldag (or on the last day of ritual idleness), the rice is stacked in the granaries without further ritual. The one who stacks the rice must be a man whose fields supply him with enough rice to last his household the entire year, since the Ifugao maintain that there is a magical bond between the rice and its stacker. The rice stacker must observe continence until after the Takdog rite, and he may not bathe because the loss of anything from his body at this time would entail the loss of some of the rice.

16. Takdog*
(August)—This rite is performed communally by villagers from the whole region at the house of the *tumoná*. Certain fields in each region are, by custom, planted first, and the owner of those fields determines for the whole valley the time of spading, repairing the dikes, and planting. In central Ifugao, where this rite precedes either the Pompon or the three days of ceremonial idleness associated with the Tuldag, several mats are spread in a shady place in the village of the *tumoná* and the idols from all the granaries of the village are placed on them, together with ritual chests and any number of wine bowls. The granary figures (*bulul*) are doused with rice wine and their faces smeared with rice cakes, the making of which is a feature of the preparations. This rite marks the ritual termination of both the harvest and the rice year in many locations in central Ifugao.

17. Kahiu*/Kahiw[†]
(August)—Performed in the home, this rite is intended to release the people from their promise to the gods, made during the Paad, not to eat aquatic foods or legumes. After this rite, the people may eat fish, shellfish, snails, and legumes. In some locations, a ritual sweeping of the house also takes place and marks the end of the Ifugao calendar, which coincides with the end of the agricultural calendar.

●

enumerated and described by Roy Franklin Barton (1946)* for various districts in central Ifugao and Manuel Dulawan (2003)† for areas in and around Kiangan and Asipulo.

What If the Bile Is Bad?

> Gibuwan da ya alan da'y manuk;
> ya yabyaban da'y mabolnat;
> gotngon da'y manuk ya ilugan da;
> hupwikon da ya tibon da'y buwa na;
> ya maphoda' abu.
>
> They finish invoking and take out the chickens;
> they fan-bless the seedlings to be transplanted;
> they slit the chickens and singe them;
> they cut them open and inspect the bile sacs
> and the signs are good. [Dulawan 2003, ll. 498–502]

Most Ifugao rites, be they agricultural or social, have common threads, such as the *mumbaki* checking the contents of the bile sac of a chicken or pig for a good omen. Back in December 1984, because he wanted to understand my culture—and perhaps his future wife as well—Gene Hettel, my then soon-to-be-husband, hailing from America's Ohio heartland, welcomed an opportunity to partake in the Moma engagement ritual. This took place in Pugo village, near Banaue, where the Ammayao family had relocated in the 1970s.

My second cousin Gambuk Ballogan—still a *mumbaki* then (see fig. 31.12) from nearby Lugu, Amganad—inspected the bile sac of a just-sacrificed pig that Gene had purchased through the traditional mediator. I still remember him asking my cousin, "What if the bile is bad?" His response was, "It means the gods do not favor your pending wedding." Now very concerned, Gene asked, "If the bile is bad, do I get another chance?" Gambuk replied, "Yes, you can buy a second pig!" Gene suspected that this could be a ploy for continued feasting on pork and rice well into the next day, but mercifully the first pig yielded "good" bile and the *mumbaki* blessed our engagement. Our relatives and friends in attendance declared, "Hiya peman, tinamtaman ta-uh chi inyali na" (Indeed, we bear witness; we tasted the pig he brought). We now had approval to carry out our planned wedding a few days later in the lowlands.

From *National Geographic* to UNESCO World Heritage Site

My husband is not alone as a foreigner with a keen interest in the Ifugao, who have been the subject of articles that date back to the early days of *National Geographic* magazine. Dean C. Worcester, then the secretary of the interior of the Philippine Islands, featured the Ifugao in a special issue of the publication, devoted entirely to the headhunters of northern Luzon. In it, he considered the Ifugao to be barbarians and excellent hydraulic engineers, as demonstrated by their marvelous rice terraces (Worcester 1912).

Nine decades later, foreigners are still fascinated with headhunting. The practice was abandoned long ago by the Ifugao, but we still have not escaped that moniker. In the book *The Last Filipino Head Hunters* (Howard 2001), we are described, along with our sister tribes the Bontoc and Kalinga, as having among our elders the last living headhunters in the Philippines. I seriously doubt that anyone now alive has ever been a headhunter.

Harold C. Conklin, a Yale University anthropologist who has devoted half a lifetime to studying the Ifugao, observed that "very few culturally distinguishable and similarly situated populations in Southeast Asia or the whole Southwestern Pacific, in general, have been written about more voluminously. The Ifugao are thus well known for their intricate ritual and legal systems; for their distinctive patterns of bilateral organizations, sex, and warfare; for their rich oral literature and other artistic achievements [such as wood carving and basketry; see Capistrano–Baker et al. 1998]; and for their skills in agricultural terracing. Ifugao is a visually impressive, remarkably pagan and culturally persistent area" (Conklin 1968, iv).

Since Conklin wrote his observations in 1968, Ifugao culture has continued to generate fascination and curiosity. Although I am not necessarily implying that this has been a good thing or even a particularly welcome circumstance, my people have continued to attract the attention of government officials, missionaries, historians, artists, writers, travelers, celebrities, and even Hollywood film directors (*Apocalypse Now*, though ostensibly set in Vietnam, was filmed using an Ifugao cast).

Throughout the 1990s and into the twenty-first century, there has been continued interest in the state of our rice terraces, which are threatened by erosion and neglect, and the related culture and rituals. In 1995 there was a flurry of activities and meetings in Manila and Banaue—some of which I attended—to formally nominate our rice terraces for inclusion in the United Nations Educational, Scientific, and Cultural Organization (UNESCO) World Heritage List as a protected cultural landscape (fig. 31.3). In one paper presented during these meetings, the Ifugao rice terraces were characterized as the very first magnificent "skyscrapers" and said to represent a unique living model of comprehensive integration of economic, sociocultural, and environmental processes (Concepcion 1995). Later that year the terraces joined the two other Philippine sites—Tubbataha Reef Marine Park and the Baroque churches of the Philippines—already on the UNESCO list.

In adding the terraces to the list, UNESCO stated: "For 2,000 years, the high rice fields of the Ifugao—which include by the way those not only in Banaue, but in Hungduan, Kiangan, Mayoyao, and Asipulo—have followed the contours of the mountain. The fruit of knowledge passed on from one generation to the next, of sacred traditions and a delicate social balance, they helped form a landscape of great beauty that expresses conquered and conserved harmony between humankind and the environment."[2] During its annual summit in December 2001 in Helsinki, UNESCO noted its continued deep concern for the rice terraces by putting them on its List of World Heritage in Danger. It stated, in part: "Despite efforts to safeguard the site by the Banaue Rice Terraces Task Force (BRTTF) and the Ifugao Terraces Commission (ITC), the BRTTF lacks full Government support and needs more resources, greater independence, and an assurance of permanence."[3]

According to provincial governor Teodoro Baguilat (fig. 31.4), the terraces' inclusion on UNESCO's endangered list only six years after being designated a World Heritage Site is an embarrassment for the Philippine government (Cabreza 2001). Indeed, it seems that local authorities have done little or nothing to implement the necessary comprehensive management plans and corrective measures to help the Ifugao save the terraces. Since the beginning of 2002, there have been mixed signals from the Philippine government. In January President Gloria Macapagal Arroyo transferred jurisdiction of the terraces' development from the Department of Agriculture to the Department of Tourism while at the same time releasing P 10,000,000 (US$200,000) to start the preservation efforts (Ventura and de Yro 2002).

Baguilat has said that he would prefer to have fewer tourists in the area to facilitate the terrace's preservation (Visaya 2002). He also stated that once the terraces are commercialized, more hotels and establishments will sprout like mushrooms. During a conversation I had with him at his Lagawe office in May 2002, he clarified that tourism could be a good strategy to help develop the rice terraces and provide additional income for the people. However, he still preferred that the terraces' development be placed under the jurisdiction of the Department of Agriculture instead of the Department of Tourism.

31.3   Women engage in the backbreaking work of transplanting rice. The Ifugao rice terraces were added to UNESCO's World Heritage List as a protected cultural landscape in 1995. Photograph by Gene Hettel, Nabyon, Banaue, Ifugao, 1991.

"Although part of the country's cultural heritage, the terraces are still primarily agricultural land," he said. He is afraid that the goals of tourism officials may not always support what is really needed to preserve our rice terraces and best serve the people. "Let's not preserve the terraces for the tourists, but for the Ifugao themselves," he told me emphatically. I agree with Baguilat that the government should focus on issues of concern to Ifugao rice farmers, including infestations of rats and golden snails as well as enhancing the irrigation systems for mountain farms. Perhaps most important of all is educating our youth to appreciate that their culture revolves around rice cultivation—and to consider staying in the region instead of moving to the lowlands to make their fortunes.[4]

In February 2002, to the chagrin of many, the president abolished the BRTTF with Executive Order 72. This development has caused great alarm in the international community, particularly the International Union for Conservation of Nature (IUCN). Archim Steiner, the IUCN director general, stated, "Mismanagement of this World Heritage Site is at a crucial stage. The abolition of its management body so early in its mandate could be a substantial threat to its future as a site of international and national outstanding universal value" ("Rice Terraces" 2002). If the Philippines continue to proceed down this path, we risk having the rice terraces removed from the UNESCO list altogether (Due 2002). In the meantime, politicians continue to discuss what to do. Some ordinary Ifugao citizens, for their part, express a wide range of feelings and are engaged in a variety of activities related to the preservation of Ifugao traditions and culture.

31.4   Ifugao governor Teodoro Baguilat has said "Let's not preserve the terraces for the tourists, but for the Ifugao themselves." Photograph by Aurora Ammayao, Lagawe, Ifugao, 2002.

"We will make do with watching your tapes on television"

Since 1995, when my husband was stationed in the Philippines as a science writer and editor for the International Rice Research Institute, we have made an effort to record on videotape the various rituals associated with the rice-growing calendar. With the help of Ana Dulnuan-Habbiling, the matriarch of the same *tumoná* family in Tucbuban for whom my late father sometimes officiated as a *mumbaki* at various rice rituals, we have been able to document many hours of ceremonies, particularly the post-transplanting rite (Kulpe) and the Ingngilin harvest rites (see fig. 31.10). We felt that we could at least show these tapes to our half-Ifugao children and future grandchildren, giving them a glimmer of understanding of what their mother's culture once was.

Some professional Filipino videographers and filmmakers—namely Fruto Corre and Kidlat de Guia—have had the same idea. Corre recently won recognition from the Film Academy of the Philippines for his ethnographic work *Ifugao: Bulubunduking Buhay*, a forty-five-minute video that documents the painful dilemmas experienced by my people today. He skillfully establishes the connections between the terraces and our traditions—indeed, how they enrich and nourish each other. The video's message is that if the terraces disappear, so will our tradition and culture. This tape has been commercially packaged and is sold in many video stores and bookstores in Manila and elsewhere.

De Guia's work debuted internationally on the Discovery Channel on December 26, 2001, as part of its *Young Filmmaker* series. In it he shows how we Ifugao ourselves can document our disappearing rituals and traditions using small, handheld video cameras. This is exactly what Gene and I have been doing since 1995, albeit as amateurs. At the end of the Discovery Channel program, the narrator states: "Up to now, uninterested youth are showing a new enthusiasm for some of the old ways and practices, even if initially only on the video screen. For the next generation of Ifugao seeing people videotaping their rituals and then watching themselves on television, they maintain their interest in the old customs and traditions. The Ifugao heritage will not just fade away in the snapshots of tourists, but might be carried on once again by a new generation of proud Ifugao. Like the ancient rice terraces that have lasted for thousands of years, the Ifugao culture will live on."

31.5   Ritual box in human form.
        Ifugao people. Luzon,
        Philippines. Carved wood,
        organic ritual contents. H:
        41 cm. FMCH X85.443a–c;
        Gift of Mrs. W. Thomas
        Davis.

31.6   Ritual box in pig form.
        Ifugao people. Luzon,
        Philippines. Carved wood.
        L: 40 cm. FMCH
        X78.L951a,b.

31.7   Imported rice-wine jar.
        Ifugao people. Kinnakin,
        Luzon, Philippines. Porcelain
        with rattan lid. H: 32 cm. ©
        Copyright 2003, Peabody
        Museum of Natural History,
        Yale University, New Haven,
        Ct., ANT.262793.
        Photograph by W. K. Sacco.

31.8   Rice-wine jar (*hinoghogan*).
        Ifugao people. Luzon,
        Philippines. Clay, rattan,
        wood. H: 16 cm. FMCH
        X76.1065a,b.

The Material Culture of Ifugao Rice Rituals

At Ifugao harvest rituals, priests (*mumbaki*) sing the appropriate epics in the presence of various ritual objects. Typical paraphernalia includes carved granary figures (*bulul,* see figs. 8.6, 26.1), ritual boxes, and rice-wine jars. The dried remains of offerings, including areca nuts and the blood or feathers of sacrificial animals, are left inside the boxes from year to year. Rice wine is always present, too, contained in imported Chinese jars or locally made vessels. The Chinese jars are highly prized and are ranked according to their value, condition, and rarity, as well as aesthetic considerations. Today, with ritual practice diminishing, many Ifugao families have elected to sell their heirloom ritual possessions.

31.9   As the matriarch of the *tumoná* family in her village, Ana Dulnuan-Habbiling is determined to continue the rice rituals. Here she reads some passages from the *Myth of the Origin of Rice* to her eldest granddaughter. Photograph by Aurora Ammayao, Amganad, Banaue, Ifugao, 2002.

Even though the production of these programs may have, in part, been motivated by profit, I think it is still a good thing that our rituals and culture are being documented for both Ifugao and the world at large. In viewing these programs, however, I could not help noticing that many of the rituals depicted appear to have been staged expressly for the camera. This is something that Gene and I tried to avoid—at least initially—in our own videotaping. In 1995, this was still possible when, in July, we taped several hours of the rice harvest ritual in Lugu, performed by local *mumbaki* Yogyog Dogapna and Buyuccan Udhuk. The ritual would have been held regardless of whether or not it was known that we were coming to record it.

Only six years later, in 2001, in an attempt to rerecord the Ingngilin ritual with better camera equipment and from different angles, I had to pay three *mumbaki* from outside Amganad to perform the ritual at Ana Dulnuan-Habbiling's family granary. If we had not come, it would have been the first time that a harvest ritual was not held in Ana's granary. That past April, Gene and I shot some good footage of the Kulpe ritual performed at the granary by Yogyog and Buyuccan. Only seven weeks later, these two *mumbaki,* both in their eighties, were too ill to come to perform the Ingngilin. The aging of *mumbaki* is not unique to Amganad but a problem across Ifugao Province in such major towns as Lagawe, Mayoyao, and Kiangan and to a somewhat lesser degree in Hungduan and Banaue.

I asked Dulnuan-Habbiling, who has been a practicing Catholic for many years, why she still persists—seemingly against all odds—in preserving the post-transplanting and harvest rituals in her family granary. She replied that it is her *tumoná* family, after all, that has been traditionally responsible for taking the lead in performing the rice rituals. "I do not want to be the one remembered for ending this centuries-old tradition here in Amganad," she said. Most likely neither Yogyog nor Buyuccan will be able to continue their *baki* duties. So, emulating de Guia's effort, she requested that Gene and I provide her with a copy of the videos of the rituals that we recorded in her granary over the years. "We will make do with watching your tapes on television—it will be better than nothing," she said. "And if someone wants to revive some of the rituals at a later date, they can use the tapes as a learning guide!" Dulnuan-Habbiling believes that Christianity and traditional ways of the Ifugao can coexist, citing the example of how her family handled the Kulpe in April 2002. "We said some Christian prayers, butchered the chickens, and then played the Kulpe tape while eating," she told me a month later.

"I will perform the rituals until my last breath"

In April 2001, just after Gene and I had recorded the Kulpe ritual in Ana's granary performed by Yogyog and Buyuccan, I chatted with Yogyog—before he became ill—about the deterioration of our culture (fig. 31.10). He knows that he is among the last of a dying breed. "It is sad," he said, "but there is nothing anyone can do." Unlike many *mumbaki* who have become Christian and given up the old ways—such as my own father and second cousin did—he has continued practicing the ways of the *mumbaki*. "I will perform the rice and other rituals until my last breath," he reaffirmed. Yogyog had come from a long line of *mumbaki,* his father and grandfather both served the people in the old ways. He is perplexed that none of his children or grandchildren have chosen to take on his family's *mumbaki* role. Interestingly, he said he has no problem with anyone becoming Christian, as many in his family have—however, he is always ready to conduct the traditional rituals if anyone gets sick.

31.10   *Mumbaki* Yogyog Dogapna (left) and Buyuccan Udhuk (see fig. 31.2) take a break during the Kulpe rituals at the rice granary of Ana Dulnuan-Habbiling to reflect on the fate of the their culture. This would be the last time that they would perform together. Photograph by Gene Hettel, Amganad, Banaue, Ifugao, 2001.

A Dichotomy in My Own Family

After my discussions with Dulnuan-Habbiling and Yogyog, I wanted to talk with some other people in Ifugao society about the status of our culture. Among members of my own family there existed an interesting dichotomy (fig. 31.11). My mother still thinks it

31.11   Ammayao Dimmangna (left) and Indanum Kinadduy raised nine children—including the author—over a marriage lasting more than fifty-five years. Nonetheless, they represented opposing views within the cultural struggle that is taking place among the Ifugao. Photograph by Jon Lee, Pugo, Banaue, Ifugao, 1999.

best to adhere to the old ways. She does not say this in so many words, but it is evident in her actions. Gene and I have some very interesting videotape of her contributing to some of the stories and debates that comprise Ifugao's oral literature (*liwliwa*) as she sat underneath Dulnuan-Habbiling's granary during the Kulpe ritual. My father, however, from around 1985 up until his death in 2002, was a faithful go-to-church-on-Sunday evangelical Christian. I asked my father why, since the mid-1980s, he no longer attended the rice rituals with my mother and, in general, shunned the ways of the *mumbaki*—or the "pagan" priests, as he now called them. He replied that he had finally seen the light and truly believed that the only way to attain true salvation was through Jesus.

Nevertheless, toward the end of his life, I think my father decided to "cover all the bases." In December 2001, perhaps sensing he did not have many more days on earth, he summoned his five daughters, four sons, and nineteen grandchildren home to Pugo village for what was essentially the Yabyab ritual. Although a Christian minister was present during certain times of the two-day affair, various activities were tied directly to Ifugao tradition, including the ritual slaying of pigs, the sharing of specific pieces of meat with the attending relatives, and the recitation of a special Yabyab blessing on the second morning. Seventeen weeks later, during his five-day wake, to the singing approval of a few of my "born-again" sisters, some of the visiting Christian ministers declared that Ammayao Dimmangna's greatest accomplishment in life was his renunciation of his ways as a "pagan" priest in order to follow the Lord. My mother remains silent on the subject.

Christianity Has Truly Changed Our Culture

My cousin Kinadduy "Jose" Binwag, a sixty-three-year-old retired rice farmer with seven sons and twenty-five grandchildren, regards the gradual disappearance of Ifugao culture with mixed emotions (fig. 31.20). A converted Christian, he uses the term "pagan" priest instead of *mumbaki*. His father was a *mumbaki*, but Binwag himself never was and none of his sons has any interest in the old ways. "I actually wanted to learn the ways of the *mumbaki* from my father, but he died suddenly in April 1966 at a relatively young age, and I lost interest after that," he said. "The introduction of Christianity is truly changing our culture. Because of the different religious sects that have recruited relentlessly since the mid-1980s, I would have difficulty teaching the old ways to my children and grandchildren even if I were so inclined," he added (personal communication, 2001).

31.12 Gambuk Ballogan, the *mumbaki* who inspected our engagement pig back in December 1984, is now a Christian and has given up officiating at rituals. "I now think having a good rice yield depends on how one tends his fields, not through protecting the crop from envious spirits by sacrificing a chicken," he said. Photograph by Aurora Ammayao, Pugo, Banaue, Ifugao, 2002.

31.13 Bognadon Bimmala and wife Rita still occasionally arrange for the performance of the Honga (health) ritual when someone in the family is very sick. "How much we can afford at the time will determine if we butcher one, three, or five pigs," Bognadon said. "I didn't pursue becoming a *mumbaki* myself because I saw what my father went through," he added. "I recall so many dos and don'ts about a specific ritual. I didn't want to deal with it." Rita mentioned that the family has decided to spend most of their limited funds on their children's education, not rituals. Photograph by Aurora Ammayao, Kiangan, Ifugao, 2002.

31.14 Tomas Liton of Mayoyao does not believe in the *baki* anymore. "Look at us now," he said. "We take a bath every day and are clean. When I was a *mumbaki* years ago, I sometimes was forbidden even to touch water, much less wash my hands or take a bath— especially if an important ritual was about to be performed." Photograph by Aurora Ammayao, Mayoyao, Ifugao, 2002.

31.15 Francisco Niwanne would like to see the *baki* continue. "But," he said, "with the old folks dying everyday and with few young people learning the proper procedures and memorizing the long epics like the *Myth of the Origin of Rice*, we rarely witness rituals of any kind nowadays." Photograph by Aurora Ammayao, Pugo, Banaue, Ifugao, 2002.

31.16   Diego Chugasna believes the *baki* itself is no longer of any use. "However," he said, "it is still useful to have *mumbaki* around to trace our ancestral lineage. When they are all gone, we will no longer know who is related to whom." Photograph by Aurora Ammayao, Mayoyao, Ifugao, 2002.

31.17   Romeo Nabannal's family operates Rita's Mountain Lodge in Batad. "We don't perform rituals anymore," he said. "We are all Christians here and have forgotten our pagan ways." Photograph by Aurora Ammayao, Batad, Banaue, Ifugao, 1999.

31.18   Mario Lachaona, member of the *tumoná* family in Mayoyao, now organizes rituals without benefit of a *mumbaki* since none can be found. "We play an audio-tape of the chants," he said. "We still butcher the pig and check the bile sac, conducting this aspect by recalling memories of the past." Photograph by Aurora Ammayao, Mayoyao, Ifugao, 2002.

31.19   Mario Hengnger from Batad practices the *baki* when he can afford it. "Pigs—even chickens—are so expensive nowadays," he lamented. "On top of this, the older folks are dying and there are few of them left who know how to properly perform the rituals. I guess for the future we need to be flexible." Photograph by Aurora Ammayao, Batad, Banaue, Ifugao, 1999.

31.20   None of the seven sons of Kinadduy "Jose" Binwag have shown any interest in the old ways. He himself calls the *mumbaki* "pagan." Photograph by Gene Hettel, Pugo, Banaue, Ifugao, 2001.

Binwag's reference to the large variety of Christian churches and sects in Ifugao Province is accurate, and all of them are vying for new members to add to their flocks. As Gene even thought the goal of increasing membership was on the agenda of all the various ministers—a Baptist, an Evangelical, a Lutheran, a Methodist, and a Catholic—who preached during my father's five-day wake.

Despite such internal tensions, some people feel that the old traditions can coexist to a certain extent. As Binwag told me, "For example, when the Christian ministers arrive for an event, the pigs are being butchered simultaneously and the meat distributed, but there is no longer any pagan chanting that goes along with it, as in the past."

During my father's wake more than twenty pigs were killed with their meat either consumed on the spot or distributed to visiting relatives and friends. Our sixteen-year-old son, Chris, even participated in the butchering when some of his cousins dared him to do so—perhaps they saw him as a soft city boy. Chris experienced his culture firsthand when he successfully sacrificed one of the pigs, stabbing the traditional sharp stick into the heart.

The Disappearance of the Rituals Is No Great Loss

In late December 2001, Gene and I visited Carlos and Maria Luglug in Lagawe (fig. 31.21). In their late seventies with eight grown children and twenty-five grandchildren, they do not believe that the disappearance of the rituals is a terrible loss to Ifugao culture. "We don't mind that most of the old rituals are no longer being performed," said Carlos.

31.21   Although their parents were *mumbaki,* Carlos and Maria Luglug do not find the disappearance of the old rituals disturbing. Photograph by Gene Hettel, Lagawe, Ifugao, 2001.

We stopped performing the rice rituals when our parents died more than twenty years ago simply because we did not believe in the old ways of appeasing a myriad of gods in the Underworld and the Skyworld. It was not the Christian way. Our parents were *mumbaki* of the highest order, but they knew we were not interested and would stop when they died. We even urged them to stop the *baki,* but out of respect we still assisted them whenever they asked us to.

Maria also explained that the way of rice planting is different now:

We were able to perform the rituals when our parents were still alive because there was only one crop—a season for planting, a season for cultivating, and a season for harvesting. But now, rice grows any time of the year. Even in our rice fields at Piwong, Hingyon, now tended by our children, we no longer have only one crop annually and so it would be difficult to time the rituals—even if we were interested.

The couple told us that the only rituals surviving in Lagawe are some that are not related to rice, such as the Moma, Tanig, and Honga—and these are now mostly Christian prayers and do not include the *baki* chanting to appease the traditional Ifugao deities. Maria also pointed out that nowadays it is simply too expensive to perform the rituals, which require chickens and pigs if they are to be carried out according to tradition. "With the economy the way it is, families cannot afford to buy a pig if someone is sick. It is less expensive to go to the hospital. I think it is best that the culture has changed in this way," she concluded.

Most of the people we talked to shared the Luglug's opinion, that the ebbing away of the *mumbaki,* the traditional rituals, and the culture they represent is of no great concern. However, two persons we interviewed—Manuel Dulawan and Juan Dait Jr.—are not content to allow Ifugao culture to disappear so easily. I had noticed their names in the Philippine press (Tarcelo-Balmes 1999A,B; Lolarga 2001) and decided to look them up where they both live in Kiangan, less than a kilometer from each other.

The Ancient Practices Are a Part of Our Identity

Gene and I found Manuel "Manny" Dulawan on his front porch during Holy Week 2001 (fig. 31.22). Now a retired high school teacher, he continues to be a student of the Ifugao culture. Among other accomplishments, he is responsible for the English translation from the Ifugao of one version of the *Myth of the Origin of Rice* and has completed an extensive work on the oral literature of the Ifugao (Dulawan 2003). Baptized as a Catholic as an infant, he was named after his maternal grandfather Dulawan, a respected *mumbaki* who attained a high social rank. "As a child, I enjoyed going along with him whenever he was invited to perform rice and other rituals in the surrounding villages," he said. "This early exposure to our socioreligious rites gave me an understanding of the rationale for their performance. I miss these rituals now as an old man, especially when our young people are so totally ignorant of our past." Dulawan had stated recently in the Philippine press (Lolarga 2001) and repeated to me that day:

> The Christian missionaries made us hate our own beliefs, telling us that these were satanic and pagan. But we're learning to question. It is interesting that wine is an element in the Catholic Mass *and* in the *mumbaki's* rituals. Our parallel culture should not be in conflict with our adopted Christian religion. Religion should not divide us. If we preserve the dignity of the *mumbaki,* the future of the Ifugao would be very well served.

Dulawan went on to say:

> Here in Kiangan, we have dispensed with many of the Ifugao rites, especially the rice rites, but we still exhume the bones of our parents, butcher chickens and pigs, and welcome friends to come to pray. We sing hymns, which are either in English or translated to Ifugao. So, in this fashion we retain our culture, but we have adopted some other practices along the way to combine with the ancient rituals. I think that there is nothing wrong or anything unchristian in this practice.

Dulawan told me that he regrets not having become a *mumbaki* himself so that he could have learned all of the rice and prestige rituals. "Even a Christian should have no reason not to learn the ancient practices because they are a part of our identity," he pointed out. He thinks that the *mumbaki* system could still be used to maintain part of our identity as a people, especially through the singing and reciting of our oral literature, such as the *Myth of the Origin of Rice.*

Putting on a Show for the Media

"Reviving Old Rituals to Preserve Ifugao Terraces" was the banner headline for a lengthy feature that ran in a February 2001 issue of the *Philippine Daily Inquirer* (Lolarga 2001). The story opened with a description of a quiet Saturday morning in Asipulo interrupted by the heavy percussion *tokotok-taktak* sounds of forty men and boys beating wooden planks and line dancing. The original intent of this old Patipat ritual—which hadn't been performed in Ifugao in almost fifty years—was to drive rats away from the rice fields. The revival of this ritual did not scare any rats but did attract a bevy of newspaper and television journalists, which was exactly the design of the man responsible for the event, Juan Dait Jr. (fig. 31.23).

31.22   Retired school teacher Manuel Dulawan suggests to the author that seldom-performed Kulpe and Ani rice rituals be taught to children in school. Photograph by Gene Hettel, Kiangan, Ifugao, 2001.

Whereas Dulawan promotes Ifugao culture through academic writing and translation projects and encouraging schools to teach our ancient myths, Dait takes a more controversial approach by attracting media attention. When he was the executive director of the ITC in the mid-1990s, he played a major role in getting the rice terraces inscribed on the UNESCO World Heritage Site list. Now at every opportunity Dait stages events aimed primarily at the Manila media. When Gene and I met with him on Easter Sunday morning in April 2001, he noted, "Resulting stories in the local press and on television written and produced by the visiting journalists are getting the message out about the plight of the Ifugao culture."

"Following the Patipat, we had the ordination of an Ifugao *mumbaki* in March [2001]," he continued. "There were eight full-fledged *mumbaki*—that we had to search high and low for in the region—who ordained this new priest—a very rare event. Seventeen journalists from Manila came and stayed the whole night from dusk to noon the following day. It was all properly documented by our local media." He pointed out that, often, foreign writers get only a superficial version of what the *baki* is all about. "They don't wait long enough to see the full picture," he said. "The perspectives of foreigners will often differ from those of the Ifugao themselves because our own interpretations come from our own feelings about how and why our rituals are done—and that is why when documents and articles come out, many aspects are often not correct."

The *Mumbaki*—Repository of the Ifugao Culture

Frank Lawrence Jenista (1987) characterizes the *mumbaki* as the repositories of Ifugao culture. He stated that the most valuable interviews for his book *The White Apos*—which details the American colonial authorities' interaction with the independent, headhunting, terrace-building Ifugao—were those with "the *mumbaki* priests with their trained memories and traditional perspectives." Since the Ifugao do not have a written language or dialect, everything is in the heads and hearts of the *mumbaki*. When I spoke with Dait in April 2001, he expressed similar admiration:

> You ask them about taboos, customs, and traditions—they know! They are also the best arbiters when there are disputes because they are respected. There may be lawyers or politicians around, but the people respect the *mumbaki* in the locality more because they are the ones who hold the culture of the people—and they cannot be bribed.

Many Ifugao now feel that *mumbaki* prayers are unchristian. Dait, however, emphasizes the ways in which the role of the *mumbaki* is not unlike that of the Catholic priest or the Protestant minister:

> When you hear the *mumbaki*, what are they praying to the gods for? They are asking for the rice fields to be blessed so that they will yield a good harvest and fill the granaries. They are asking that no sickness will afflict the people. And they are asking that the children will grow up to be good and that the wicked will be punished. So, now I ask, are these not the prayers of Catholic priests and Protestant pastors? So, I think that it is certainly a wrong impression to give to our children that an Ifugao in a G-string is ignorant or maybe even backward.

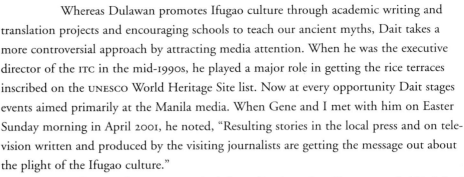

31.23   A promoter of traditional Ifugao culture, Juan Dait Jr. proudly displays a prized family *bulul*. Photograph by Gene Hettel, Kiangan, Ifugao, 2001.

Dait feels we should practice what he calls *inculturation*, that is, mixing the good values of the Church with the good values of our traditional culture. "I think this is a perfect combination," he said. "We should not turn our backs to either one. To what extent we should preserve the Ifugao culture is a very debatable issue. But a dialogue should continue among what I call the integrated indigenous peoples—or Ifugao like myself who embrace Christianity but at the same time feel that the values of our ancestors should not be lost." Dait believes that with the proper education of Ifugao youth, including exposure to the old rituals, it will be possible to preserve the Ifugao culture for future generations.

Back in Tucbuban, if Yogyog butchered a pig today and then checked the bile sac for an omen to determine if there is a future for our culture, I hope that he would declare the bile to be good. It will be interesting to see what the next twenty years will bring.

Epilogue, a Museum Experience Twenty Years Hence, April 2023

Ifugao culture will of course continue to evolve in ways that no one can foretell. I fear that twenty years from now, perhaps the only experience of our rich heritage of Ifugao rice traditions that any future grandchildren of mine might have may be during a visit to a museum. I have a sad future scenario in my mind's eye: My two little grandsons and I amble down a path along a museum diorama that artistically depicts the breathtaking landscape of the terraces. Up ahead, an Ifugao hut—built at about three-fourths actual scale to fit the cramped museum setting—is silhouetted against this artificial scene. From inside the hut we hear the recorded voice of *mumbaki* Yogyog Dogapna chanting a portion of the *Myth of the Origin of Rice:*

> Kimmali ama da'n hi Tad-ona: "Mu ya ngana dakayu ke ya nadijyunan di nige yu? Haj-oy ke ya kawwiliyok di bagbaguwan." Wada da'y hintulang kimmali da: "Mu maid di inanup mu page. Man dakami ke ya inanup mi'y page." Wada ama da'n hi Tad-ona ya kanana: "Ukaton yu." Ukaton di hintulang di inanup da'n pagen' di I-Kabunyan. "Ay, hiya peman," an kanan ama da'n hi Tad-ona.

> Their father Tad-ona speaks:
> "Why has your hunting taken so long?
> In my time I always brought home my quarry early."
> The brothers answer:
> "But you never hunted rice.
> In our case we have hunted rice."
> Their father Tad-ona says:
> "Bring it out."
> The brothers bring out what they hunted
> which is the rice from Kabunyan.
> "Ay, it is so," says their father Tad-ona. [Dulawan 2003, ll. 244–54]

The kids run ahead to the hut and then climb up the rickety ladder to peer inside at Yogyog of long ago as he prepares to sacrifice a chicken as part of the Kulpe ritual. Ironically, the luminesence of the television monitor, showing the video clip of Yogyog, adds a realistic glow to the hut's interior—much like real flames in the fireplace of bygone days.

I had seen Yogyog perform this very same ritual many times before. The last time was nearly twenty-one years ago when Gene videotaped the rites in Lugu. We linger a few more moments at the entrance to the hut to watch some more of the ritual on the videotape; then we move on to let other museum visitors lined up behind us have their turn to look inside to see what once was.

•

32

Mbok Sri Dethroned:
Changing Rice Rituals in Rural East Java

Rens Heringa

The rural traditions of Java are often viewed as attenuated versions of antecedents originating in the courts of Central Java.[1] In reality, however, rural expression may communicate basic concepts that tend to be veiled by elaborate aestheticism or abstraction in the court environment. This misconception concerning courtly origins—harbored by elite Javanese and the colonial government alike during the early twentieth century—has contributed to the gradual disappearance of distinct regional forms of expression. In more recent times, comparably myopic views have informed the development programs initiated by the Indonesian government. The official disregard for the relationships between differing methods of performing tasks and distinct underlying notions has inadvertently produced adverse social and economic effects on the populace in areas where such traditions have endured. In spite of efforts to bridge the gap between new ways and old concepts, the villagers' search for meaningful solutions has often been frustrated. The subdistrict of Kerek on the northeast coast of Java is one such example of a remote pocket where, until the mid-1990s, a local body of myths presented the model for the division of activities related to the yearly rice cycle (fig. 32.2). While the villagers succeeded in fusing several decades of gradual change into the existing conceptual mold, recent changes in planting and harvesting methods are perceived as the cause of social and cosmic chaos.

Kerek

A single-access road leads to Kerek, an arid valley enclosed by the foothills of the Northern Limestone Mountains, twenty-five kilometers inland from the ancient harbor of Tuban. Between the eleventh and seventeenth centuries, the abundant supplies of rice, fruits, palm wine, and meat delivered to the harbor by villagers of the then-prosperous area made the roadstead of Tuban a favored anchorage in the trade network linking the archipelago to the Asian mainland and China. The densely forested mountains provided timber for the local shipbuilding trade and, in later times, for Dutch houses in Batavia. By the mid-nineteenth century, the resulting deforestation had turned the region into a barren backwater. Since that time, the population of Kerek has been known for its fierce insistence on customs and beliefs that contrast with the standard Javanese cultural mold. An antiquated variant of spoken Javanese and strong autochthonous Kejawen beliefs, sprinkled with a minimum of Islamic expression, set these villagers apart.[2] Their claim

32.1   Mbok Karni's harvesting outfit, consisting of a dark blue and red hip wrapper (*jarit pipitan*) and dark blue-black shoulder cloth (*sayut irengan*), marks her as a fully mature, middle-aged woman. The shoulder cloth is used to hold the women's basket for carrying Sri. Only the wide-brimmed bamboo hat is a recent innovation. Photograph by Rens Heringa, Gendong, Kerek, 2000.

32.2   Two earthenware pots (left) personify Mbok Sri and Sedana, the mythical Rice Goddess and her husband, enthroned in the rice shed. In this instance, prior to a wedding, two separate cloths are used to cover the pots, whereas a single cloth is used at most other occasions. Photograph by Rens Heringa, Gendong, Kerek, 1989.

32.3   Once the first rains have flooded the rice field, a second plowing takes place. Photograph by Rens Heringa, Gendong, Kerek, 1989.

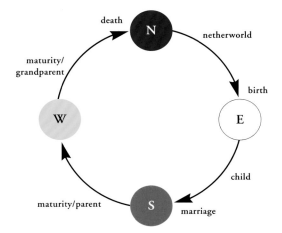

32.4   Space and time fuse into a color cycle that starts in the east with white (birth) and ends in the north with blue-black (death). Red and yellow in the south and west stand for increasing maturity. Along the path, each color gradually changes, uniting midway with the next, while all colors come together in the center. The blue-and-white (or gray) section between north and east represents the indefinite period between death and renewed life. Black denotes spirits and demons who find their place in the lower world, while white corresponds with the ancestors and gods resting above.

to the ritual guardianship of the forests and their sacred water sources causes frequent conflicts with representatives of the forestry department and with government officials in charge of regulating the limited water supply.[3] In an effective response to the ecological conditions, maize and tubers have been chosen as daily staples, while the single yearly crop of rice from the rain-dependent fields is reserved for ceremonial purposes. The rice cycle is initiated with the first rains in October or November and lasts from four to six months (fig. 32.3). Modern, low-growing strains of both regular and glutinous rice have dominated the crops since the late 1960s. Officially, the cultivation of old rice strains has been prohibited; the *pare menjeng* (tall, "voluptuous rice" believed to personify the Rice Goddess) and the ritually purifying *ketan ireng* (black glutinous rice) are only sown in secret among the *pare gogo* (dry rice) crop in upland fields.

Agricultural labor alternates with a range of craft activities, some of them specialized according to gender. Men and women each plait specific basket types. While women are accomplished weavers and batik makers,[4] men specialize in carpentry—a remnant of the old foresting tradition—and raising livestock. Sturdy textiles and carved wooden objects play a significant role in the rice-cycle rituals; their form and decoration, and particularly their color, serving as expressions of the persistent Javanese need to create a balanced whole. Everything in the universe is seen as part of an ever-moving, living continuum, with each entity's temporary position marked by a specific color set out on horizontal and vertical planes (fig. 32.4). Like human beings, material things and ritual activities are considered living entities, always evolving and, therefore, essentially ephemeral. Textiles, carved wooden containers, pots, and baskets are only intended to have a short lifespan; their "children" and "grandchildren" constantly replace older generations of objects. Immaterial activities, such as the utterance of a ritual invocation, are similarly considered descendants of earlier versions, which makes exact reproduction irrelevant. Symbolic value rather than costly materials or elaborate effort determines whether objects are chosen for ritual use and thereby acquire a mostly transient degree of sacredness. Devotion to the precepts handed down by the ancestors and the act of transforming ingredients into a harmonious assemblage is of central importance.[5]

32.5   Gifts of steamed rice representing the body of the Rice
Goddess are distributed by
the mature women to all
married men. Photograph
by Rens Heringa, Gendong,
Kerek, 1990.

32.6   Young boys wait in suspense
for a chance to heap their
teak leaves with the leftovers
of the rice gifts. Standing to
the left of the boys is an elderly villager with the
difficult task of keeping
them from fighting over the
leftover rice. Photograph by
Rens Heringa, Gendong,
Kerek, 1990.

The community's distinct character is also expressed through concepts regarding rice. In the Javanese courts, the cultivation of rice is attributed to Dewi Sri, a goddess with Hindu characteristics. Addressed as Mbok Sri (Mother Sri) in Kerek, the deity shows qualities related to pre-Hindu concepts encountered among other groups elsewhere in Indonesia. Mythical accounts establish the goddess's close relationship with the human community; in fact, the main tale presents Mbok Sri as the initiator of general moral and social order. By offering herself as "food" for the ritual period, she effectively put an end to a time of chaos when groups of men used to take women by force or, in east Javanese dialect, *mangan* (eat) them. The goddess thereby transformed herself into the regenerative medium for all, while also initiating the proper relationship between mature men and women by setting an example for all wives to "offer" themselves to their husbands (fig. 32.5). In contrast, adolescent boys, as yet incomplete members of the community, are excluded from the regulated exchange. They must *ngroyok* (vie for) a share of rice and, implicitly, a chance for regeneration (fig. 32.6).[6]

While the first tale is known to all villagers, a second narrative forms the restricted property of the local elite of landowners as it is related to the distinct local form of landownership. In contrast to Central Java, where all land is considered the property of the ruler, agricultural land in Kerek is the individual property of men and is inherited from father to son. The myth, in addition to introducing the elite's virilocal marriage system,[7] presents an individual example of the impetuous behavior typifying young men. A landholder's son, Jaka Tarub (which may be translated as "Young Man of the Circumcision Shed"), takes away the "flying cloth" of the sky nymph Nawang Wulan (Resemblance of the Moon) while she bathes in a forest spring and hides it, thus tricking her into marriage. Miraculously, she feeds her family on a single stalk of rice until her husband breaks the spell and, contrary to her instructions, lifts the lid of the rice steamer and curiously peeks into it. Now the store in the rice shed gradually diminishes, and once the goddess recovers her cloth that is hidden underneath, she returns to her father's abode in the sky.[8] The villagers have had to toil in their rice fields ever since that day.

The first tale recounts Sri's lasting social pact with the human community. The second story refers to the exclusive partnership imposed on Nawang Wulan by Jaka Tarub, thus further legitimizing the high social status of his descendants as hereditary owners of rice land. The village chief, foremost among them, is accorded the use of a plot of communally owned prime wet-rice land in return for his services. These fields and the village chief's family house serve as the ritual focus for the initiation of each phase of the yearly rice cycle. Once the family of the village head and their helpers have set each phase in motion, individual owners repeat the performance in varying degrees of complexity in their own fields. Before fulfilling their respective tasks, all those involved—male and female—are required to purify themselves with water in which black glutinous rice has been boiled.

An intricate sequence of ceremonies moves the cycle forward and in its totality forms a metaphor for a woman's life. The human participants, either as a group or individually, stage specific kinship relations to the goddess and thereby reconstitute the community. The focal phases related to Sri's regeneration—planting and harvesting—are enacted by mature men and women in their reproductive phase, while postmenopausal women perform the preparatory and concluding tasks. Women may act as grandmother, mother, or aunt to Mbok Sri; men fulfill the role of her father or uncle, as well as that of sexual partner. Finally, the whole community partakes of her transformed body, the offerings of steamed rice exchanged on the graves of the ancestors. Throughout the cycle steamed rice in various forms, accompanied by raw and cooked food, is exchanged among social groups and offered to a wide range of spirits, ancestors, and gods.[9] Each individual phase is initiated by the *tukang tanduk*, the male agricultural officiant, with the recitation of Javanese *donga,* or prayers (fig. 32.7). In the central invocation, pronounced when harvest—the last phase of the goddess's life—is imminent, he addresses Mbok Sri by the six epithets evoking the separate stages of her life journey. Several of the colors encountered in the life continuum also mark specific phases of Mbok Sri's life (see fig. 32.4). The epithets are as follows:

Mbok Sri Ngemanti	Mother Sri "in anticipation"
Mbok Sri Sedana	Mother Sri (and her husband) Sedana
Mbok Sri Abang	Mother Sri the Red One
Mbok Sri Kuning	Mother Sri the Yellow One
Mbok Sri Ayu	Mother Sri the Beautiful
Mbok Sri Puteh	Mother Sri the White One

Each of these stages of Sri's life journey indicates the different roles assumed by the villagers in the traditional process described below. The effects of recently imposed changes on the focal phases of planting and harvesting are also considered in order to define the extent of their social impact.

Mbok Sri Ngemanti—Initiation

Sri's first phase in life is that of *anti-anti,* or anticipation, when she is a nubile young girl soon to be given in marriage. Like all Javanese brides, she must remain in ritual confinement in a replica of the womb before being reborn in her new phase of life (Achjadi 1989, 152). To this end an elderly female officiant rinses the bride with fragrant floral water so that she will resemble the sky nymph (Bratawidjaja 1985, 66). Analogously, the senior woman of the household enters into the rice storage in the inner reaches of the house and sprinkles the seed of the *pare menjeng,* the symbolic embodiment of Sri, with floral water. Since the new rice strains are not considered subject to ancestral precepts by the people of Kerek, their seed is taken without the preliminary sprinkling to the location considered to be the "female space" of the house—in front of the kitchen—where the mature, married women set out all of the earthen and brass pots in the house. Each womblike pot is filled with seed, covered with tepid water up to the brim, and left to germinate for two days. The wet kernels are then removed two handfuls at a time and laid out on a bed of rice straw—remnant of the goddess in her previous year's incarnation. The straw absorbs all excess fluid, although the period of rest is specifically believed to enable the seed to grow long, strong roots. The effect of the treatment is referred to as *atus* (dripped dry), a term that also alludes to a person of noble descent. The typical Javanese wordplay indicates that at this point the noble ancestress acknowledges her new descendant by bestowing "strong roots." While the two previous treatments have infused the seed with the female ability to regenerate, the male complement must necessarily follow. The seed, therefore, is tightly packed between layers of rice straw in bamboo baskets that are lined and topped off with fresh teak leaves (fig. 32.8). An extra stone weighs down the basket that contains the *pare menjeng* seed, to keep it from "flying off like the nymph" before the process of germination is completed. Another day and night must pass before the senior woman's early

32.7   The male ritual specialist (*tukang tanduk*) invokes the protection of the ancestors by sending up the black smoke of incense burned at the northeastern corner of the rice field before plowing can begin. Photograph by Rens Heringa, Gendong, Kerek, 1996.

morning activity in which she carefully separates the rice straw from the seed that has been spread out on large mats. The "male" baskets—made and used by men to carry manure or grass to feed their livestock—complement the "female" pots. Similarly, the leaves from a small forest of teakwood trees near the ancestral graves "complement" the straw in order to heat and more quickly germinate the seed and infuse it with *sari* (ancestral life essence). The complementary gendered objects used in the preparation for a *kemanten* (wedding) are instrumental in transforming the seed into a fertile couple: Sri and her bridegroom, Sedana.[10]

Mbok Sri Sedana—The Wedding

The preliminary phase of the wedding takes place in the shelter of the village head's house, an appropriately secretive setting for processes related to the generation of fertility. The next phase—the public celebration of the wedding of Sri and Sedana—is staged for all to witness, in imitation of a long-defunct village tradition. Since at least 1822 to as late as the 1950s, the elite village bride was dressed in festive garb and carried by her paternal uncles around the village in a palanquin en route to being brought to live in her husband's house. She was accompanied by a procession of family members bearing her belongings and was followed by her groom on horseback.[11] In the agricultural context, the dressing of the bride finds its equivalent in the task of the grandmother of the house who arranges "Sri and Sedana"—the pre-germinated regular rice and the glutinous variety—in separate ceremonial baskets befitting their gender. The grandmother also provides in each basket a package wrapped in banana leaf containing a small mirror, fragrant yellow ointment, and coconut oil intended to beautify the couple, as well as medicinal roots to guard against illness. Also included are ingredients for a betel quid, the basic offering for the establishment of a marital union (Jordaan and Niehof 1988, 170). When everything is ready, the mature men of the house fulfill their role as uncles by carrying the female and male baskets—the seat of the bridal couple, Sri and Sedana—to the fields, followed by helpers bearing sacks of modern seed rice.

Until a few years ago, the pre-germinated seed was broadcast directly on the rain-inundated fields.[12] The method, referred to as *gogo rancah*, did not require transplanting and produced a yield sufficient for ritual purposes, in line with Mbok Sri's injunction. The planting sequence was executed by men and followed a set pattern that found an analogy in the successive steps executed by the mature women when filling in the format of a batik hip wrapper (fig. 32.9). First, several young helpers walked around the field to mark its perimeter, evening out the loose soil and the weeds near the banks with their bare feet. Then, the two main officiants would sow wide borders of broad-leafed white and black glutinous rice—representations of Sedana—along the short ends of the field where rows of palms and fruit trees form a natural boundary. The "male" plants were meant to cool and protect the female, shoulder-height *pare menjeng*—the representation of Mbok Sri—which used to be sown in the central area (fig. 32.10). Together the two men traced a spiraling path through the inundated field, broadcasting the seed (*nampek*) with wide sweeps of their right arms (fig. 32.11). The older man walked in front, carrying the seed openly visible in a man's basket as a sign of his legitimate union with the goddess. As a further indication of his married state, he used to wear sturdy tan shorts and a matching jacket woven by his wife especially for the planting season. A red batik sarong draped over his shoulder completed the outfit. A younger

32.8   At the front is one of the womb-like copper pots that held the rice seed during the first, "female" phase of germination. The seed is here being transferred to "male" baskets where, heated between layers of rice straw (the dry remainder of the previous year's crop) and protected with teak leaves, it will undergo the final, "male" phase of germination. Photograph by Rens Heringa, Gendong, Kerek, 1990.

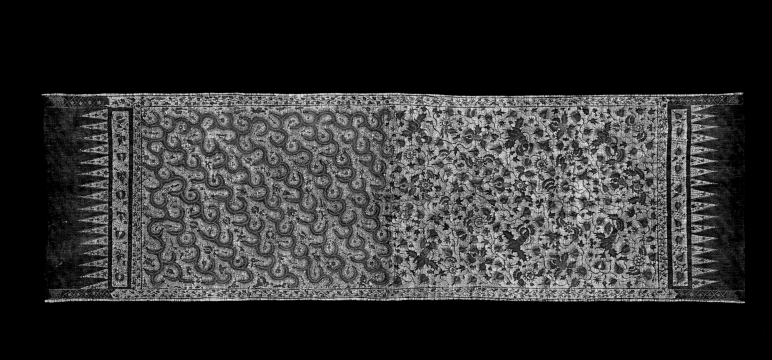

32.9   Hip cloth (*jarit*). Kerek, East Java, Indonesia. 1979. Handwoven cotton, natural indigo dye; hand-drawn wax resist (batik) technique. L: 293 cm. Collection of the author.

The pattern on the cloth represents a rice field, and the woman who applies the wax resist to the cloth works in a manner analogous to the way men plant rice fields (see fig. 32.10). A central field (*pelemahan*) is plotted out by first outlining a surrounding "bank" (*pinggir*) and two end sections, each consisting of a rectangular section (*bogeman*; place for valuables), surrounded by a small border referred to as a drainage ditch (*glontor*) and closed off by a row of isosceles triangles (*pucuk rebung*; "bamboo sprouts"). The two halves of the central field bear contrasting patterns. One features a bold pattern of waterweeds (*ganggeng*; *Hydrilla verticillata*), a plant that trails its spiky offshoots in the inundated rice fields.

The other has a floral and bird pattern. The small dotted blue pinpricks (*coblosan*) stand for the young seedlings that will soon spring up in the rain-flooded rice field. Hip cloths of this type are worn by mature women for the harvest (see figure 32.1). A young mother would wrap the cloth so that the water-weed pattern shows on the outside, while a grand-mother would show the more restrained floral and bird pattern.

32.10 End borders of "male," glutinous rice are sown before and harvested after the "female" rice in the center of the field, thus "cooling" and protecting it. Until a few years ago a small plot of the traditional *pare menjeng* was grown in the very center of the field, surrounded by the modern low varieties. Palms and fruit trees guard the boundaries, while edible herbaceous creepers sprout along the banks that enclose the field. Key to diagram: (a) dike (*galengan*); (b) drainage ditch (*glontor*); (c) glutinous rice (*ketan*); (d) traditional varieties of rice (*pare menjeng*); (e) modern varieties of rice (*pare*); (f) palm and teak trees.

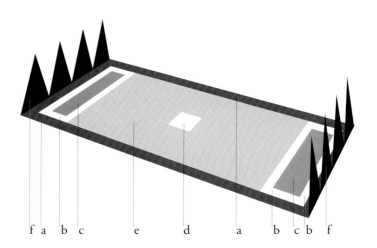

lad impersonating Jaka Tarub followed closely behind. He carried the seed in a woman's basket hidden inside a woman's red batik shoulder cloth, which is said by the villagers to be a descendant of the mythic cloth stolen long ago from Nawang Wulan (fig. 32.12). The red color of the men's clothing was a fitting representation of the beginning of the rice cycle. The maturity of the older man was marked by the reddish tan of his suit and the deep red of his sarong, both of which contrasted with the brighter red shade of the younger man's sling.[13] While the sowing of the fields proclaimed the male reproductive role, the waxing of batik patterns served the same purpose for mature women.

Although the time-honored *gogo rancah* method was eminently suited to the region's infertile soil and low precipitation, government determination to make Kerek adjust to modern ways was relentless. In the late 1990s, government officials decided that the weeding needed to make the old, "primitive" system succeed was too costly and labor-intensive, and they admonished villagers to convert to the gendered division of tasks that has been common since time immemorial in irrigated areas elsewhere on Java. The men were instructed to install *winen* (seedbeds) at the first rains, while women were given the task of transplanting the seedlings once the torrential second rains had broken through. As one might expect, the new system proved to be an ecological failure. Young plants in the seedbed need twenty to twenty-five days to develop fully. If the second rains arrive sooner, the plants will not be ready for replanting. If the rains take longer to come, the seedlings' roots start rotting away. In either case, the crop fails and, not surprisingly, complaints about meager results abound. The principal victims of the new system were the ritually required traditional rice types. A small patch of *pare menjeng* continued to proudly raise its head in the center of the fields as long as the new low rice strains continued to be sown. For a while this modification was upheld even though the combination of the sowing and the new transplanting system were felt to be an extra burden. But the conviction that the new procedure was in discord with the goddess's erstwhile precepts finally made the villagers decide to cease sowing her image in the center of the field. No longer do the *tuan tanah*, the owner of the land, and his younger brother stage their mythically ordained union with Mbok Sri and Nawang Wulan.

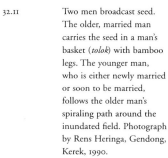

32.11   Two men broadcast seed. The older, married man carries the seed in a man's basket (*tolok*) with bamboo legs. The younger man, who is either newly married or soon to be married, follows the older man's spiraling path around the inundated field. Photograph by Rens Heringa, Gendong, Kerek, 1990.

32.12   The younger of the two broadcasters carries the seed rice in a bamboo woman's basket (*senik*), wrapped in a red batik carrying cloth (*sayut*). These cloths, normally worn by women, are considered to be descendants of Nawang Wulan's mythic flying cloth. Photograph by Rens Heringa, Gendong, Kerek, 1990.

Mbok Sri Abang—Pregnancy

During the months that follow the sowing phase, water, sun, and moon are said to aid the landowners to make the union prosper. The weeding, once done by Sri's "mothers" and "siblings," has now, however, been made redundant by a profuse use of chemical fertilizers. After almost seven weeks the reddish tinge of the flowering rice plants reveals the next step in Sri's life called the *abang* (red) phase. Again, the bright color indicates beginnings—at this point making reference to the new life hidden inside the flowering plant. Now the time has come for *tingkeban* (the main pregnancy ritual), which consists of the mature women preparing *rujak ceprot* (a fruit concoction) for the well-being of their "daughter" Sri and her "offspring." The gift consists of three different components quite similar to those prepared early in the seventh month of a human pregnancy, when the fetus is considered viable and ready to engage in its journey toward the human world during the remaining months of the pregnancy. Seven types of unripe raw fruits of various colors and a similar number of different tubers are cut into thin strips and seasoned with tasty spices believed to ensure the fruition process. The seven fruits and tubers stand for the period of time that has passed, while their unripe, raw state is in accordance with the still unfinished condition of Mbok Sri's fruit. The final ingredient consists of tiny *uler-uler* (worm-shaped lumps of white rice porridge, part of which have been dyed yellow and green). The white (*puteh*), yellow (*kuning*), and green (*gadung*) worms are said to accompany the red worm—Mbok Sri's offspring—in analogy to the *dulur papat* (four spiritual siblings) who accompany every human being on the journey through life.[14] Once all the ingredients have been mixed together in an earthen pot, a mature woman carries it in a red shoulder cloth to the village head's rice field. During a solemn circumambulation of the field, making as little noise as possible in order not to frighten Mbok Sri, she splashes (*ceprot*) the mixture among the rice plants.

Mbok Sri Kuning—The Harvest

Once the rice has turned a deep yellow (*kuning*), Mbok Sri is ready for the next phase of her life—the harvest. Symbolically, yellow is the color of full maturity and being reaped is a metaphor for giving birth. In the human world, the pregnant woman's family arranges a new plaited mat and pillow on the floor of the house and prepares the appropriate offerings. The traditional midwife places a broom made from the central ribs of the coconut palm's leaves at the pillow's end and decorates it with medicinal roots, onions, garlic, and chili peppers to guard against untoward influences (Mayer 1897, 1: 282). After guiding the delivery, the midwife cuts the umbilical cord with a sharp bamboo knife, smears the baby's navel with yellow turmeric to fight infection, and rubs the young mother with yellow medicinal paste (Koentjaraningrat 1984, 352). In the field, a male specialist cuts the first rice stalks using a traditional *ani-ani* (woman's rice knife). During the days preceding the harvest, the women of the house have prepared elaborate offerings in which rice forms the main ingredient. Small containers folded from banana leaf are filled with multiple ingredients and set out on a large offering tray fashioned from the central ribs of coconut palm leaves. The combination of complementary gendered ingredients, the form and color of the cookies and flowers, turns the offering into a metaphor for the cosmos (fig. 32.13). Early in the morning of the appointed day, family members—both male and female—together with the agricultural officiant, carry all offerings to the northeastern corner of the field, the point in the cycle where life starts anew. The tray is arranged at the edge of the field, next to a new

32.13   Five sets of small figurines (*bekakak*): a man and a woman, livestock (*raja kaya*), and fowl (*raja brana*), molded from rice dough, represent "the villagers and their helpers." Photograph by Rens Heringa, Gendong, Kerek, 2000.

32.14   At the northeastern corner of the field, the *tukang tanduk* presents the offerings to Mbok Sri, imploring her not to be frightened when he cuts her with his *wesi pulesani* (enticing weapon), the ritual name for the rice knife. Photograph by Rens Heringa, Gendong Kerek, 2000.

mat and a pillow shaded by an umbrella, which will serve as a bed for Mbok Sri during her confinement. While purifying incense burns, the officiant implores Mbok Sri not to be frightened, promising her that she will be made to feel sweet and fragrant after he cuts her with his *wesi pulesani* ("enticing weapon"; fig. 32.14).[15] Searching out fully developed stalks of the same length, he cuts two small bundles of four stalks each and binds them together.

In the past, the newly cut stalks of *pare menjeng* were anointed with yellow ointment and decorated with flowers before being carried home. Currently, all participants merely share a quick communal meal of rice, after which the small bundle is laid on the bottom of the now-empty basket and carried home without further ado. While the bunched stalks are generally referred to in literature as the *pare penganten* (the rice couple) Sri Sedana (van der Weijden 1981, 33–34), the villagers of Kerek stress that the freshly cut stalks are conceived of as the *penganten* (waiting pair), the new generation who will not become a bridal pair until the next planting season. Nevertheless, birth and marriage seem to flow together. The use of yellow ointment is an analogy for the treatment given to a young mother and her baby but also for that given to a bridal couple. The elaborate offerings to Mbok Sri Kuning bring to mind the festive gifts (*sasrahan*) offered by the elite to fasten the ties to a prospective daughter-in-law.

Once the ritual cutting of the stalks is complete, the other villagers are allowed to start harvesting their own fields. Those who have no *sawah* of their own assist those who do in exchange for part of the yield. Traditionally, it was the province of married women to carefully cut each separate stalk with an *ani-ani* (harvesting knife) in the form of a flying bird. The women used to dress in a dark blue batik hip wrapper, and the newly cut rice—their baby—was safely stored in a matching batik shoulder cloth (fig 32.1). The cloths, with a floral pattern in dark blue with red-and-black accents, are called *pipitan* (close together) in reference to the mature woman's position close to her husband and also to the baby, carried close to her body in the sling. The second, ritual meaning of *pipitan* is "yellowed," an allusion to the resemblance between Mbok Sri Kuning and mature women, both of whom are engaged in the western phase of procreation.[16]

In recent years, the harvest—like the planting ritual—has been subjected to a labor-saving process that has again brought about a reversal of gendered tasks. Now male hired hands, rather than the women of the family, cut stalks by the handful using an *arit* (sickle) in swift, almost aggressive motions (fig. 32.15). Only a decade ago, newly harvested rice was brought to the rice storage on the stalk; today, the crop is threshed directly on the field. Sometimes the threshing is done by hand but more often a simple machine is used. Finally, the *gabah* (rice grain) is poured into sacks and taken home unceremoniously on the back of a bicycle. The men have taken over from the women and again the new working divisions have upset the original kinship tie with Sri. No longer do women lovingly carry home their new "rice baby" for shelter in the rice shed in anticipation of the next cycle.

32.15   An *arit* (sickle) is used by the hired hands in the modern harvesting mode. Photograph by Rens Heringa, Gendong, Kerek, 2000.

Mbok Sri Ayu—The Rice Shed

In the human world, a new daughter-in-law is received in her husband's family house by an elderly female expert who prepares and dresses her in formal attire, so she and the groom can be displayed on a *kobongan* (bridal seat) and introduced to the community in the central room of the house (fig. 32.16). In the agricultural simile, aspects of marriage and birth once again overlap. From the field, the young mother Mbok Sri Ayu (Mother of

32.16   During the public part of a contemporary wedding, the bridal couple appears in a modern version of Central Javanese court-style dress. A pair of chairs just visible behind the couple comprise the *kobongan* on which they will sit in state. Photograph by Rens Heringa, Luwuk, Kerek. 1990.

32.17   The boat-shaped rice mortar (*lesung*) used to carry Mbok Sri on her journey toward the otherworld. Photograph by Rens Heringa, Gendong, Kerek, 1981.

32.18   Mbok Sri Puteh—now a
          layer of steamed rice—is laid
          out on her funeral bier by
          Ibu Petinggi, the village
          head's wife. Photograph by
          Rens Heringa, Gendong,
          Kerek, 1990.

32.19   A "roof" of "moon crackers"
          (*rengginang*), made from
          glutinous rice, provides extra
          protection for Mbok Sri
          Puteh on her way to the
          grave. Photograph by
          Rens Heringa, Gendong,
          Kerek, 1990.

32.20   A *slendang pati* is unfolded
          over the bier of Mbok Sri
          Puteh. Photograph by
          Rens Heringa, Gendong,
          Kerek, 1990.

32.21   The *dhondhang* on its way
          to the graveyard on the
          shoulders of two young
          "grandsons" of Mbok Sri
          Puteh. Photograph by
          Rens Heringa, Gendong,
          Kerek, 1990

32.22   Palanquin (*dhondhang*). Kerek, East Java, Indonesia. 2000. Carved and painted teakwood. L: 135 cm. FMCH X2002.9.1.

Dhondhang are used to carry the "body" of Sri in state at her "funeral" held after the harvest but also during a wedding, the pinnacle of the human life cycle. A bamboo pole is run through the holes to allow the palanquin to be carried. The colors of regeneration—yellow and green—that were hidden in the cloth worn during the harvest are fully visible on the *dhondhang*. Some red is still visible, but most of it is now hidden on the inside of the wooden container. Blue has disappeared but is vaguely visible in the silk *slendang* used as a shroud to cover the *dhondhang* (see fig. 32.23).

Radiant Beauty) and her offspring are brought home to be installed in the *gladhak* (elevated rice shed) in the inner reaches of the house.[17] In analogy to the bridal seat two *pedaringan* (earthen jars) filled with seed rice are set out near each other (see fig. 32.2). A single white cloth covers the mouth of the jars and a skein of handspun yarn joins the two. Two small oil lamps representing the sun and the moon shine on the jars day and night, while a *gendhi* (earthen water container) stands nearby to quench their thirst. Traditionally, rice could not be removed from the shed for a period of forty days, as Sri and her baby needed much rest—not unlike a young mother confined to the house after giving birth. Once the period of seclusion was over, the necessary quantity of rice was removed for pounding in a *lesung* (boat-shaped rice mortar; fig. 32.17).

A mechanical rice huller has been installed in the local market town for the past twenty years and *wos* (milled rice) is available in the shops throughout the year. Today most of the rice is sold as *ijon*, or "in its green state," quite a while before it will be harvested. Gradually, the rice shed has been rendered redundant. In the search for an appropriate use for the sacred space, various modifications have been attempted. Some traditionally minded villagers decided to use the former rice shed as the main bedroom, in an effort to retain its symbolic position as the house's center of fecundity. Others, considered more modern, turned the sacred shed toward Mecca to serve as prayer house in reponse to Muslim prosetylizing efforts. Nevertheless for the duration of each ritual occasion, the two earthen jars are set out anew, flanked by the required paraphernalia. Sometimes the old set has been kept for the purpose, but a newly bought pair is considered equally acceptable.

32.23 Heirloom shoulder cloth (*slendang pati*). Pati, north coast of Java, Indonesia. Early twentieth century. Silk; wax resist (batik) dye technique. L: 293 cm (half shown). FMCH LX74.270.

The cloth's name refers to the district of Pati where such cloths were formerly made by commercial Chinese batik shops, but it also indicates that it is a covering "for the death" ([*pe*]*pati*). A *slendang pati* is used to cover the funeral bier of Mbok Sri Puteh. The dark, bluish-brown patterns contrasting with the light background signify the otherworld that Sri will soon join.

32.24   The contents of the *dhon-dhang*—Mbok Sri's dead body—are laid out in state for all to see before being divided. Photograph by Rens Heringa, Gendong, Kerek, 2000.

Mbok Sri Puteh—Regeneration of Social Ties

The goddess is conceived of as a human being in the last phase of her life—a wife contributing to the undertakings of her husband's family by fulfilling her side of the social pact and nurturing her husband and family. As human life necessarily must come to an end, Sri's life in this world ends during the *manganan besar* (big eating), a communal meal held on the ancestors' graves.[18] During this period, the goddess is referred to as Mbok Sri Puteh (Mother Sri the White or Pure One), fulfilling her mythical role by serving as ritual sustenance to the whole community. While the "funeral" regenerates social and cosmic unity, it also forms the link between two rice cycles. In human life, relatives by marriage are given the task of preparing deceased parents-in-law for the funeral. Similarly, the wife of the village head acts as Sri's daughter-in-law when she lays out her body—a layer of steamed rice—on a bed of teak leaves at the bottom of a colorfully decorated wooden funerary palanquin (*dhondhang*) and supplies her with cosmic gifts made of rice (figs. 32.18, 32.22). First come symbols of different aspects of life on earth. A bunch of bananas, flanked by two slices of *jenang* (thick pudding made from rice flour, coconut milk, and brown sugar), forms a metaphor for the fusion of the male and female principles. A variety of *panganan* (colorful sweet snacks), each in multiples of five, tops the "terrestrial set" and acts as a symbol of the elderly generation. After a long bamboo pole has been inserted through the round holes in the handles of the *dhondhang*, a "roof" of the heavenly bodies is added. These include the sun, which consists of round yellow rice cookies (*kucur*), and the moon, large, pale disks made of glutinous rice (*rengginang*; fig. 32.19). Once the contents have been covered by a protective layer of teak leaves, a shroud consisting of an antique silk batik cloth is carefully spread out and tied to the four corner posts of the wooden chest (figs. 32.20, 32.23). Carried on the shoulders of two of her classificatory grandsons, Sri's bier is taken to the graveyard, where villagers have already gathered (fig. 32.21). Women have brought their own, more modest containers of steamed rice, bananas, and cookies to complement the elaborate offering of the wife of the head of the village. Together, the women divide all the food, distributing it among the mature men (fig. 32.24). The boys—as it is expected of them—make a show of fighting for a share of the rice and snacks, hiding the spoils away in the teak leaves they have brought for the purpose (see fig. 32.6). In the presence of the ancestors, all the men eat Mbok Sri Puteh, who is offered by all mature women. Beyond death and, according to the poetic phrase, hidden "in the depths of the golden storehouse" (a metaphor for the stomach), the goddess contributes to the continuity of the community.[19]

•

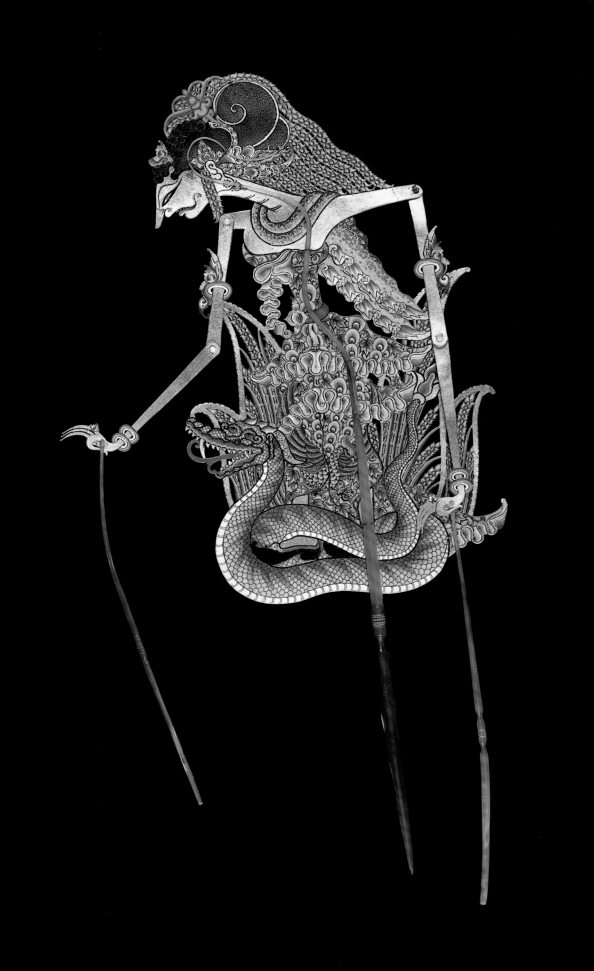

33

The Descent of Good Fortune and Material Wealth: A Contemporary Javanese Shadow Puppet Play

Roy W. Hamilton with Kik Soleh Adi Pramono

One of the best-loved stories in the repertoire of the Javanese shadow play theater (*wayang purwa*) is that of Dewi Sri, the Rice Goddess, and the origins of rice. In traditional Javanese agricultural communities, this play was supposed to be performed at least once annually, generally at harvesttime, as a reminder to humans of the divine origins and sacred nature of their most important crop. As every shadow play, and in fact every performance, is a unique creation of the puppet master (*dhalang*), there have undoubtedly been countless versions of this play performed in Java over the centuries. Featured here is a recent version created by Kik Soleh Adi Pramono, a puppet master currently practicing in the town of Tumpang, a few miles outside the city of Malang in East Java (fig. 33.4).

Soleh's version of the story is entitled *Mudhuné Réjeki lan Kumalané Raja Brana* (*The Descent of Good Fortune and Material Wealth*). *Mudhuné* implies a descent from Heaven. *Réjeki* (good fortune) is an alternate name for Sri herself. *Kumalané* is the material wealth that is the symbol of Sri's consort, Jaka Sedana, referred to here by his alternate name Raja Brana. Thus the title could equally well be translated as *The Descent of Sri and Sedana*. To the people of the villages surrounding Tumpang, who regard their area as the point of origin of rice, this is the story of *their* rice, and the origin of *their* sacred crop in *their* landscape. While this belief may run contrary to archaeological findings regarding the ultimate origin of the domestication of rice, it is quite literally true when understood in the sense of the development of localized varieties unique to a specific area.

Though the play itself is ancient, and Soleh essentially follows a traditional story line, his version is decidedly contemporary in several important ways. In the past, while many shadow play characters had distinctive, immediately recognizable forms, not all did. Any set of puppets would contain many generalized puppets (a princess, a soldier, an ogre, etc.) that could be put to use as necessary. Despite her fame, Dewi Sri typically fell into this category, and many a puppet master has selected a rather small, nondescript female puppet of modest character to represent Sri. Soleh has brought his creative genius to bear in creating new puppets not only for Sri but for the entire cast of characters in the play, including a puppet representing rice itself (fig. 33.3). Soleh's Sri is uniquely the goddess of rice, with her hair in the form of golden strands of rice decorated with a Javanese rice-cutting knife (*ani-ani*) as her hair ornament (fig. 33.1). Soleh himself is instrumental in the design of his puppets, providing his talented puppet maker, Daniel Mulyana, with drawings of the effects he wants. The result is a richly imaginative landscape whether

33.1   Dewi Sri. By Daniel Mulyana. Tumpang, East Java. 2002. Leather, paint, buffalo horn. H: 66 cm. FMCH X2002.16.1.

Dewi Sri's sacred vehicle is a rice field snake, which has many important symbolic roles in Javanese agriculture and mythology, in part because it protects the growing rice from rats and other pests.

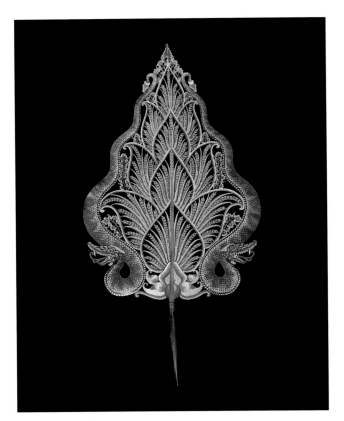

33.2   Kala Singa Mulya. By
Daniel Mulyana. Tumpang.
East Java. 2002. Leather,
paint, buffalo horn. H: 78
cm. FMCH X2002.16.5.

Most of the demons that
pursue Sri and Sedana transform themselves in the climactic scene of the play into
the pests that attack rice
crops. These puppets thus
represent the vermin and
diseases that are tremen-
dously important to Javanese
farmers in their daily experience. Kala Singa Mulya
(Great Lion) is an exception,
as he protects the rice.

33.3   Gunungan Padi. By Daniel
Mulyana. Tumpang, East
Java. 2002. Leather, paint,
buffalo horn. H: 82 cm.
FMCH X2002.16.8.

Shadow puppets of this
shape, called *gunungan*,
appear at key moments,
such as the opening and
closing of scenes, and are
also used as scenery. Soleh
uses them frequently to
"cover" a scene where one
character is magically transforming into another; there
are many transformations of
this type in his Dewi Sri
play. This *gunungan* bears
rice stalks heavy with grain,
and Soleh uses it to represent the rice plant at the
moment Sri creates it (see
fig. 33.10).

viewed in black and white on the shadow side of the screen or in full color on the puppet master's side (fig. 34.2). Soleh and other leading puppet masters of his generation have in this manner reinvigorated the visual artistry of *wayang*.

Another unique aspect of Soleh's play is that the female singer (*pesinden*) who accompanies the performance is not Javanese, but an American who performs under the name of Nyi Karen Elizabeth Sekar Arum. Together the two have formed the Mangun Dharma Arts Center (Padepokan Seni Mangun Dhara) in Tumpang. This collaboration is emblematic of the growing globalization of the arts. *Wayang*, can now frequently be enjoyed by audiences in Europe or America, while at its point of origin in Java, it often loses out to a host of imported forms of entertainment such as television or Western popular music.

Not only the features of Soleh's play itself but also the context in which it has been performed and the meanings that it has for its audience have strongly contemporary aspects. Soleh was first commissioned to perform this play in 1998 by a group of environmental and agricultural activists based in Malang. The group was concerned that rice lands were being damaged by heavy reliance on chemical fertilizers and pesticides, that valuable biogenetic diversity was being lost due to the ongoing replacement of traditional varieties with a much smaller number of modern ones, and that the balanced relationship between humans and the environment that had previously prevailed in Java's rice lands was being rapidly destroyed.

Puppet masters traditionally have enjoyed a unique role in Javanese society, in part based on their ability to breathe life into their cast of characters. The *dhalang* is not only an entertainer but also a philosopher, a spirit medium for the deities, a shaman (*dukun*), a spiritual adviser, and a social critic. To Soleh, the problems threatening the rice fields were as much political as agricultural:

> In the past rice was grown without synthetic fertilizers, chemicals, pesticides, etc. Natural remedies were adequate. Then in the 1970s, there were instructions to pull out all the indigenous Javanese varieties, to change from varieties that matured in four and a half months to those that matured in three and a half. The farmers couldn't use the natural methods anymore, but they had chemicals. It turned out that those varieties were imported from abroad, and the chemicals too were imported through the government run village Co-ops. So, the livelihood of the farmers became dependent on government bureaucrats. It was great for the bureaucrats, but the farmers who sweat in the sun every day have increasingly had enough of the Co-op credit system. They are continually squeezed by it. In the end they can't even bring their harvest home—it is taken directly from the field to the Co-op. And then what? The earth needs to rest, but there is no time anymore. Straight away it is plowed again, and worked over and over, with more and more chemicals. In the end, something that was natural has become totally ruined, and the crop is no good anymore.

As a *dhalang*, Soleh sees spiritual aspects to the crisis as well: "At the same time there were no longer elders who wanted to teach the spiritual aspects [of rice production], beginning with the natural planting process, the meditative process of beseeching [the rice spirits]—with prayers said over the seeds and the holy water given to the seedlings so that they are free of disease and grow well."

As he performs his play in East Javanese villages, Soleh is raising awareness of these issues in a way that touches his audience as no other form of public education can. They hear and see, in their own language and treasured medium, the story of their very own material wealth and good fortune. To restore the abundance of their rice lands, they are urged to begin by restoring respect for their goddess Sri and all for which she stands.

•

33.5   The serpent Anantaboga is about to swallow the heavenly gemstone Retna Dumilah, which has dropped to earth through one of the holes that have inexplicably appeared in the hands of all of the gods. Anantaboga's magical venom will turn the gem into the beautiful Retnowati.

33.4   Kik Soleh Adi Pramono performs *The Descent of Good Fortune and Material Wealth.* Photograph courtesy of Mangun Dharma Arts Center, Tumpang, East Java, 2001.

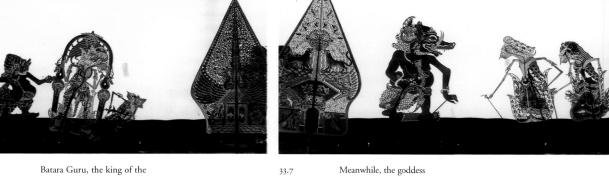

33.6   Batara Guru, the king of the gods, has fallen in love with Retnowati and wants to marry her. Because she has already promised herself that she will give herself to all the people on earth, she refuses him. She pulls a hair from her head, turns it into a dagger, and stabs herself to death. Narada, Batara Guru's prime minister, reminds his master that Retnowati was not a real goddess, but a bewitched gemstone, and was not meant for him.

33.7   Meanwhile, the goddess Dewi Sri and the god Wisnu (in the form of Sri's consort, Jaka Sedana), have been sent to earth to produce a new kind of food. They meet the wild boar Celeng Srenggi, who lusts after Sri. The boar tries to overpower Sri and Sedana until, with the help of the other gods, they kill him with a special bamboo spear.

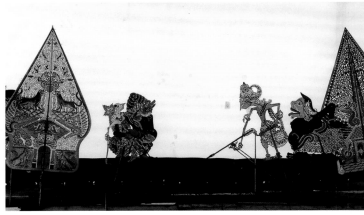

33.8   At the moment of his death, Celeng Srenggi's body parts are transformed into all manner of demons (*kala*), among them the rat-like Kala Tikus Jinadha (center left). His ears become the six-eyed Walang Sangit (right), his blood becomes Kala Srunthul (top), and his head becomes Kala Rau (flying off at upper left), who swallows the moon to cause eclipses. Sri and Sedana flee to the land of Medhang Kemulan, hotly pursued by the demons.

33.9   Batara Guru has ordered Narada to bury his beloved Retnowati in the fragrant soil of Medhang Kemulan. Siman Mikukuhan, the king of Medhang Kemulan, grants Narada's request for a plot. The king plunges his magical sword, forged in the shape of a rice stalk, into the ground and makes a fiery hole. The great corpulent Javanese god Semar (far right) looks on as Narada lays Retnowati to rest in the hole.

33.10   At that moment Sri and Sedana, still running to escape the demons, come upon the scene and leap into the hole with Retnowati. When the grave is covered over, a new plant rises out of it. Narada names this rice (*padi*). Sri ascends from the grave, proclaiming that she has become rice.

33.11   The demons arrive and begin to goad Sri, who is by now accompanied by Sedana astride a *kolang-kaling* palm, his symbol of material wealth. Sri announces that whoever wants to marry her must take a new form. The demons transform themselves into pests that prey on rice. Walang Sangit becomes the stinky locust Lembing Walang Sangit (red, at top) who munches on rice plants. Kala Tikus Jinadha becomes a rat. One demon, Kala Badhug Basu, becomes a tuna so that people will have something to eat with their rice! In the end, Semar, in his role as shaman of agriculture, expels the pests with a song so that the crops may flourish.

Figures 33.5–33.11: Photographs by Onny Tatang, courtesy of Mangun Dharma Arts Center.

34.1   After the harvest, the sacks
of grain are loaded on a
cart for transport to the village or the nearest cooperative. Photograph by
Kwang-Kyu Lee, near Pusan,
South Gyungsang Province,
early 1970s.

34

Social and Agricultural Change in Korea's Rice Farming Communities

Kwang-Kyu Lee

Korean society traditionally has been supported by two main pillars: Confucianism and agriculture, especially rice cultivation. Rice has been the main dish of the Korean people since the Neolithic period. Not only flat plains but also mountain slopes were cultivated as rice fields. Flat rice fields were watered from dikes, and the terraced paddies on the slopes, looking like stairs to heaven, from mountain springs.

In the past as well as in recent periods, the criteria for a good or a bad overall harvest depended solely on the rice production. Even a poor yield of other grains and vegetables was not considered a bad harvest if the rice production was good. The Korean government never intervenes in rice production even though many other sectors of Korean agriculture are now forced to compete with producers in other countries due to international trade agreements.

Many aspects of traditional Korean culture have their roots in the rice cycle. Rice itself was considered to be a spiritual or divine entity, preserved in a small jar that was placed in the main hall of a house. There is a proverb, "one rice seven pounds," meaning that seven pounds of peasant sweat went into a single grain of rice. This emphasizes the high value of rice. Besides rice as grain, rice straw, too, was needed for many material uses around the house and fields. For example, straw shoes, straw roofs, and many different kinds of straw mats indicate the great importance of rice straw in rural areas.

During the past half century, Korea underwent rapid socioeconomic change as it developed from an agrarian society to an industrial one. A key indication of this change is the ratio of agricultural workers to industrial workers. This was formerly 7 to 3 but has been reversed; by 1986 the ratio was approximately 2 to 8. Nowadays less than 20 percent of Korea's total population of 45 million works in the agricultural sector. This demographic shift has brought tremendous transformations in agriculture, including in the rice cycle. Social changes are also apparent in families and in rural communities, including enormous alterations in the value system and worldview.

The Modernization of Korean Rice Farming

After liberation from the Japanese occupation in 1945, Korean industrial production dropped to one-fifth of its 1940 level. South Korea recovered consumer production by 1950, but the economy again sank into chaos during the Korean War (1950–1953). Afterward, aid from the United States promised one means of economic reconstruction,

34.2   A farmer's band, joined by
villagers for *jisin balgi*,
the worship of the village
god, marches to each house
to invoke good luck.
Photograph by Kwang-Kyu
Lee, near Pusan, South
Gyungsang Province,
early 1970s.

34.3   A straw shelter covers a large storage jar (*onggi*) filled with *kimchi*, which has been lowered into the ground for optimal climate control. The straw structure in the background, called *toeju*, marks the location of the house god. Photograph by Kwang-Kyu Lee, South Gyungsang Province, 1968.

but this aid was planned by the givers and not the receivers. Using the slogan of "Let's Overcome Poverty," the Third Republic under Chung-Hee Park launched a new five-year economic plan in 1962, promoting scientific technology and the founding of industries. This plan was designed to boost production of export items and was dependent on foreign capital and foreign technology. South Korea's real industrialization phase began to set in during the second and third consecutive five-year plans, establishing heavy industrial complexes for steel, mechanical, chemical, electronic, and naval industries that produced for both domestic and foreign markets. More export-oriented consumer products, such as televisions, VCRs, automobiles, clothes, and household items, were produced during the 1970s and 1980s. This period also saw Korea's first exports of software and capital goods. Taken all together, these achievements earned Korea the epithet of the "Miracle on the Han."

While South Korea's industry prospered rapidly during the 1960s and 1970s, Korean farmers had to wait for development and export-oriented production. There was, however, growing demand for food in the cities. In 1955, South Korea's total population was 21 million, a figure that climbed to 43 million in 1990. The rural-urban population ratio was 77 percent to 33 percent in 1971; by 1990, 75 percent of the population lived in urban areas.

The development motto "Escape from Poverty" addressed the government's goal of reaching national self-sufficiency in food production, especially rice. A new high-yield variety of rice developed by the International Rice Research Institute, known in Korea as "Unification Rice," was introduced in 1971. The government started an extensive propaganda campaign and enforced the cultivation of the new grain by farmers. At the same time rice production necessarily became mechanized due to the shortage of human power in rural areas. Instead of humans and farm animals, small multipurpose tractors were used not only for cultivating the land but also for transport and as a mobile power plant for irrigation, harvesting, and threshing. Mechanical methods were even developed for transplanting seedlings. New production technologies—such as the use of vinyl row covers, fertilizers, and agricultural chemicals—have shortened growing times for rice.

One of the most important aspects of agricultural change in Korea is the use of fertilizers. In traditional society, animal dung and human excrement were considered "good fertilizers." Weeds and ashes were also worked into the rice fields. During the early part of Korean industrialization, several companies launched large-scale production of chemical fertilizers, which were among the top export commodities during the 1970s. Traditionally, farmers had to weed at least three times after transplanting the rice into the fields. Nowadays, special chemicals take care of weed and pest control, disinfection of the seed, and sterilization of the fields. The introduction of these agrochemicals has, however, had a cost in terms of worker health, loss of soil biodiversity, and changes in the biosystem.

Recently two new methods for rice cultivation have appeared, "direct seeding" and the "duck method." Neither technique is yet uniformly applied nationwide. The former involves directly seeding the rice with airplanes and large seeding machines in open areas that are large enough to obviate the need for transplanting. The latter consists of letting ducks into wet-rice fields to weed and organically fertilize around the plant stalks after transplantation.

The tractor first appeared in Korea in 1959, when it was imported from the United States and Japan. In 1962, the Korean farm supply industry produced a remodeled, multipurpose tractor adjusted to Korean circumstances. Since the advent of the new tractor, the use of oxen as draft animals has disappeared. Tools such as plows, hoes, spades,

34.4   The rice loop game (*jajun-nori*) is a harvesttime amusement in which two loops—one male and one female—have to be linked together. The contest symbolizes the interplay between nature and humans, which may either produce a bumper crop or result in failure. Photograph by Kwang-Kyu Lee, near Andong, North Gyungsang Province, early 1970s.

34.5   A farmer performs the spring plowing with oxen in the early 1970s, just prior to the widespread adoption of mechanized plowing. Photograph by Kwang-Kyu Lee, South Gyungsang Province.

sickles, and weeding tools—many of them uniquely Korean—have also vanished from the Korean farm. New farming tools and machines are now manufactured according to the needs of mechanized farming, including cultivators, soil tillers, large and small tractors, water pumps, transplanters, sowing machines, combines, winders, and threshers. Nowadays villagers own some of the larger machines collectively.

Social Change in Korea's Rural Areas

In the past, the basic rice production unit in a rural community was the household. However, members of a village carried out many farming activities jointly. The village was therefore the organic unit representing community life, including farming. All male members of a village would carry out work organized as a *durae*, a common or cooperative task force. For example, a *durae* could be organized to cultivate all the land in the village. One representative from each household had to participate not only in transplanting rice for all *durae* members but also in communal work such as repairing village roads and paddy dikes. Formerly, many rural *durae* owned their own "farmers' flag" and had their own musical band. Today there is no longer an organized system of mutual help in the traditional sense of the *durae*, although sometimes several people may pool assets to buy large machines like harvesters.

Individual labor exchanges, called *pumashi*, are another form of mutual aid that villagers engaged in. The work involved ranged from farm labor to a variety of family affairs, including the execution of funeral tasks and helping to prepare for wedding celebrations.

One of the most popular types of associations in rural areas in the past was the financial association called a *kye*, organized according to village members' needs. A widespread form of *kye* was that organized by villagers who had elderly parents. When a member's parent passed away, all of the association members shared the expenses of the funeral and helped the grieving family in all aspects of the arrangements, including the preparation of the portable bier. Farmers typically participated simultaneously in several *kye* for different purposes. There are still many *kye* in rural Korea today, but they tend to be commercialized associations for the purpose of saving money.

Above all, the traditional village in Korean agrarian society was protected by a particular village deity that symbolized the common fortune of the entire community. The people of the village not only worked together but also shared feelings about their common fate. The worship of a village god entirely disappeared, however, after the government introduced the "New Community" movement (Saemaul Undong) during the 1970s.

Important demographic changes have also occurred in the family life of rural areas. The average family size has decreased from 6.5 persons in 1963 to 2.8 persons in 1990. The most common family type of traditional rural Korea was the extended family, but today it is the nuclear family. The extended family was ideal for rural conditions because married sons lived with their parents, strictly oriented toward the paternal line. Practically speaking, the real family was a stem family, i.e., the first son lived together with the parents after his marriage, while the second and younger sons had to leave the parents' home. Change began when many oldest sons decided to move to the city or form new nuclear families, meaning that the son who lived with the parents was no longer the oldest son. Nowadays many young rural Koreans do not want to live with older people anymore, and many old couples can be found living by themselves. Only when one has died does the other move in with a son.

34.6   Threshing was formerly
accomplished by pulling the
grain-bearing stalks through
a metal comb. This tradi-
tional activity, performed in
the village, was subsequently
replaced by mechanical
threshing directly in the
fields. Photograph by
Kwang-Kyu Lee, 1966.

34.7 Another old method of threshing, using a foot-pedal-driven threshing wheel, was also accomplished in the yard of a farmer's home. It was the advent of mechanized threshing machines, which cooperative members lined up in the fields to use, that lead to the relocation of threshing away from the village. Photograph by Kwang-Kyu Lee.

34.8 Men repair a belt-driven threshing machine, a piece of equipment that represented advanced technology in the 1960s. Photograph by Kwang-Kyu Lee, near Pusan, south Gyungsang Province.

Beginning in the 1970s, many new consumer goods could be found in Korean homes. First the electric rice cooker appeared, followed by the radio, then the sewing machine, the record player, the refrigerator, the TV set, the telephone, the washing machine, and finally the car. Today, more than half of all people under the age of forty own a car. This clearly reflects the elevated standard of living in South Korea's rural areas, where the generally low average annual income has increased steadily from the 1960s until today. Village housing standards have changed as well; the traditional houses thatched with rice straw are being replaced by apartment-style houses with slate roofs, similar to those found in urban areas. The fuel used for heating has changed from wood or straw to smokeless coal, and then to oil and gas.

Rural social change was accelerated by the Saemaul. This wide-ranging government policy was launched in a top-down manner in 1970 with the goal of increasing community diligence, self-help, and collaboration. Because of strong governmental funding, the movement soon spread throughout the country. In each community a Saemaul leader was nominated to work beside the village head. New community songs, village flags, and uniforms were devised. There were also Saemaul women's associations. New community halls and libraries were built, and in the morning, one would wake up early to the sound of the new community song, broadcast over the village with a loudspeaker system. All villagers were supposed to work together day by day to improve their community and the adjacent environment.

According to Saemaul policies, each household had to change first its roof, then the toilet, followed by the kitchen. Step by step, the entire house and yard were to be renovated, including the construction of larger and better storage facilities for agricultural products. In the 1980s, the straightening and widening of roads and the building of bridges altered the village infrastructure. In some cases entire villages were removed where they posed an obstacle to progress. The income of farming villagers gradually increased, which in turn supported the buying power of the Korean countryside.

In the 1990s, the government granted additional financial aid to model communities, after classifying rural areas into either highly successful, successful, or less developed villages. At the same time the government encouraged the rotation of vegetable crops planted next to rice.

Farming has become increasingly commercialized since the 1980s. Landowners now employ laborers for the various stages of rice cultivation, including seeding, transplanting, applying chemicals, and harvesting. In the past these tasks would have been taken care of through the mutual aid practices of *durae* and *pumashi*. The landowner used to invite his neighbors as labor partners and bring quantities of food to the fields for them. The sharing of hard work and food served as a unifying force and a source of enjoyment for the whole village. This nostalgic scene has disappeared. In modern business-oriented farming, the landowner pays a wage for daily labor, while the workers bring their own lunches.

There are now farm managers and produce brokers who manage vast areas of the Korean agricultural market. They may conduct research to find the best quality of agricultural chemicals, select the most qualified laborers, and rent the least expensive machines. During the busy season of rice transplantation, groups of specialized workers from urban areas, including both men and women, offer skilled labor. These are people who gained agricultural skills in their youth before they moved to the cities and can now capitalize on them during periods of high demand for agricultural labor. The groups are contracted by the farm manager, who manages the entire enterprise from an office room.

34.9   This ancient tree was considered to be a protector of the
village and its environs.
Villagers worshiped the tree
as their god by bringing food
sacrifices year round. At the
base of the tree lay a sacred
rice-straw bundle, a symbolic
manifestation of the same
god (see fig. 34.10).
Photograph by Kwang-Kyu
Lee, Kyunggi Province, 1977.

The relocation and remodeling of entire villages created a very different atmosphere in today's rural South Korea. A renovated house does not provide a place for rooftop vines to grow, for example, in the way that pumpkins were once cultivated on roofs thatched with rice straw. Neither cows nor cow stables are tolerated anymore near modern houses, nor do grazing cows dot the landscape. There are no more tools for cutting the rice straw to feed the cows. In the past, cows were raised to support major family needs, such as a wedding, or a child's education, or the sixtieth birthday celebrations of the parents. During the summer, boys would be given the task of fetching the family cow home from a nearby hill in the evening. Then the cow would be fed with rice chaff and corn, which were cooked together in large iron pots. Even the pots are no longer found among the utensils of a modern household.

Another by-product of the Saemaul has been the great damage done to religious life in the countryside. At home there usually was a god for the head of the household, a god for fire and the kitchen, a god to protect the buildings, a god of the house site, a god of wealth, a god of the toilet, and a god for the entrance that was worshiped by the housewife. But now there are no gods worshiped in the home.

On the village level, there was a village god, a god for the front gate of the village, and a road god symbolized by a heap of stones. The single most important Saemaul objective for the government was the destruction of the village gods. The worship of village gods was frowned upon, and shaman rituals conducted for family welfare and for rain gradually also disappeared.

The annual array of celebrations was once closely aligned to the seasonal cycles of rice cultivation tied to the lunar calendar. Some of the most important holidays in the past were the Full Moon Day (Daeborum) in the first lunar month (approximately January 15), the Day of the Weeding Hoe Cleaning (Baekjung) on the full moon of the seventh month (about July 15), and the Full Moon Day (Ch'usuk) in the eighth lunar month (around August 15). With the advent of Saemaul, however, the government enforced the usage of the solar calendar. In an act of compromise with rural people and their traditional system of holidays, New Year's Day (Sulnal) was pronounced a folk holiday in 1986. It took until 1991 for Lunar New Year's Day and August 15 to be declared national holidays.

Many games that had sprung from the rice cycle have not survived Saemaul innovation. The *yut* game, the Moon-House Burning (Daljiptengi), Weeding Hoe Cleaning (Homishikgi), Stepping-on-Earth Deities (Jishinbalgi), all of which were performed on the Full Moon Day of the first lunar month, are only a few examples of traditional Korean games that have disappeared. Last, but not least, folk songs have not fared well since the advent of radio because they, too, were not thought to be in accord with progress.

Women in traditional society were expected to be obedient and submissive to the family and parents-in-law. Nevertheless, a rural housewife in old Choson (1392–1910) was accustomed to managing a fairly large arena with many activities related to the household and cooking. Usually she worked around the house and commuted between the well, the outdoor platform where storage jars (*onggi*) were kept, and the kitchen—even in the cold of winter. In a modern house, her realm has literally shrunk to such narrow limits that she may well think of herself as being a caged bird. If she then compares her circumstances with those of the well-dressed ladies she sees on TV, the result will be complaints against rural life and her husband.

34.10   A closer view highlights the
sacred rice-straw bundle seen
in figure 34.9. The symbolic
meaning of the bundle was
based on the eternally sustain-
ing power of rice, interpreted
as an implicit indication of
the multigenerational inter-
dependency of humans,
ancestors, spirits, and nature.
The tree symbolically protected the village, resulting in
the yield of rice, while the
villagers in turn sustained
the tree by building a protective wall around it. Photograph by Kwang-Kyu Lee,
Sungdock Ri, Gangha
Myun, Yongpyong Kun,
Kyunggi Province, 1977.

Ongoing innovation in the agricultural sector has put farming in a financial and social position subordinate to that of industry. Many farmers face heavy debts from agricultural cooperatives for expenses accruing from inorganic farming and the purchase of heavy machinery. But the price of rice, for example, which is fixed by the government, is always lower than the costs invested in production. Moreover, modern mass culture has created a negative image of rural lifestyles. Rural people, in turn, often feel uncomfortable when confronted with modern culture. At the same time, traditions and traditional ways of thinking have no place in the new ways of urban life. The result is often a certain chaos and disarray in the consciousness of rural people, the consequences of which may be seen in increasing divorce rates and lax moral attitudes on the part of younger people in rural areas.

Rice in Modern Korea

Even though there has been extensive agricultural reform followed by vast social changes in rural areas, rice has always remained among the primary interests of Korean people. Polished twice to become white rice, it is a symbol of prosperity. Recently even unpolished rice has achieved a new status as a health food. Other new rice-based health foods are now advertised, such as ginseng rice, which consists of unpolished rice mixed with ginseng powder. The nutrient-rich by-products of rice milling are used as ingredients for vitamins A and B and mineral supplements. Many new kinds of rice cakes are on the market, catering to the tastes of urban people and particularly young people. Rice wine products are strategically marketed to compete with Western alcoholic beverages such as whiskey and grape wines.

In the process of becoming a modern industrialized nation, South Korea seems to have stolen the reputation and pride of its rural people and their orientation toward rice agriculture. Younger generations in urban areas, for example, no longer eat rice for breakfast but toast and coffee. Nevertheless, many Koreans still have three meals a day with rice as the main component. Korean farmers have no apprehensions as long as they have a good rice harvest, even if their harvest of other crops is poor. Korea imports around 70 percent of its food, with the exception of rice. Currently the country produces a surplus of rice. As long as this continues, Koreans will be satisfied, even though they may have to import almost all other crops. In an attempt to satisfy Korean mainstream culture, rice farmers are now competing in a new environment and trying their best to survive in a commercialized and commodified society.

•

35

Rice in South Korean Life: The Transformation of Agricultural Icons

Michael Reinschmidt

Introduction

If one were to search the traditional Korean rice cycle for symbolic and material manifestations that express the essence of Korean life, one would not have to search long. There were four realms—religion, economy, society, and the land itself—that together formed one's sense of identity and purpose in a life surrounded by rice. A similar search conducted today would require more subtle observational methods, but it would provide a rich yield nevertheless. Some material objects used on a daily basis may follow us into the remotest corners of our existence (for example, the laptop computer or the cell phone), and we may find ourselves adopting the symbolism of such things—at least in part—to define ourselves, our lives, and how we relate to our societies (e.g., a laptop may symbolize progress, a solid work ethic, and personal mobility). Such self-definitions through material things incorporate the social values that have been attached to various goods in a way that most members of a society agree upon. The gaining of status, enhanced feelings of importance, higher pay, and membership in a distinguished social class or age group may be among the social gains we expect from socially defining material objects.

In South Korea the above-mentioned four realms, the focus of the present essay, are still imbued with the presence of rice, but its overall visibility has faded into the background of modern life. There are still peasants—called farmers today—making a living in rice agriculture, but their number has dwindled significantly since the 1950s. Overall rice consumption has dropped in favor of rising food imports from the West. Even so, hardly a day passes during which a South Korean would not eat rice at least once, while North Koreans, for different reasons, would consider themselves lucky to have a guaranteed once-a-day rice meal. Due to intensive urbanization, however, most South Koreans do not conduct their lives according to the once-so-prominent traditional rice cycle. In which ways—subtle or prominent, real or symbolic—does rice still find a place in South Korean life today?

Rice in the Life of Today's South Koreans

Today's South Koreans encounter rice mostly in commercial settings. If a rice consumer doesn't obtain bulk supplies from the rice merchant at the corner or directly from a befriended farmer, she can go to any supermarket or department store where she will find household-size sacks of rice surrounded by a large array of modern rice products, such as processed rice in drinks, snacks, instant dishes, and cosmetics. She would not, however, find items such as rice-straw shoes, handcrafted rice-straw baskets, and window blinds densely woven with the finest rice-straw strands. Such items, along with artsy books about

35.1   Despite fierce competition from other foods, rice maintains its hold in a whirl of change. The proprietor of this rice store displays some of his attractive products: sweet fried rice cookies (*ssal gangjong*, the whitish snacks right in front of him) and sweet rice and chestnut pieces (*yagshig*, the display in front of the *gangjong*), among a variety of other traditional foods. Photograph by Michael Reinschmidt, Andong City, Korea, 2000.

Korea's glorified past, can be found in the store's (at least in the trendy department stores in Seoul) folk section that usually attracts more foreigners than locals. In the dining area of the very same department store she can enjoy modern rice-based food hybrids that reflect the clashes of local and global cuisines (fig. 35.2).

After this modern culinary experience with a traditional touch, our homemaker may pay her bill with one of the many and varied credit cards issued by the financial branch of Nonghyob, the largest national agricultural association that has evolved directly from the old rice world into a modern Jaebol (a term that has entered many world languages, usually referring to a family-owned or founded Korean conglomerate) of its own kind. She also may have conducted her shopping at one of the countless Nonghyob supermarkets and associated *sintoburi* (lit., "body and earth cannot be separated"; figuratively, since the Korean body cannot be separated from its native soil, one should buy Korean foods from Korean soil) stores from which she may carry away bags full of rice products. She needs those in the evening to entertain her female friends who may convene at her residence—on a thin rush mat that has replaced the old rice-straw mat called *mongseok* (see fig. 35.10)—to have a *kye* meeting, while an occasional rice commercial may flash by on prime-time TV. The *kye* was—and is—a significant means of providing for instant cash if large, important purchases must be made. To the dismay of her Nonghyob bank, she, like so many other Koreans, still engages in this traditional credit-sharing practice.

While the women discuss *kye* business, their husbands may gather in a small restaurant (*sigdang*) around the corner for a hearty rice-based dinner to debate political issues such as the protection of the domestic rice market from foreign rice imports (figs. 35.3, 35.4). They also may frown upon their children, who are lured into a modern pop culture that encourages youngsters, among other things, to eat more "unhealthy" international foods and less "good" Korean rice. (That is one reason why their wives endorsed the advent of domestically grown and processed instant rice, which helps to ensure at least a minimum daily intake of Korean rice.) The doctor among them tells his friends about the undeniable increase in the percentage of obese children during the last ten years. In the old days this wouldn't have happened, due to a diet largely based on rice and the hard physical work in the fields. Until recently, most mothers spent much time and effort preparing rice-based lunch boxes for their children. They sacrificed their dawn hours of sleep, for the freshly made food was considered fundamental for a child's health, growth, and mind. Automated food commerce has entered schools in South Korea, as elsewhere in the industrialized world. If the young people today do consume rice at all, so the men in the restaurant may lament, then it happens in a cultural realm that used to be dominated by chopstick etiquette but today is corrupted by pop, neon, and hip-hop and void of any—if they ever existed—good old table manners. Finally, with increasing consumption of vodka-like rice liquor (*soju*), rice wine (*makkoli*), and American beer containing rice to lighten its color, they may nostalgically grieve the fact that they somehow were forced to leave behind their rural homes, their aging parents, and, most importantly, their family farms and paddies perhaps snugly sited in a beautiful rural home-land-scape. At last, one of the men who studied abroad for several years insists they should not worry too much about all that rice stuff, namely the relationship between rice and young people and between rice and modern Korea. In the end, he says, the youngsters, many of whom dream of attending foreign universities, will appreciate rice after they have lived in a foreign country. He himself came to appreciate the distinct taste of Korean rice and its overall value mostly because of his foreign experience. However, he may state that it is important to keep a thumb on the government so that it comes up with concrete ways to protect domestic rice growing, provides strong incentives to maintain the central role of rice in Korean culture, and avoids the nutritional mistakes of the West.

35.2   The Korean rice burger, a late 1990s creation by the Lotteria Corporation, enjoys great popularity among young urban South Koreans. Served in the hip neon ambiance of Lotteria's restaurants, this hybrid burger contains both meat, which caters to young peoples' wish to identify with a "modern" lifestyle, and rice, which alludes to the "traditions" of the nation's old rice culture. Photograph by Michael Reinschmidt, Seoul, Korea, 2000.

Material Objects in Cultural Contexts

I will bolster and append this fictionalized account with a materialist ethnography emanating from fieldwork observations and archival research conducted in the Republic of Korea between 1997 and 2002. To understand traditional Korean rice life, it is helpful to examine the material culture of rice farming; tools, utensils, furniture, and dogmatic signs (reinforcing the existing agricultural ideology) provide a rich representation of traditional Koreans living with rice, reflecting many of their ideas, needs, and hopes. This excursion into Korean rice traditions will also explore the multiple transformations of rice within South Korea's modern world. While the text will focus on material objects, my digressions into subjects only lightly tied to materialism provide a necessary backdrop that reveals cultural change by contrasting views of the past with those of contemporary developments in South Korean popular rice culture.

Material objects surrounding the traditional Korean rice complex can be sorted into five major categories: farming implements, household utensils, special rice-straw creations, storage containers, and semiotic objects (such as flags carrying messages in support of agriculture) directly speaking from the rice cycle. All can be seen as strong icons that symbolize Korea at national, intranational, regional, local, and personal levels and have significant physical, psychological, linguistic, and cultural effects on the lives of Koreans. Aesthetics and refinement of objects, thus providing deep insights into aspects of politics, the economy, religion, language, social class, gender, and age, can range from creative states of crude to highly sophisticated, depending on functional and representational needs. For reasons of brevity, the five areas are introduced here through a selection of representative objects.

35.3   Hoe (*homi*). Korea. Circa 2000. Metal, wood. L: 32 cm. FMCH X2000.30.12.

Farming Implements

Farming implements had to be highly practical because they helped produce the crops needed to sustain people, animals, and the land in a harsh climate. Although farming tools are similar in many respects all over the world, varying local climates, soil conditions, and tool-handling customs account for regional differences. The Korean hoe (*homi*), for example, counted among the most important and versatile implements in farming and was mainly used for weeding and soil treatment in the rice fields (fig. 35.3). It could also be used for gardening, yard work, trail construction, plowing, and defense during times of confrontation. The Korean *homi* is unlike hoes of the neighboring peoples. The Chinese, for example, largely used comparatively heavy hoes with thick, left-right symmetrical and round-edged blades. But increasingly small, curved, and pointed *homi* could be found in Korea the farther south one moved into the peninsula away from Chinese influence, with the smallest ones occurring on Jeju Island. What causes such differences in shape and handling? In regions with little rainfall, we find hoes with wide bases and rounded tips, whereas areas with high precipitation feature small hoes with pointed tips. Since the roots of weeds don't grow as deep in hard subsoil as they would in a soft ground, a broad hoe-blade is sufficient for uprooting the weeds, whereas a hoe with a small and pointed blade helps the user to lift the deeper roots from the wet subsoil (Nonghyob Agricultural Museum 1997).

Rice cultivation gradually entered the peninsula until about the fourth century B.C.E. and with it came the need for increasingly refined hoes. Antler, stone, and wooden hoes were superseded by iron hoes during the fifth and fourth centuries B.C.E. (Nelson 1993, 58–110, 174). The Korean hand hoe consists of an iron piece with the blade on one end and a pointed rod, onto which a wooden handle piece is driven, on the other. Wooden handles could be short, medium, or long. Otherwise unadorned, the wooden piece is sometimes replaced by, or densely wrapped with, braided rice-straw rope to facilitate a better grip on the tool when the hands are sweaty. These hoes are called *nonhomi* with the particle *non* (lit., "rice paddy") indicating that they were preferably used in the fields during wet periods, where the rice-straw rope would absorb not only sweat from the hands but also ground moisture if the tool fell into the muddy paddy or, more likely, got rained on. The hand hoe required—as does a sickle—excellent handling skills in order to avoid injury to the user because of the potential danger that could be wrought with the inward-turned blade (Lee 1994, 25; Nonghyob Agricultural Museum 1997, 38, 39, 84).

The *homi* remains a famously unique symbol of the countryside although it is needed as a "rice-winning" tool neither by millions of urban office, high-tech, or industrial workers nor by farmers who nowadays weed their paddies largely with mechanical or chemical means. Everything that distinguishes Korea and Koreans from the rest of East Asia, and especially from the West, however, is gladly upheld in order to emphasize Korean uniqueness. Today *homi*—together with other traditional farming tool artifacts—are used to point out this uniqueness to thousands of annual visitors to the folk villages scattered across the country. Among the traditional artisan workshops of these villages are blacksmith shops that manufacture *homi* and many other iron tools originally needed for farming. I bought a *homi* at the Suwon Folk Village in 2000 after I watched it being wrought by the local blacksmiths. There are hundreds of such traditional icons of the past sold every day for token prices to buyers who take them home mainly with a desire to make a connection with a bygone way of life. They also might wish to be reassured of their distinct identity as a people. Having said that, I am convinced that quite a few people will also actually use the *homi* in their small suburban backyard gardens.

Another Korean farming implement worth mentioning here is the Korean carrying frame (*jigae*). This backpack rack was simply called an "A-frame" by United Nations soldiers during the Korean War (1950–1953) for it has the shape of an *A* (fig. 35.6). The *jigae*, with its typical A-frame, can be found only in Korea and thus, like the *homi*, can be used to represent a part of the national identity. Unlike the *homi*, however, the *jigae* still plays a vital role in most areas of the contemporary transportation sector. Carrying large and bulky amounts of crops or other heavy loads on the shoulders with a *jigae* is admittedly difficult but nevertheless practical and energy efficient in the often winding and terraced terrains of the rice paddies. A *jigae* was usually made of two oak saplings each of which had a branch protruding backward. Either a cargo platform or a basket is supported by these branches. To help soften the load, a cushion made of rice straw was inserted in the midsection between the saplings. Two thick rice-straw braids served as shoulder straps, which demonstrates the enormous strength of rice straw. When put on the ground, the frame was supported by a cane with a forked head so that the device looked like a tripod from a distance. In this position, the *jigae* could be loaded or unloaded. Almost anything movable has been carried by men with a *jigae*, from plows to manure barrels, from living livestock to a tower of ceramic *onggi* jars, as well as firewood, furniture, construction materials, rice-straw bundles, sacks of rice and other grains, and even ammunition and weapons during the Korean War (Nonghyob Agricultural Museum 1997, 40, 62, 116; Knez 1997, 147). There are still many traditional *jigae* used in agriculture and by porters around most markets in Korea. In the big cities one can see short-distance carriers hauling all kinds of cargo, including industrial materials, appliances, raw textiles, finished clothing, sacks of raw rice, and boxes of instant rice foods. Most saliently, however, the *jigae* occupies a special place in the "national museum" of Koreans' minds because of its widespread use during the Korean War, when thousands of fleeing families tried to rescue the bare minimum of their possessions by loading them tower-high on their *jigae*.

Interestingly, the *jigae* has undergone a modern-day adaptation in Korea. Cargo bicycles and especially heavy motorcycles (seen in many of the mercantile areas in Seoul, such as in the Yongsan electronics market, Hwanghakdong market, and the big markets at the old city gates) are fitted with a tall device constructed of steel pipes in the A-shape resembling the old *jigae* form. The structure is mounted right behind the driver's seat so that boxes or sacks can be piled up on a platform and tied against the A-frame. Drivers never have to wait long around the markets to get hired by shoppers who want to take home large bulky items, including TV sets, refrigerators, large rolls of wall-to-wall carpet, or even the largest pieces of furniture imaginable. It is breathtaking to see the kinds, shapes, and sheer quantity of loads transported this way, by drivers skillfully maneuvering their vehicles at high speeds through thick traffic. The mentioning of such adaptations may seem irrelevant to the topic at hand, but I have spoken to a few identity-conscious Koreans for whom the sheer sight of cargo bikes serves as an instant opportunity to link themselves with the lives of their ancestors from the rice cycle.

Household Utensils

Since rice is present everywhere in Korean food culture, there are many household utensils and devices that have become intimately associated with rice. One of these symbols is the *ssal duiju*, or rice storage chest (fig. 35.8). The *duiju* probably assumed highest rank among all furniture in the house because of its powerful appearance and its central role and location in the household. If the contents of the *duiju* could not be replenished, it literally meant the household was bankrupt. The fact that a woman was considered a household's senior female only after she achieved undisputed access to the stored rice and to the condiments bay (*jangdokdae*) also attests to the significance of the *duiju*. Moreover, the inherent

35.4   "Dear fellow citizens, we are concerned about your health; therefore, [buy] rice from Kangwha Island" (subway advertisement). Photograph by Michael Reinschmidt, Seoul, Korea, 2000.

35.5   "The most famous name under the sky is Pochon rice wine [*makkoli*], made by the Idong Company" (department store advertisement). In addition to their economic functions, signs like these help people form positive opinions about the regions where rice products come from or where they were born and raised, and thus form part of their identity. Photograph by Michael Reinschmidt, Seoul, Korea, 2000.

power conferred by rice was buttressed by impressive rice storage facilities such as the *duiju*, the *onggi* jar, and the *narakduiju*, the latter being the traditional Korean outdoor silo for grains (fig. 35.9). Finally, these containers were never allowed to go empty lest a decline in "rice power" result in famine for the family or the entire community.

Rice chests were generally made of thick elm, pine, or zelkova wood and measured around eighty centimeters high, seventy centimeters wide, and sixty centimeters deep. If they were ornamented, it was only with the oil-finished wooden grain features most visible on the front boards. Another type was ornamented with black iron fittings, especially in the north of the country; such were preferable to shiny brass because of the latter's tendency to tarnish more easily. A chest was usually placed in the *mareu* section of the house, a part-porch and part-entrance area under rice-straw eaves that provided shelter and ventilation for the stored rice grains. Straight and sturdy square legs lifted the case portion of the chest about fifteen centimeters off the ground to help keep the contents ventilated and free of rodents. The lid of the chest consisted of two panels. The front panel was either hinged to the hind part or, if not hinged, fitted with two short shafts attached to its underside. The lid's shafts could be positioned to prevent unauthorized opening when the front of the chest was locked. Up until the 1960s, *duiju* could be seen in middle-class and upper middle-class homes, while poorer households used *onggi* jars for outside rice and food storage. When plastic *duiju* became available during the 1970s, they were modified for modern, smaller households and also featured a grain release near the bottom of the chest so that things could be stored on top of the box. During the 1980s, plastic chests replaced most of the traditional *duiju* (Knez 1997, 136), although in the 1990s new wooden versions reentered households as part of a national wave of nostalgia for Korean traditions.

Large *duiju* of the traditional kind (see fig. 35.8) were commonly used in the Gyeonggi region near Seoul, whereas in the southern provinces more refined versions could be found, including elaborate two-story chests. The upper level held the rice while bottles, pots, small baskets, and ceramic bowls were kept in the bottom compartment. Despite some occasional ornamentation, *duiju* were meant to be functional and not necessarily beautiful. They were located around the kitchen, a place of dirt and hard work, where no one was supposed to rejoice in the appearance of beautifully made furniture (Wickman 1978, 84–88; Wright and Pai 2000, 81, 133).

But that's what the *duiju* is: a highly dignified household object commanding respect due to its impressive appearance. It demands inclusion in a group of symbols that inspire feelings of belonging and pride in Koreans even as they go about their lives in the modern world. While the large, old, wooden *duiju* no longer fits in the small living spaces of urban high-rises, there is a fresh demand for this object, which symbolizes rice culture like almost no other object from traditional Korea. This demand has led to the production of a smaller *duiju*-like item called a rice box (*ssaldong*), available in the common furniture markets. It is a sort of miniature *duiju* that can be stored beside electric kitchen appliances or even placed on the kitchen counter next to the microwave oven. The *ssaldong*'s size suits the needs of the modern Korean nuclear family. It also mirrors the decreasing per capita rice consumption of Koreans. Nevertheless, as a symbol the *duiju* still claims a central place in the image that Koreans have of themselves and of their homeland.

Rice-Straw Creations

Rice-straw creations played an indispensable role in rural Korea. There was an entire rice-straw complex that increasingly flourished since the introduction of rice during the second millennium B.C.E. Rice straw was used for thatching, insulation, animal fodder, fertilization, decoration, and ritual practices on one end of the spectrum, and for utility objects such as rush mats, fine curtains, baskets, bags, ropes, strings, threads, capes, storage

35.6   Carrying frame (*jigae*).
Korea. Early twentieth
century. Wood, rice straw.
H: 127 cm. American
Museum of Natural History
70/13289.

35.9   The traditional granary (*narakduiju*) is a sophisticated structure whose walls are woven with bamboo slats. The granary is elevated on a clay pedestal and covered with a rice-straw roof. The inside is tightly sealed with clay to prevent rice from leaking out or rodents from getting in. Photograph by Michael Reinschmidt, Suwon Folk Village, Korea, 1999.

color is nothing without black-and-white contrasts, Korean food would be nothing without the stabilizing presence of rice.

Depending on the occasion, next to the rice bowl there is always a bowl of soup. Rice and soup are surrounded by three, five, seven, nine, or twelve side dishes. A table set with twelve side dishes is called *surassang*, which means "table set for a king." Tables are set according to two basic groups of requirements, one for the ordinary setting (*bansang*) and the other for the special purpose setting (*juansang*). A *bansang*, for example, consisting of three or more side dishes, even today could be a standard everyday lunch. It is served on the *soban*, the low individual table, and contains a bowl of rice as the main element, surrounded by a bowl of clear soup, a bowl of kimchi, and a bowl of meat or fish mixed with vegetables (fig. 35.7). No matter how small a meal may be, it is always nutritionally as varied as possible. Koreans appreciate that their dietary culture emerged in large part to suit special human needs related to the unusually varied climate of the peninsula. For example, a Western-style meal of meat, potatoes, and bread is too heavy to be an appropriate meal in the hot and humid Korean summer, when meals must be light and filling, yet still enable a person to have enough physical power to undertake often very hard work. Adherents of traditional Korean food therefore argue that food imports from the West will never be able to nourish Koreans to the fullest extent because they lack nutritional variety.

The *soban* served the practical purpose of being a food tray and table in one. These tables were loaded in the kitchen and then carried to the dining area and placed directly in front of the recipient. Usually the oldest male in the household received the first prepared table. The *soban* also reflected the importance of the "resting culture" in Korea, a pride in being "down to earth" that is seen in Koreans' love of sitting or lying on the ground or floor instead of chairs or beds. To be near the floor was ideal because in the winter one was near the subfloor ducts of the *ondol* heating system that extended throughout the house, and in the summer near the cool ground.

Soban were manufactured by skilled craftspeople who carefully selected boards with decorative openwork grains. The top board, which could be made of round, polygonal, or rectangular shapes, served as a tray with a slightly raised rim. Legs were short and straight or curved and about thirty to forty centimeters long (Lim 1997, 210).

A *juansang* meal for a number of persons required the setting of a large dining table called *gyeojasang*. *Juansang* was rather unusual among ordinary people during the Yi dynasty (1392–1910), but today it is a common feature of traditional restaurant culture.

35.10   Sandals (*chipshin*). Korea.
Circa 2000. Braided rice
straw. L: 25 cm. FMCH
X2000.30.9a,b.

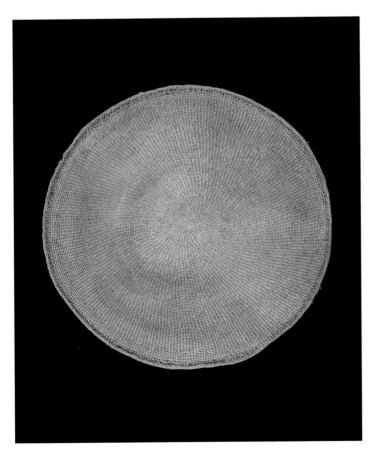

35.11   Rice-straw mat (*mongseok*).
Korea. Early twentieth
century. Diam: 139 cm.
FMCH X2000.30.2.

White rice, a coveted luxury not readily available to the average person, and soup were the main dishes among the nobility (*yangban*), but it was largely the wine that dictated the arrangement of side dishes. Hot soups—like ones made with noodles, rice cakes, or stuffed dumplings—combined with seasonal raw vegetable and seaweed side dishes were all considered suitable. The *gyeojasang* was further loaded with deep-fried fish, boiled meat slices, freshly grilled and/or seasoned meat or fish, and steamed vegetables. The meal was sealed with a dessert consisting of various rice confections, sweet beverages, a small array of freshly cut fruits, and perhaps a little more wine or liquor (Han 1993, 43), which traditionally could have been *makkoli* (especially popular among commoners), *soju*, ginseng wine, azalea wine, and the longevity-granting wolfberry wine (Yoo 1997, 228–237). What is more, imagine the festivity of a *juansang* crowned with the presence of expressive wine ewers from the Yi era or even further enhanced by the stunning company of bottles from the Goryeo period (918–1392).

A description of a set dining table is not complete without mentioning the role of spoons (*sugarak*) and chopsticks (*jogarak*). The use of chopsticks and spoons in Korea goes back many centuries; the earliest evidence dates from the late fifth century. Upscale Korean-style spoons are oval in shape and made of silver—the most popular and "touristy" ones feature the blue and red yin-yang motif in the center. One uses the spoon for scooping rice, stew, and soups, and chopsticks for eating rice and side dishes. Forks and knives are not needed because larger pieces of meat, vegetables, fruits, or rice cakes are cut and prepared in the kitchen. Spoons are also used to eat mixed grain dishes, in which the rice, joined with other cereals and beans, no longer holds together well enough to be picked up by chopsticks. Because a large part of Korean traditional food is still consumed in many varieties of soups, the Korean spoon is thus the busiest one in East Asia (Park 1993, 26–27).

Once spoon and chopsticks have been taken up, they are not supposed to touch the table until the meal is over. They are seen as complementary work companions and must not be separated from their intended modes of either use or rest, i.e., chopsticks handle yang-natured solids and spoons yin-natured liquids (Lee 1994, 81). When not in use, both are placed at rest on either the rice or the soup bowl. According to a recent explanation, "Koreans are trained to use the spoon and the chopsticks correctly from childhood. Using both utensils at the same time is considered bad manners" (Korean Overseas Culture and Information Service 1998, 470). Given what one sees in every restaurant today, however, Koreans seem to be constantly violating this traditional etiquette requirement.

Conclusion

Rice used to be a highly visible and undisputed cultural icon at the center of the "traditional Korean agricultural complex." In the highly intricate modern complex of rice, however, it has been thrown into an arena of competition within the Korean and international corporate food worlds. Symbols of the old rice culture such as the *mongseok*, however, remind Koreans that they have lost the community spirit of the old world. Rice used to be synonymous with life, dignity, and pride in social status, no matter what it may have been. While today status can accrue by means of education, achievement, and wealth, rice seems to be needed only for its nutritional benefits and—to no small degree—for the satisfaction of the culinary habitus. This term—*habitus*—conceptually used by Aristotle and reintroduced by Bourdieu (1990, 52–65), refers to an individual's habitualization of objective social norms, including food habits. Habitus, thus, generates subjective practices that tend to reproduce the regularities existing in the objective conditions of a society.

35.12   Platform for jars (*onggi*)
behind a restaurant.
Although they are not solely
rice storage containers, *onggi*
often serve in storing large
quantities of raw rice (*ssal*),
rice wine, and rice liquor
(*soju*). Traditionally they
were kept in backyards or a
specially designated condiments bay. Today one can
see them also on rooftops or
cemented into the concrete
foundation of town houses.
Photograph by Michael
Reinschmidt, Yoju Town,
Korea, 2002.

35.13   Ceramic storage jar (*onggi*).
Korea. Circa 2000. H: 46
cm. FMCH X2000.30.1a,b.

As South Korea comes to terms with its import obligations as stipulated in international trade agreements (such as GATT and the so called Uruguay Round), the country is exploring new methods of rice cultivation with an eye toward realizing higher and more cost-effective yields. Improvements in science and production are also intended to increase per capita rice consumption through greater publicity and heightened levels of awareness regarding personal and public health and nutrition. In the meantime, rice is reemerging in many material, ideational, and culinary forms. These symbolic forms range from Korean-style sushi and the Lotteria rice burger to the rice varieties listed on the menus of American-style fast-food chains, and from the old and new *jigae* to the new mini rice chest. The *mongseok* has undergone a metamorphosis from the rice-straw versions found in the old farm setting to plastic ones in the modern living room. The *kye*, with its roots in the old rice cycle, is still around to compete with modern financial institutions, one of which is Nonghyob, a Jaebol that grew directly out of its rice origins and in part still tries to revive or at least remind people of the old culture. This preservation attempt is most visible in Nonghyob's impressive Agricultural Museum near Seoul's Sodaemun subway station. The old rice order and its symbolic and visual imagery was successfully tapped during the 1980s to accelerate the country's struggle toward democratization through the creation, display, and powerful protest of *minjung* art. Moreover, the old forms of baskets and storage containers are still found in modern households, and they are internationally highly regarded as fine or folk art, thereby playing a role in vital businesses such as the arts and tourism and globally representing Korean culture and commerce.

According to anthropologist Victor Turner (1967), one symbol can stand for and signify many things, e.g., food as a symbol can have multiple, highly charged meanings. Although rice is not as strong an economic pillar as it used to be, it is still a core feature of Korean identity. The sociologist Pierre Bourdieu uses the phenomenon of "symbol" as a case in point in order to specify how it is used as an important tool to establish sociopolitical function and social solidarity: "Symbols are the instruments *par excellence* of 'social integration': as instruments of knowledge and communication... [and]... festivity, they make it possible for there to be a *consensus* on the meaning of the social world, a consensus which contributes fundamentally to the reproduction of the social order" (1991, 166).

Rice in Korea, being synonymous with life itself, stands for a wide variety of aspects of persons *and* ancestors, including language, politics, religion, economy, ecology, social class, nation, gender, and age. Although diminished in its quantitative status as a daily staple food, rice yet maintains its position as a factor that indisputably allows all Koreans to distinguish themselves as Koreans at various points in their lives and even at death. Unfortunately, taking advantage of the nutritional benefits and the symbolic forces of rice does not seem to be a current priority in the northern part of the peninsula. In the South, however, in the wake of modern agriculture, food industries, gastronomy, and households, rice has successfully adapted to contemporary needs. While still a significant carrier of traditional symbols, values, and meanings, rice has embraced modernity and thus found a new qualitative status.

•

Notes to the Text

Chapter 1 (Hamilton)

1. Madagascar is a special case, as its cultural heritage is in part directly inherited from Southeast Asian ancestors. Rice is of central importance for the island's highland peoples, who have the highest per capita rate of rice consumption in the world.

2. Detailed descriptions of these beliefs and practices can be found in the classic descriptions of rice agriculture (including Freeman 1970, Conklin 1975, and Anuman Rajadhon 1961) and in a continuing series of more recent treatments, including Visser 1989, Ohnuki-Tierney 1993, Klopfer 1994, Trankell 1995, and Helliwell 2001, as well as a great number of individual articles scattered through the literature (see References Cited).

Chapter 2 (Hamilton)

1. This is one of eight rice-planting songs at <http://www.myanmar.com/gov/per-spec/2001/8-2001/ric.htm>.

Chapter 3 (Smutkupt and Kitiarsa)

1. The "Tai" are an ethnolinguistic group dwelling in Mainland Southeast Asia, whereas "Thai" refers to the citizens of the modern nation-state of Thailand, who belong to many ethnocultural groups.

2. Translated by Suriya Smutkupt and Pattana Kitiarsa.

3. This essay is based upon a series of separate short-term field notes made about Isan rice farmers in Nakhon Ratchasima and Ubon Ratchathani in 1995, 1996, and 1998. We have also relied on findings from the anthropological studies of rice culture in this area that we have conducted since 1980, primarily in the provinces of Kalasin, Khon Kaen, Maha Sarakham, and Sisaket. In early 2000, we conducted further ethnographic fieldwork relating to contemporary rice festivals in the Borabue District of Maha Sarakham and the market town of Roi-et.

4. Similar twelve-month rituals are observed in other parts of the country such as Lanna, or northern Thailand (see Premchit and Dore 1992; Phayomyong 1994), and at the royal court in Bangkok (Chulalongkorn 1963).

5. For an overview of the Isan twelve-month rituals, see Priichaa Phinthong (1991, 57–154).

6. An interview conducted by research assistant Chantana Suraphinit with a fifty-two-year-old woman, Nakhon Ratchasima, Thailand. April 6, 1995.

7. An interview conducted by research assistant Silpakit Teekhantikul with a sixty-year-old woman, Nakhon Ratchasima, Thailand. April 5, 1995.

8. An interview conducted by research assistant Silpakit Teekhantikul with a fifty-two-year-old woman, Nakhon Ratchasima, Thailand. April 6, 1995.

9. An interview conducted by research assistant Chantana Suraphinit with an eighty-two-year-old woman, Nakhon Ratchasima, Thailand. April 11, 1998.

10. *Phii*, meaning "ghost" or "spirit," is sometimes spelled *phi* in other systems for transliterating Tai.

11. An interview conducted by research assistant Silpakit Teekhantikul with a fifty-eight-year-old woman, Nakhon Ratchasima, Thailand. April 26, 1995.

12. An interview conducted by research assistant Silpakit Teekhantikul with a seventy-five-year-old woman, Nakhon Ratchasima, Thailand. April 26, 1995. An interview conducted by research assistant Chantana Suraphinit with a fifty-seven-year-old man and a seventy-five-year-old man, Ubon Ratchathani, Thailand. April 17, 1995.

13. For more details on the Rice Mother, see our unpublished paper of 1998 on the subject, "Myths of Mae Phosop (the Rice Mother)."

14. An interview conducted by research assistant Chanthana Suraphinit with a seventy-five-year-old woman, Nakhon Ratchasima, Thailand. April 5, 1998.

15. An interview conducted by research assistant Silpakit Teekhantikul with a seventy-one-year-old man, Nakhon Ratchasima, Thailand. January 16, 1996.

16. In parts of southern Thailand, farmers acknowlege the contribution of water buffalo in the rice cultivation process by making a kite shaped like the head of a buffalo and flying it in the sky over the vast golden rice field. This kind of kite is widely known in the region as the "buffalo kite" (*wow khwaai* or *wow na khwaai*; see Smutkupt and Kitiarsa 2000a, 186–37.

17. Students of Thai culture since the 1960s have approached the twelve-month rituals in Thai rural villages from agroecological, symbolic-structuralist, and functionalist standpoints (Anuman Rajadhon 1987; L. Hanks 1972; Kirsch 1977; Smutkupt et al. 1998, 1991a, 1991b; Tambiah 1970).

Chapter 4 (Krauskopff)

1. The Tharu wilderness was a refuge for the small Hindu kingdoms in the Himalyas following the Muslim "invasion" of India.

2. This situation occurred primarily before the eradication of malaria in the 1960s and varied considerably from one area to another depending on the state of deforestation and the administrative and economic situation of the local Tharu farmers vis-à-vis the central government (Krauskopff 2000).

3. The Tharu population bears a genetic mutation of the coding for alpha-hemoglobin resulting in alpha-thalassemia, which disrupts the proliferation of the plasmodium and gives an extremely high protection against malaria, an endemic and virulent disease in Terai (Modiano et al. 1991).

4. See Krauskopff 1999. The association of rice cultivation and fishing is a very old pattern of subsistence in the Middle Ganges Valley as illustrated by excavations of Neolithic sites in the area of Gorakhpur in India dating from the tenth century B.C.E. (Singh 1994).

5. In 1940 William Archer assumed the post of superintendent of Bihar in India close to the Nepalese border. The photographs that he took in Champaran, a Terai district of Bihar, during his early years in this position reveal his interest in artistic expression. They also confirm that dance was certainly the strongest aesthetic component of the Tharu culture. India Office Library and Records, European Manuscript, Mss Eur. F 236/26.

6. In Nepal, the day devoted to planting rice has long been celebrated, as illustrated in an essay written by the Nepali poet Lakhsmi Prasad Devokta and entitled "The Fifteenth Day of Asar" (Asar starts June 15 and ends July 15); see Treu 1993.

7. *Phulwar* (garden) is the term applied to all sacred songs of origin (Krauskopff 1996).

8. *Dahara daphul* refers to doves playing (*daphul*) in a pond (*dahara*). "Doves playing" is a metaphor for sexual intercourse. It is actually a core motif that defines a song as a *phulwar,* or a mythical song of origin (Krauskopff 1996). For a summary of these myths, see Krauskopff 1989.

9. The Tharu do not make material reproductions of this dimension of fertility; instead, performance and singing are valued. In other societies of eastern India such material creations do exist. In Jain (1989, 43) there is a reproduction of a bronze ring of the Muria of Bastar in Madhya Pradesh, which depicts a farmer with a pair of bullocks and a rich harvest. These rings are presented to the bride at the time of marriage and are worn for seed-sowing ceremonies.

10. See, for instance, the rice dances and courtship customs of the Limbu of far western Nepal in a Himalayan context (Jones and Jones 1976).

11. Archer notes that the Nawalpurya Tharu, as their name indicates, migrated from Chitwan (Nawalpur) three generations before. On the significant and socializing role of women's dances among the Tharu of Chitwan in Nepal from an outsider's point of view, see also Bjork Guneratne 1999.

12. There is Holi for the Rana Tharu, Dasain for the Dangaura Tharu, Tihar (Devali) for the Chitwanya or Nawalpurya Tharu. Holi is not really performed among the Dangaura Tharu while it lasts two months for the Rana. On the Rana dance tradition, see Srivastava 1948–1949; 1958. See also Korvald 1999.

13. For more details about the Tharu rice granary, see chapter 9 of this volume.

14. The film is entitled *Mahabharata: The Barka Naach as Sung and Danced by the Tharu of Jalaura, Dang Valley, Nepal,* a Deependra Gauchan film from Rusca Productions (1998).

15. As noted by William Sax, the *Mahabharata* or *Pandav Lila* in the Kumaon Garwhal Himalayas "is not a text, much less a book, but rather a tradition of performance" (1994, 131). It is interesting that, as in the Dang Valley of Nepal, the hero is not always Krishna but is at times the villain of the epic, Karna (or Kanha as he is known among the Tharu; Sax 1996). The parallel between the *Mahabharata* in Dang and in the Kumaon Gharwal is striking and should be further explored in relation to the Kanphata Yogi influence.

Chapter 5 (Bahadur)

1. The villagers of Sugnu and Tangjeng bring pinewood; those of Nungu, pine bark; those of Chairel and Thongjao, three earthen pots; those of Wangoo, rice and pineapples; those of Kakching Khunou, iron tools for clearing ground; those of Thongam, a basketful of unhusked rice; those of Waikhong, salt and chilies; those of Kumbi, molasses and milk; those of Ithai, two *sareng*; those of Elangkhang-pokpi and Langmeidong, areca nut, betel leaves, and *kabok* (rice sweets). The Kom tribe from Sagang and Tonshem villages brings one pig, one white cock, and one egg.

Chapter 6 (Headley)

All illustrations featured in this essay are taken from Leiden University Oriental Ms. 12.542. Research funding to work with this collection was generously provided by the Getty Grant Program. The collection contains twenty stories of rice field purification rituals. They have been numbered one to twenty, and their successive illustrations (about ten per story) are consecutively numbered as well, thus 1.3, 2.6, and so forth. The story and illustration numbers appear within parentheses at the end of the captions for the illustrations.

1. According to Soetrisno Santoso, the grandson of Widi Prayitna, his grandfather, in addition to writing up hundreds of stories himself during the 1930s, also taught Cermodiyasa, a *dhalang* of his own age, how to compose such texts. Cermodiyasa came from the village of Dewan (west of Maguwa and east of Yogyakarta). Surprisingly Soetrisno said that Cermodiyasa produced better tales than his grandfather.

2. Moens assembled approximately 170 volumes of oral literature from diverse areas of south Central Java (Leiden University Oriental Mss 10.886–10.974 and Leiden University Oriental Mss 12.507–12.577). They remained in his possession until he died in 1954 in Jakarta at which point they were sold privately. Eventually, however, most of them were purchased by the Leiden University Library, which reassembled the collection in Holland. See Hosein (1955, i–iv).

3. *Manikmaya* (1981, 1:16–17).

4. The ninth illustration of story number eight contains such a phallus/whip.

5. On the cult of Bima in ancient Java, see W. F. Stutterheim (1956, 107–43). Another of the Sentolo stories relates Bima's rape of Durga. In that story the ogre Kala Jamus turns Bima's penis into a magical javelin so that he may rape/murder his adversary Durga.

6. Leiden University Oriental Ms. 12.542, 1,368 pp. See Pigeaud (1980, 66–67 [= Moens Kris Planken, vol. 2; the first volume has been lost]). The third tale appears on pages 141–208, and the illustrations begin on page 167. It was transcribed in 1990 by Soetrisno Santoso, grandson of *dhalang* Widi Prayitna.

7. In this collection of tales *aking* is used (as opposed to *angking*) as a synonym of *tlawingan.* The word *aking* is sometimes written *akik,* which means an "agate," but unless we take this choice to indicate that the board is a wooden talisman, as opposed to one made of semiprecious stone, the reason for it remains unclear. According to Poerwadarminta (1939, 609), *tlawingan* means a *plawungan,* or a board for hanging spears or kris blades. Such boards are often carved and decorated with the same sort of motifs used on calendrical counting boards (*blabag pétangan*). Otherwise, the word *aking* means "dried out" (as applied to wood or rice) and may be an opposite of *teles,* or wet. The opposition between the

body as a wet book of living symbols and the Koran as a dry book of written symbols (Beatty 1999, 161) does not seem to apply here. In the context of rice field exorcisms, the *aking* or *tlawingan* is dry wood (not part of a living tree), but perhaps more germane, it is personified and given a name. Thus it is Kyai Barung that expels the Baju Barat demons from the rice fields.

8. Pigeaud (1994, 268) translates *mèl* as "muttered" (for instance, a proverb). Prawiroatmojo (1981, 344) translates *mèl* as "mantra" or "prayer" (*doa*), and Poerwardarminta (1939, 301) gives *donga* or *rapal,* which amount to the same.

9. See Poerwardarminta (1930, 46, 57).

10. Demons can best be seen at night when one is naked; in broad daylight they are considered to be invisible.

11. This is a corruption of *Bismilah salla'llahu 'alaihi wasallam.*

12. Soetrisno Santoso translated this as "working on the pests that are under his leadership."

13. Perhaps this is planet Johar, or *al-zuhara* (Arabic for the planet Venus). This planet is also identified in Java with a part of the body in the macro/microcosmic parallels of the masseurs (Headley 1996). There are two other possibilities: to read *jahar* or *jahr,* which means "aloud"; or *jahir be zahir,* that is, *lahir,* meaning "outside."

14. For lack of a real translation, I give the interpretation of Soetrisno Santoso.

15. *(Am)bleg* may be translated as "collapsed"; *bleg* refers to the sound of a fall; *gudu* is the equivalent of *dudu* (not).

16. Since *borang* means "pitfalls" or "mantraps," I have read this as *bareng-bareng,* or "at the very same time."

17. This may be intended as *bi-llâhi,* with Allah.

18. This ritual sequence is equivalent to what is called: "the four directions, the five relationships"; *keplat papat, lima pancer.*

19. This is the name of one of the exorcism prayers (Headley 2000, ch. 5), also used in the Maésa Lawung buffalo sacrifice (Headley 1979) and in the coronation anniversary rituals in the Surakarta palace, See Nancy Florida (1993, 128).

20. The meaning of *enter* also shades off to that of *possession.*

21. In the version of the *Manikmaya* published in 1852 by J. J. de Hollander, the grandson of Guru and "son" of Kala—Kala Gumarang—personifies the rice parasites. When he is wounded by Wisnu's (Vishnu's) arrow, the blood spurting from his wound produces insects and diseases that attack ripening rice.

22. Although nothing in the myth says that Uma, Guru's consort, is menstruating, the taboo against intercourse during menstruation does exist in urban Java. Whether it forms part of peasant taboos in Sentolo is a moot point.

23. See Headley's study (2000) of the cosmologies in this collection of origin myths. Another version is to be found in Ranggawarsita's *Serat Pustakaraja Purwa* (1993–1994).

24. See Scubla (1999, 160).

25. For the Balinese version of this covenant, see Hooykaas (1974, 75).

26. Chains of "self-structuring concepts" take us even further afield for, upon inspection, the birth of Kala interacts locally with the

"broken grain" myths. Known in China, Mainland Southeast Asia, and the insular areas of the Pacific Rim stretching from Java to as far north as Japan, these myths have been studied among the Black and White Tai, the Atayal, the Bunun and the Paiwan of Taiwan, and the Vietnamese. A single inexhaustible primordial grain of rice loses it heavenly qualities when it descends to earth and is metaphorically "broken." Henceforth, mankind must depend on all-too-human labor to cultivate cereals on earth. The descent of rice into the human world is a new creation, but it is not without the loss of its paradisiacal qualities.

27. Pigeaud (1994, 51) says *boma* or *boman* is a decorated wooden beam that may be the same as the *sunduk* (Frick 1994, 8) that unites the frame resting on the king posts or *saka guru*. Bhauma (in Sanskrit) means "coming from the earth," but some Javanese dictionaries imply it is to be derived from the Sanskrit *wyoma*, or sky, ethereal emptiness (cf. Poerwardarminta 1930, 94–95: s.v. *boma: awang-awang*). The word as an epithet is most commonly used to characterize a terrifying Balinese monster head with hands.

Chapter 7 (Vi and Crystal)
This essay is the result of a nascent research collaboration between the UCLA Fowler Museum of Cultural History and the Vietnam Museum of Ethnology in Hanoi. With grant funds supplied by the Getty Grant Program, the authors, with the assistance of Trần Thị Thu Thủy, undertook research on Tai traditional culture in Sơn La and Nghệ An Provinces. This contribution specifically relates to two rice harvest rituals observed in Nghệ An in fall 2000. Each ceremony celebrates the spirit of rice, demonstrates the cognitive link between rice and fertility in Tai culture, and involves the graphic depiction of the spirit of rice in the midst of ritual harvest activity.

1. The term *Tai* encompasses the majority populations of Thailand and Laos, the inhabitants of the Shan State in Burma, and Black and White Tai mountain people in the mountains of Vietnam and Laos. There are also substantial Tai populations in Yunnan and elsewhere in southwestern China. The term *Thai*, therefore, refers specifically to the citizens of the kingdom of Thailand. Tai refers to members of the Tai-Kadai language family in East and Southeast Asia.
2. Vi Văn An has circulated a research memorandum at the Vietnam Museum of Ethnology discussing the dialectical and cultural differences between Black and White Tai in Nghệ An Province. His paper is entitled, "A Contribution to the Discussion about Black Thai and White Thai in Western Nghệ An."
3. Coauthor, Vi Văn An selected his natal village, Bản Đốc, in Chi Khe commune in the Con Cuông District for initial fieldwork.

Chapter 9 (Meyer and Deuel)
1. Even with the most careful construction, a certain amount of grain loss occurs due to dampness and insects. This loss has been magnified by the increase of population density as the hill dwellers moved into the Terai. Preserving grain in the tropical Terai requires different strategies than those used in the drier and colder hills; rice storage techniques that worked well in the hills cannot be adapted to the Terai climate. Another factor is the growing use of the new high-yield varieties of rice. The new strains appear to be less resistant to insects than the old, time-tested local grains.

Chapter 11 (Hamilton)
The author thanks the Getty Grant Program, which provided funding for the research that forms the basis of this chapter.

1. The author thanks Ramon Villegas for pointing out this clue to dating various depictions of Isidro.

Chapter 12 (Stuart-Fox)
1. See photographs 275–78 in Goris and Dronkers (1952).
2. There are several possible transliterations of "Bhatara" and the female equivalent "Bhatari." Elsewhere in this book "Batara" and "Batari" have been used as consistent with the system of transliteration employed by the author of the chapter.

Chapter 13 (Hamilton and Kanteewong)
The authors thank the Getty Grant Program, which provided funding for the research that forms the basis of this chapter.

1. The northern kingdom of Lanna (valley of a million rice fields), with its capital at Chiang Mai, has a history quite distinct from that of the central Thai kingdom of Siam. The Khon Muang, the majority population, grow rice in irrigated or rain-fed bunded fields and are politically dominant over various mountain-dwelling minority groups who grow rice in dry hill-side fields.
2. A second, smaller crop of rice may be grown in the dry season from March to June if there is adequate moisture, but this crop is not considered important and does not entail the rituals of the main crop. The fields may also be used for other types of crops in the off-season.
3. The authors thank Praá Learn (the former abbot of Wat Paá Daád) and the congregants of Wat Paá Daád and Wat Yang Lúang for their hospitality and cooperation.
4. This is one of many Buddhist uses of the nine-strand cotton string, called *fai sai sin* (*fai*, cotton; *sai*, line or string; *sin*, power). One of the most spectacular examples is during the ceremonial recitation of the Jataka stories, when a grid of strings is run from the Buddha image to the head of each participant, as if to give them a symbolic "charge."
5. The arrangement varies from monastery to monastery. At Wat Yang Lúang, the community gathers in the *wihăn* and the uncooked rice is heaped into two conical piles of loose grain at the feet of the Buddha, one of husked white rice and the other of unhusked brown rice. At Wat Paá Daád, the community is too large to gather

in the original *wihăn* and assembles instead in a larger adjacent meeting hall (*săalaa-baat*) that also houses a giant Buddha image. In front of this Buddha, they pile husked rice contained in plastic sacks.
6. Standard religious texts for specific rituals such as this are widely available in Thailand in commercially published versions. They are normally written in Pali.

Chapter 14 (Gerson)
The majority of this text is based on the author's original research, drawn from direct interviews in the Thai language with Dan Sai residents. Personal interviews that yielded valuable material were conducted in July 1999 with Apichat Kamkasem, the chief mask maker and artist in Dan Sai, and in July 2000 with Thanong Suwannasingh, Governor's Office, Loei Province.

1. The master mask maker of Dan Sai stated that this change took place about three hundred years ago (personal communication 1999), but there is no independent verification of this fact.

Chapter 15 (Crystal)
1. The Ma'bugi' trance ritual is depicted in the film *Ma'bugi: Trance of the Toraja* (1973), produced by Eric and Catherine Crystal and Lee Rhodes and distributed by Extension Media Center, University of California at Berkeley
2. A good discussion of the two types of Bua' festival found in western Toraja districts may be found in Waterson (1984, 23).
3. An extensive analysis of the Bua' Kasalle feast may be found in Nooy-Palm (1986, 10–61).

Chapter 16 (Vũ and Hamilton)
The authors thank the Getty Grant Program, which provided funding for the research that forms the basis of this chapter.

1. The eighteen kings of the Hùng Vương dynasty (2879–258 B.C.E.) are known only by numbers, not individual names.
2. See chapter 29 regarding the important pairing of rice agriculture and silk production in Chinese cultural history.
3. The Lê court honored her with the epithet "*Hậu hậu chi thần phú quốc an dân*," perhaps meaning "[she] will be a deity for the salvation of the people" (various translators could not agree whether this phrase in classical Vietnamese was literal or indirect). As it is considered disrespectful to refer to a deity using her real name, in daily usage the people of Trám call their goddess Hậu Hậu Chi Thần.
4. In 1993, 1997, 1998, 2000, 2001, and 2002.
5. The formal name for this ritual is Lễ Mật. *Lễ* means ritual. The normal meaning of *mật* in Vietnamese is "secret," but in this case there is a more complex meaning derived from Chinese, expressing the fertility aspects of the ritual.
6. According to the village elders, *nõ* derives from *lỗ*, the tip of a penis. *Nường* is a derivative of *nàng*, vulva.
7. *Vật* are tools of production, or in this case reproduction. *Linh* indicates their use in religious rituals that mimic sexual intercourse.

8. There is no literal translation for these words, but their sound, with the stress placed on the last word, more or less sums up the meaning!

9. In practice, the rice contained in the bundle is no different from the ordinary crop of modern high-yield varieties, as it has been replaced with these varieties in recent times.

10. This text is an excerpt from a longer song, as are the texts that follow. The full songs were published in Vietnam (Dương 1974) at the time when the festival had gone into abeyance. They had presumably been collected as oral history from elders who still recalled the last performances in the 1940s. When the festival was revived in the 1990s, the performers relied on the published texts, making only minor changes. The texts used in this essay are those taken from the publication with corrections provided in interviews with village elders. They were translated into English by Nguyễn Anh Hiếu. Additional translations of the published versions were provided by Dung N. Tran.

11. This is the opinion of Professor Phan Đăng Nhật, expressed at a Trò Trám seminar in Tứ Xã Commune.

Chapter 17 (Sano and Hamilton)

Funding to support the research that provides the basis for this chapter was generously provided by the University of California Pacific Rim Research Program.

1. Also called Ideha Sanzan Jinja, or just Dewa Jinja.

2. Also called the Toshiya festival. The most thorough English-language description of the festival and of Haguro Shugendō can be found in Earhart (1970), which served as an admirable guide for this documentation of the festival. Useful descriptions in Japanese include Togawa (1969) and Ito (1996).

3. The concept of the five sacred grains was adopted in Japan from China, where it is also still found. It is an abstraction that represents all grains. The identity of the individual grains varies with time and place but is generally considered to consist of rice, wheat, two types of millet, and beans. In modern Japan, rice is of course by far the most important of these. Some Japanese displays of the "five grains" consist of rice and four types of beans.

4. <http://scarab.msu.montana.edu/history-bug/scrub typhus.htm>; <http://www.ciesin.org/docs/001-613/001-613.html>.

Chapter 18 (Hamilton)

1. There are many possible transliterations of "Batara" and the female equivalent "Batari." Elsewhere in this book Bhatara and Bhatari have been used as consistent with the system of transliteration employed by the author of the chapter.

2. The other is Lara Kidul, Queen of the Southern Ocean. Depending on the context, both Dewi Sri and Lara Kidul are considered symbolic consorts of the sultan (see chapter 26).

3. The major temple to this deity is located at Tirunelveli (in Tamil, *tiru* means "god," *nel* is "rice," and *veli* is "a fence or field boundary"; thus, the rice field deity).

Chapter 19 (Krishna)

1. In Muttavakkam, Thirumalper, and Madipakkam, there is no tradition of animal sacrifice. The priest at those Ponniyamman temples where sacrifice is not performed is a Brahmin, while the temples that practice animal sacrifice have priests of the dominant local caste.

2. Both the groves and the potter-priests are also associated with the Tamil village guardian god Ayyanar, but the Ayyanar cult is not directly connected with rice.

Chapter 20 (Stuart-Fox)

1. For a full study of Pura Besakih, see Stuart-Fox (2002), for a summary discussion, see Stuart-Fox (1996; 1998).

2. The Usaba Buluh gets its name from a variety of bamboo (*buluh*), a piece of which every family must bring to the temple. The subsidiary branches of each piece are decorated with colored leaves, flowers, palm-leaf ornaments, and chains of young coconut leaves.

3. Usually a *dangsil* is like a small slender multiroofed *meru* or pagoda. The *dangsil* used in Usaba Ngeed is a tall and narrow cylindrical offering composed of various ingredients and with many rectangular rice-dough cookies on the sides. The *penek* is a kind of rice offering, together with various other ingredients, arranged on a footed tray and crowned by an ornament of young coconut palm leaf.

Chapter 21 (Hamilton)

1. Translation by Fumiko Cranston, in Rosenfield (1979, catalog 31).

2. If grain containers are difficult to interpret, the "wine" containers are even more problematic, as alcoholic beverages have been made from many ingredients in China; the most likely candidate in ancient times, at least in northern China, is millet beer.

Chapter 22 (Ray)

1. The city of Calcutta has recently been renamed Kolkata.

2. I collected several editions of the text from different women. This is a summary translation that integrates all the versions.

3. *Sindur* is the local term for the vermilion that married women put in the parts of their hair as a sign of marriage.

Chapter 23 (Brinkgreve)

1. The basic data for this article were collected during field research in Bali in 1988–1989, financed by the Programme of Indonesian Studies, the Netherlands, and sponsored by the Indonesian Institute of Sciences (LIPI) and Universitas Udayana, Denpasar. I am very grateful to my Balinese informants and friends. See Brinkgreve and Stuart-Fox (1992) for more general information on, and numerous photographs of, Balinese offerings.

2. See Stuart-Fox (1974, 75–76) for more information on names and ways of making *jaja*, especially in Karangasem, East Bali.

3. In Brinkgreve (1993, 141) I give a further analysis of the meaning of the *cili*.

Chapter 24 (Fujita)

1. The term *kagami-biraki* refers to the cask-breaking ceremony and also to the ceremony of breaking *mochi* cakes at the New Year (see p. 299 of this volume). *Kagami*, which means "mirror," applies to the lid of the sake barrel in the one instance and the *mochi* cakes in the other. The word *biraki* comes from the verb *hiraku*, which usually means "to open," but in each of these ceremonies, means "to break."

Chapter 26 (Hamilton)

1. The inspiration for the title of this part of the book and this chapter came from Emiko Ohnuki-Tierney's seminal study, *Rice as Self: Japanese Identities through Time* (1993).

2. The equivalent ritual in Bali, consisting of the making of the Nini Pantun prior to the harvest, is still widely performed (see fig. 2.24).

3. Personal communication, Stephen C. Headley.

4. Personal communication, Kik Soleh Adi Pramono

5. I thank Rens Heringa for discussions that developed this idea.

Chapter 27 (Krishna)

1. Ahamasmi prathamaja rtasya. Purvam devebhyo amrtasya nabhih. Yo ma dadati sa ideva ma'vah. Ahamannamannam-adantamadmi (I am the first progenitor of *yajna*. It is I who, at the earliest times, became the nucleus of the nectar. The first *yajna* was born of me. I am the god of rice). *Taittiriya Brahmana* 2.8.8.1–8. *Yajna* is the sacrifice, the fire rituals, and offerings to the gods for their blessings. It is performed only by an accomplished priest, who follows strictly prescribed rules and observes the necessary precautions.

2. Brahman is not a god but is above all gods. It is in Brahman that the gods, as all existence, have their being. All creation emanates from Brahman. However, Brahman is a mystical all-pervading spirit, without form, and is not to be confused with Brahma, the creator god of Hinduism.

3. The descriptions given here are neither exhaustive nor do they cover all the rituals performed in a *samskara*. They are restricted to the use of rice in each *samskara*. I have tried to document as many caste and tribal customs as I could find, but there are so many that it is not possible to survey and document all of them.

Chapter 28 (Sano)

I would like to thank Nasu Hisayo and Kojima Sayoko for their assistance in collecting material in the early stages of planning this essay.

1. The sample books in question are reproduced in Yamanobe (1974a; 1974b; 1974c; 1974d) and Imao (1972a; 1972b).

2. Depending on the word in the pattern, the fish could be identified either as *koi* (carp) or as *tai* (sea bream). I believe this is *koi*, judging from its shape.

Chapter 29 (Bray)

1. In the daily ritual of honoring the ancestors, in southern China a bowl of cooked rice and a cup of tea were customarily laid out before the ancestral tablets on the domestic altar (and indeed they still are in some homes today, especially in Taiwan; Wolf 1974). But the point of this minor sacrifice is to offer the ancestral spirits a portion of the staple food, and in the north some other grain would be substituted for rice. Similarly in funeral rituals in the south, rice plays an important symbolic role, but not so much as rice in its own right, as in the dyad of grain and flesh (symbolizing male and female kin ties, respectively), represented in this instance by rice and pork (Thompson 1988).
2. Chang (1977) is still the standard account of the earliest emergence of regional cultures in China. Two very readable general histories of China from the Bronze Age through the imperial dynasties are Ebrey (1996) and Shaughnessy (2000).
3. Vivid illustrations of standardization during the Qin and Han dynasties are provided by Ledderose's discussion of the mass production of iron tools and weapons, luxury lacquerware, and the famous clay warriors discovered in the tomb of the First Emperor (2000). On the establishment of the imperial university and the standardization of education, see Powers (1991).
4. For a general account of the historical development of rice cultivation techniques, field types, forms of irrigation, and rice varieties in China and other parts of East and Southeast Asia, see Bray (1986; 1984, 477–510).

Chapter 30 (Hamilton)

1. Exceptions after 1975 would include Cambodia, decimated by the Khmer Rouge in the late 1970s, and isolated North Korea (primarily a wheat-growing country) at the turn of the twenty-first century.
2. See <http://www.citypopulation.de/cities.html>. The ten cities are Calcutta, Delhi, Dhaka, Jakarta, Manila, Mumbai, Osaka, Seoul, Shanghai, and Tokyo.
3. I thank Ruaraidh Sackville Hamilton, IRRI, for providing details about the International Rice Genebank.

Chapter 31 (Ammayao with Hettel)

1. In some versions of the myth, it is Balituk instead of Wigan; in others, it is Wigan only, without a brother.
2. <http://whc.unesco.org/sites/722.htm>.
3. <http://whc.unesco.org/archive/repcom01.htm#riceterraces>.
4. Year 2000 census data show there are approximately 161,000 people—most of whom are Ifugao—in what is now Ifugao Province (up from the 123,000 estimated by Worcester in 1912). The small increase in population over the last nine decades is, I believe, indicative of the large outward migration from the province.

Chapter 32 (Heringa)

Research in this area has been carried out intermittently since 1977. Funding was provided in 1989–1990 by the Program of Indonesian Studies (PRIS) at Leiden University, The Netherlands, under the auspices of the Indonesian Research Council (LIPI). A grant from the Getty Grant Program enabled research in Indonesia during February and March 2000. The author would like to express her appreciation to all agencies involved.

1. See Alan Dundes on this "outmoded theory" (1994, vii).
2. The term *Kejawen* refers to the complex of specifically Javanese beliefs.
3. The situation resembles that of the Badui and the Tenggerese, other marginal groups dwelling in mountainous areas in Java. Oral tales further indicate a possible ancestry from the Kalang, a group of nomadic forest dwellers first mentioned in Javanese sources in the fourteenth century (Lombard 1990, 3: 33, 131).
4. Locally made textiles are mentioned in the report of the first Dutch visit to Tuban in 1599 (Keuning 1839/1849, 3: 34, 36). The contemporary function of textiles as symbolic expression of the social structure in Kerek has been analyzed in Heringa 1989; 1991; 1993; and 1994.
5. See Brinkgreve and Stuart-Fox (1992, 47), for comparable ideas in Bali.
6. The cannibalistic aspect—which does not form part of the acknowledged historical Javanese tradition—seems to relate this tale to a much earlier background.
7. Like the sky nymph, wives of landowners move into their husband's family house. Commoners generally follow a matrilocal residential pattern with husbands moving into the compound of their wife's mother.
8. The rice goddess's relation to the sky and the moon is encountered throughout the archipelago. See Kruijt 1935 for the Toraja of Central Celebes.
9. Limited space does not allow further elaboration; Heringa (forthcoming) gives an extensive analysis of ritual food exchanges.
10. In written versions of the myth, Sri is generally presented as Sedono's elder sister (Ranggawarsita 1993–1994, 3: 36), suggesting that the match was not only incestuous but also countered the Javanese stipulation that the bride must be of a younger line than the groom. As the point has little relevancy in the context of this essay, it will not be elaborated.
11. A series of drawings of a village wedding procession in a nearby area forms part of a manuscript compiled in 1822 by an official of the Dutch Colonial Government, Cornets de Groot, "Statstiek van de residentie Gresik."
12. The system used to be widespread in dry areas throughout Java and is generally described as an intermediate method between dry rice (*gogo*) and irrigated *sawah* cultivation (Paerels 1913, 57).
13. Locally woven male dress fell into disuse several decades ago, although the preferred color of the shop-bought fabric that is used at present is still a rusty brown.
14. Details of a human pregnancy ritual are given in Heringa (forthcoming).
15. *Wesi pulesani*, literally, "the enticing iron," is the ritual name for the rice knife (*ani-ani*).
16. In modern Javanese, *pipitan* is derived from *pipit*, or close together. The probably much older use of the ritual term *pi-pit[a]-an* is a compound from Old Javanese *pita* (yellow), which stands for yellowed.
17. *Sri* may be translated in several ways, although it is generally considered to be the name of the Rice Goddess. In its original Sanskrit and Old Javanese forms, it stands for "radiant" or "magnificent." An alternative translation of Mbok Sri Ayu, therefore, may be "Mother of Radiant Beauty."
18. Elsewhere on Java, the ritual is referred to as the purification of the village (*bersih desa*; see Geertz 1960; Mayer 1897; Pemberton 1994).
19. "Wis neng jeroh gedhong kencana" may be translated as "already enclosed in the innermost reaches of the golden storage room."

References Cited

Abelmann, Nancy
1996 *Echoes of the Past, Epics of Dissent: A South Korean Social Movement.* Berkeley: University of California Press.

Achjadi, Judi
1989 "Batiks in the Central Javanese Wedding Ceremony." In *To Speak with Cloth: Studies in Indonesian Textiles,* edited by Mattiebelle Gittinger, 151–61. Los Angeles: Museum of Cultural History, University of California.

af Edholm, Erik
1984 "Canda and the Sacrificial Remnants: A Contribution to Indian Gastrotheology." *Indologica Taurinensia* 12: 75–91.

Allen, Jane
1997 "Inland Angkor, Coastal Kedah: Landscapes, Subsistence Systems, and State Development in Early Southeast Asia." In *Indo-Pacific Prehistory: The Chiang Mai Papers,* vol. 3, edited by Peter Bellwood and Dianne Tillotson, 79–87. Bulletin of the Indo-Pacific Prehistory Association 16. Canberra: Australian National University.

Amirthalingam, M.
1998 *Sacred Groves of Tamilnadu—A Survey.* Madras: C.P.R. Environmental Education Centre.

An Zhimin
1999 "Origin of Chinese Rice Cultivation and Its Spread East." Translated by W. Tsao. Edited by B. Gordon. <www.carleton.ca/~bgordon/Rice/papers/zhimin99.htm.>

Anderson, Charlotte
2001 *Rice and Rituals.* New York: Japan Society.

Anderson, E. N.
1988 *The Food of China.* New Haven, Conn.: Yale University Press.

Antlöv, Hans
1986 "Tradition and Transition: Harvest and Social Change in Rural Java." In *Rice Societies: Asian Problems and Prospects,* edited by Irene Norlund, Sven Cederroth, and Ingela Gerdin, 151–70. London: Curzon Press.

Anuman Rajadhon, Phya
1961 *Life and Ritual in Old Siam: Three Studies of Thai Life and Customs.* Translated and edited by William J. Gedney. New Haven: HRAF Press.
1987 *Some Traditions of the Thai.* Bangkok: Thai Inter-Religious Commission for Development and Sathirakoses Nagapradipa Foundation.

Apffel Marglin, Frédérique
1985 *Wives of the God-King: The Rituals of the Devadasis of Puri.* Dehli: Oxford University Press.

Ashkenazi, Michael
1993 *Matsuri: Festivals of a Japanese Town.* Honolulu: University of Hawaii Press.

Aung-Thwin, Michael
1990 *Irrigation in the Heartland of Burma: Foundations of the Pre-Colonial Burmese State.* Monograph Series on Southeast Asia. DeKalb: Center for Southeast Asian Studies, Northern Illinois University.

Averbuch, Irit
1995 *The Gods Come Dancing: A Study of the Japanese Ritual Dance of Yamabushi Kagura.* Ithaca, N.Y.: East Asia Program, Cornell University.

Babb, Lawrence A.
1970 "The Food of the Gods in Chhattisgarh: Some Structural Features of Hindu Ritual." *Southwestern Journal of Anthropology* 26: 287–304.

Bagley, Robert, ed.
2001 *Ancient Sichuan: Treasures from a Lost Civilization.* Princeton, N.J.: Princeton University Press.

Bajaj, J., and M. D. Srinivas
1996 *Annam Bhu Kurvita.* Madras: Centre for Policy Studies.

Barraud, Cécile
2001 "De la distinction de sexe dans les sociétés: Une présentation." In *Sexe relatif ou sexe absolue,* edited by C. Alès and C. Barraud, 23–99. Paris: Editions de la Maison des sciences de l'homme.

Barton, Roy Franklin
1922 "Ifugao Economics." *University of California Publications in American Archeology and Ethnology* 15, no. 5: 385–446.
1946 "The Religion of the Ifugaos." *American Anthropologist* 48, no. 4: 1–219.
1955 *The Mythology of the Ifugaos.* Philadelphia: Memoirs of the American Folklore Society.

Beatty, Andrew
1999 *Varieties of Javanese Religion: An Anthropological Account.* Cambridge: Cambridge University Press.

Bell, Mrs. Arthur
1904 *Lives and Legends of the English Bishops and Kings, Medieval Monks, and Other Later Saints.* London: George Bell and Sons.

Bethe, Monica
2002 "The Staging of Noh: Costumes and Masks in a Performance Context." In *Miracles and Mischief: Noh and Kyōgen Theater in Japan,* edited by Sharon Sadako Takeda, 176–227. Los Angeles: Los Angeles County Museum of Art.

Bishop, Isabella B.
1970 *Korea and Her Neighbors.* Western Books on Korea, no. 4. Seoul: Yonsei University Press.

Bjork Guneratne, Katherine
1999 *In the Circle of the Dance.* Ithaca, N.Y.: Cornell University Press.

Björnsen Gurung, Astrid
N.D. "Storage Management and Food Security in Gobardiha Village, Deukhuri-Dang: Case Study in Nepal." Zurich: Department of Environmental Sciences, Swiss Federal Institute of Technology. Unpublished manuscript.

2000   *The Sacred in Storage Pest Management in Rural Nepal.* Zurich: Department of Environmental Sciences, Swiss Federal Institute of Technology.

Bourdieu, Pierre
1990   *The Logic of Practice.* Palo Alto, Calif.: Stanford University Press.
1991   *Language and Symbolic Power.* Cambridge, Mass.: Harvard University Press.

Bratawidjaja, Thomas Wiyasa
1985   *Upacara Perkawinan Adat Jawa.* Jakarta: Penerbit Sinar Harapan.

Bray, Francesca
1984   *Agriculture.* Vol. 6, pt. 2 of *Science and Civilization in China,* edited by Joseph Needham. Cambridge: Cambridge University Press.
1986   *The Rice Economies: Technology and Development in Asian Societies.* Oxford: Basil Blackwell.
1997   *Technology and Gender: Fabrics of Power in Late Imperial China.* Berkeley: University of California Press.
2000   *Technology and Society in Ming China (1368–1644).* Washington D.C.: American Historical Association.

Brill, Maria
1999   "Ghosts in the Streets." *Sawaddi* (American Women's Club, Bangkok; 1st qtr.): 33–36.

Brinkgreve, Francine
1993   "The Woven Balinese *Lamak* Reconsidered." In *Weaving Patterns of Life: Indonesian Textile Symposium 1991,* edited by Marie-Louise Nabholz-Kartaschoff, Ruth Barnes, and David J. Stuart-Fox, 135–53. Basel: Museum of Ethnography.
1997   "Offerings to Durga and Pretiwi in Bali." *Asian Folklore Studies* 56: 227–51.

Brinkgreve, Francine, and David J. Stuart-Fox (photographer)
1992   *Offerings: The Ritual Art of Bali.* Sanur (Bali): Image Network Indonesia.

Brohier, R. L.
1934   *Ancient Irrigation Works in Ceylon.* Colombo: Ceylon Government Press.

Brook, Timothy
1998   *The Confusions of Pleasure: Commerce and Culture in Ming China.* Berkeley: University of California Press.

Bryde, John A.
1986   "The Decline in Paddy Cultivation in a Dry Zone Village of Sri Lanka." In *Rice Societies: Asian Problems and Prospects,* edited by Irene Norlund, Sven Cederroth, and Ingela Gerdin, 81–116. London: Curzon Press.

Buahapakdee, Apinan
1988   "Festival of Mischief, Ghosts, and Phallic Symbols." *Saen Sanuk* (Bangkok; June): 53–56.

Bui Thiet
1996   *Vietnamese Festivals Dictionary* (in Vietnamese). Hanoi: Cultural Publishing House.

Bui Xuan My, Bui Thiet, and Pham Minh Thao
1996   *Vietnamese Ceremonies and Customs Dictionary* (in Vietnamese). Hanoi: Culture and Information Publishing House.

Cabreza, Vincent
2001   "Group to Raise Funds for Rice Terraces." *Philippine Daily Inquirer* (December 20): A12.

Caldwell, Sarah
1996   "Bhagavati: Ball of Fire." In *Devī: Goddesses of India,* edited by John Stratton Hawley and Donna Marie Wulff, 195–226. Berkeley: University of California Press.

Cameron, Elisabeth L.
1985   "Ancestors and Living Men among the Batak." In *The Eloquent Dead: Ancestral Sculpture of Indonesia and Southeast Asia,* edited by Jerome Feldman, 79–100. Los Angeles: UCLA Museum of Cultural History.

Capistrano-Baker, Florina H., et al.
1998   *Basketry of the Luzon Cordillera, Philippines.* Los Angeles: UCLA Fowler Museum of Cultural History.

CECAP (Central Cordillera Agricultural Programme) and PhilRice (Philippine Rice Research Institute)
2000   *Highland Rice Production in the Philippine Cordillera.* Banaue (Philippines): CECAP.

Cederroth, Sven, and Ingela Gerdin
1986   "Cultivating Poverty: The Case of the Green Revolution in Lombok." In *Rice Societies: Asian Problems and Prospects,* edited by Irene Norlund, Sven Cederroth, and Ingela Gerdin, 124–50. London: Curzon Press.

Chang, Kwang-chih
1977a   *The Archaeology of Ancient China.* New Haven, Conn.: Yale University Press.
1977b   *Food in Chinese Culture: Anthropological and Historical Perspectives.* New Haven, Conn.: Yale University Press.

Chatterjee, Tapanmohan
1948   *Alpana: Ritual Decoration in Bengal.* Calcutta: Orient Longman Pvt. Ltd.

Chin, Lucas
1988   *Ceramics in the Sarawak Museum.* Kuching (Malaysia): Sarawak Museum.

Chulalongkorn, King
1963   *Phraraadcha prapheenii sip song dyan* (The royal twelve-month traditions). Bangkok: Prae Phittaya.

Chung, Yang-Mo
1998   "The Art of the Korean Potter: From the Neolithic Period to the Choson Dynasty." In *Arts of Korea,* edited by Judith G. Smith, 220–49. New York: The Metropolitan Museum of Art.

Clunas, Craig
1996   *Fruitful Sites: Garden Culture in Ming Dynasty China.* Durham, N.C.: Duke University Press.

Collet, Octave J. A.
1925   *Terres et peuples de Sumatra.* Amsterdam: Société d'édition Elsevier.

Collier, William L., et al.
1973   "Recent Changes in Rice Harvesting Methods." *Bulletin of Indonesian Economic Studies* 9, no. 2: 36–45.

Concepcion, Rogelio N.
1995   "The Ifugao Rice Terraces Today." Paper presented at the Regional Thematic Study Meeting on the Asian Rice Culture and Its Terraced Landscapes, organized by the UNESCO World Heritage Center, Banaue, Ifugao, Philippines, March 28–29.

Condominias, Georges
1986   "Ritual Technology in Mong Gar Swidden Agriculture." In *Rice Societies: Asian Problems and Prospects,* edited by Irene Norlund, Sven Cederroth, and Ingela Gerdin, 28–46. London: Curzon Press.

Conklin, Harold C.
1968   *Ifugao Bibliography.* New Haven, Conn.: Yale University Press.
1975   *Hanunóo Agriculture: A Report on an Integral System of Shifting Cultivation in the Philippines.* Northford, Conn.: Elliot's Books.
1980   *Ethnographic Atlas of Ifugao.* New Haven, Conn.: Yale University Press.

Cornets de Groot, Adriaan David
1822   "Statistiek van de residentie Gresik." The Royal Institute of Linguistics and Anthropology (KITLV) Leiden, The Netherlands, Manuscript DH379.

Cort, Louise Allison
1992   "Whose Sleeves?: Gender, Class, and Meaning in Japanese Dress of the Seventeenth Century." In *Dress and Gender: Making and Meaning,* edited by R. Barnes and J. B. Eicher. Oxford: Berg Publishers.

Covarrubias, Miguel
1936   *Island of Bali.* New York: Alfred A. Knopf.

Czaja, Michael
1974   *Gods of Myth and Stone: Phallicism in Japanese Folk Religion.* New York: Weatherhill.

Diamond, Norma
1988   "The Miao and Poison: Interactions on China's Southwestern Frontier." *Ethnology* 27, no. 1: 1–25.

Directorate of Economics and Statistics, Government of Manipur
2001   *Statistical Abstract of Manipur, 2001.*

Do Phuong Quynh
1995   *Traditional Festivals in Vietnam.* Hanoi: The Gioi Publishers.

Due, JoJo
2002   "World-Famous Banaue Rice Terraces Facing Removal from UNESCO List." *Today* (Manila; January 26): 3.

Dulawan, Manuel
2003   *Oral Literature of the Ifugao.* Manila: National Commission for Culture and the Arts.

Dumont, Louis
1999   "A Folk Deity of Tamil Nadu: Aiyanar, the Lord." In *Religion in India* edited by T. N. Madan. Oxford: Oxford University Press.

Dundes, Alan
1994   "Foreword." In *The Folk Art of Java,* by Joseph Fischer, vii–ix. Kuala Lumpur: Oxford University Press.

Dương Văn Thàm
1974   "Trò Trám." *Tạp chí dân tộc học* (Ethnographical studies) 4: 94–102.

Dutt, Gurusaday
1990   *Folk Arts and Crafts of Bengal: The Collected Papers.* Calcutta: Seagull Books.

Earhart, H. Byron
1970   *A Religious Study of the Mount Haguro Sect of Shugendō: An Example of Japanese Mountain Religion.* Tokyo: Sophia University.

Eberhard, Wolfram
1968   *The Local Cultures of South and East China.* Leiden: Brill.

Ebrey, Patricia Buckley
1996   *The Cambridge Illustrated History of China.* Cambridge: Cambridge University Press.

E'ertai, comp.
1742   *Shoushi tongkao* (Compendium of works and days). Imperial compilation.

Eichinger Ferro-Luzzi, Gabriella
1977   "The Logic of South Indian Food Offerings." *Anthropos* 72: 529–56.

Eliade, Mircea
1957   *The Sacred and the Profane: The Nature of Religion.* New York: Harcourt-Brace.

Elvin, Mark
1973   *The Pattern of the Chinese Past.* Stanford, Calif.: Stanford University Press.

Fernandez, R.
1999   "Rice Terraces Face Destruction." *Philippine Star* (May 17).

Fischer, Joseph
1994   *The Folk Art of Java.* Kuala Lumpur: Oxford University Press.

Fischer, Joseph, and Thomas Cooper
1998   *The Folk Art of Bali: The Narrative Tradition.* Kuala Lumpur: Oxford University Press.

Fisher, Robert E.
1988   *The Earth and Soul: The Evans Collection of Asian Ceramics.* San Bernardino: California State University San Bernardino, University Art Gallery.

Florida, Nancy
1993   *Javanese Literature in Surakarta Manuscripts,* vol. 1. Ithaca, N.Y.: SEAP.

Fontein, Jan
1990   *The Sculpture of Indonesia.* Washington, D.C.: National Gallery of Art.

Foulston, Lynn
2002   *At the Feet of the Goddess: The Divine Feminine in Local Hindu Religion.* Brighton: Sussex Academic Press.

Fraser-Lu, Sylvia
2000   *Burmese Lacquerware.* Bangkok: Orchid Press.

Freeman, Derek
1970   *Report on the Iban.* London School of Economics Monographs on Social Anthropology, no. 41. London: Athlone Press.

Frick, Heinz
1994   "Rite de Passage of House and Man in Central Java." Paper for the Seventh Annual Workshop of the European Social Science Java Network, London, April 21–22.

Fruzzetti, Lina
1975   "Conch-Shell Bangles, Iron Bangles: An Analysis of Women, Marriage and Ritual in Bengali Society." Ph.D. diss., University of Minnesota.

Geertz, Clifford
1960   *The Religion of Java.* Glencoe, Ill.: The Free Press.
1963   *Agricultural Involution: The Processes of Ecological Change in Indonesia.* Berkeley: University of California Press.

Gerson, Ruth
1996   *Traditional Festivals in Thailand.* Kuala Lumpur: Oxford University Press.

Gittinger, Mattiebelle, and H. Leedom Lefferts Jr.
1992   *Textiles and the Tai Experience in Southeast Asia.* Washington, D.C.: The Textile Museum Washington.

Goris, Roelof
1939   "Het groote tienjaarlijksche Feest te Selat." *Djawa* 19: 94–112.
1969   "The Decennial Festival in the Village of Selat." In *Bali: Further Studies in Life, Thought, and Ritual,* edited by J. Van Baal, 105–29. The Hague: W. van Hoeve. (Translation of Goris 1939).

Goris, Roelof, and P. L. Dronkers
1952   *Bali : Atlas kebudajaan = Cults and Customs = Cultuurgeschiedenis in Beeld.* Jakarta: Pemerintah Republik Indonesia.

Graham, Penelope
1991   "To Follow the Blood: The Path of Life in a Domain of Eastern Flores, Indonesia." Ph.D. diss., Australian National University.

Grist, D. H.
1965   *Rice.* London: Longmans, Green, and Co.

Guang Fan
1959   *Bian min tu zuan* (Collection of pictures for the convenience of the people). Beijing: Zhonghua shu ju. (First published 1502; facsimile of 1593 edition).

Hà Văn Tấn
1993   *Buddhist Temples in Vietnam.* Hanoi: Social Sciences Publishing House.

Haas, Mary
1964   *Thai-English Student's Dictionary.* Stanford, Calif.: Stanford University Press.

Han, Pok-Chin
1993   "Table Settings and Cookery." *Koreana—Korean Art and Culture Journal* 7, no. 3: 36–43

Hanks, Jane Richardson
1960   "Reflections on the Ontology of Rice." In *Culture in History: Essays in Honor of Paul Radin,* edited by Stanley Diamond, 298–301. New York: Columbia University Press.

Hanks, Lucien M.
1972   *Rice and Man: Agricultural Ecology in Southeast Asia.* Arlington Heights, Ill.: AHM Publishing Corporation.

Harrisson, Barbara
1986   *Pusaka: Heirloom Jars of Borneo.* Oxford in Asia Studies in Ceramics. Singapore: Oxford University Press.

Hawkes, David, trans.
1959   *Ch'u Tz'u: The Songs of the South.* Oxford: Clarendon Press.

Hawley, John Stratton
1996   "The Goddess in India." In *Devī: Goddesses of India,* edited by John Stratton Hawley and Donna Marie Wulff. Berkeley: University of California Press.

Headley, Stephen C.
1979   "The Ritual Lancing of Durga's Buffalo in Surakarta and the Offering in the Krendowahono Forest of Its Blood." In *Between People and Statistics: Essays on Modern Indonesian History Presented to P. Creutsberg,* edited by Francien van Anrooij et al., 49–58. The Hague: M. Nijhoff.
1983   "Le lit-grenier de la déesse de la fécondité, rites nuptiaux?" *Dialogue,* special issue *"Le lit,"* no. 82: 77–86.
1996   *"Notes sur les types de soignants à Java."* In *Soigner au pluriel: Actes du colloque de Baume-les Aix organisé par Amades,* edited by Jean Benoit, 225–50. Paris: Harmattan.
2000   *From Cosmogony to Exorcism in a Javanese Genesis: The Spilt Seed.* Oxford Studies in Social and Cultural Anthropology. Oxford: Oxford University Press
2001   *"The Rice Goddess in a Javanese Court."* Unpublished manuscript.

Helliwell, Christine
2001   *"Never Stand Alone": A Study of Borneo Sociality.* Borneo Research Council Monograph Series 5. Phillips, Maine: Borneo research Council, Inc.

Heppell, Michael
1992   *Masks of Kalimantan.* Melbourne: Indonesian Arts Society.

Heringa, Rens
1989   "Dye Process and Life Sequence: The Coloring of Textiles in an East Javanese Village." In *To Speak with Cloth: Studies in Indonesian Textiles,* edited by Mattiebelle Gittinger, 151–61. Los Angeles: Museum of Cultural History, University of California.
1991   "Textiles and the Social Fabric on Northeast Java." In *Indonesian Textiles: Symposium 1985,* edited by Gisela Völger and Karin von Welck, 45–53. Ethnologica, n.s., vol. 14. Cologne: Rautenstrauch-Joest Museum für Völkerkunde.
1993   "Tilling the Cloth and Weaving the Land: Textiles, Land, and Regeneration in an East Javanese Area." In *Weaving Patterns of Life: Indonesian Textile Symposium 1991,* edited by Marie-Louise Nabholz-Kartaschoff et al., 155–76. Basel: Museum of Ethnography.
1994   *Spiegels van Ruimte en Tijd: Textiel uit Tuban.* The Hague: Museon.
1997   "Dewi Sri in Village Garb: Fertility, Myth, and Ritual in Northeast Java." In *The Divine Female in Indonesia,* edited by Robert Wessing, 355–77. Asian Folklore Studies 56, no. 2. Nagoya (Japan): Anthropological Institute, Nanzan University.

Forthcoming
    "Reconstructing the Whole: Seven
    Months Pregnancy Rituals in Kerek,
    East Java." In *The Alimentary
    Structures of Kinship*, edited by
    Monica Janowski and Fiona G.
    Kerlogue.

Héritier, Françoise
1979   "Symbolique de l'inceste et de sa
    prohibition." In *La fonction symbol-
    ique: Essais d'anthropologie*, edited by
    Michael Izard and Pierre Smith,
    209–45. Paris: NRF Editions
    Gallimard.
1981   *L'exercice de la parenté*. Paris: Hautes
    Etudes-Gallimard, Le Seuil.

Higgins, J. G.
1998   *Notes on Meithei (Manipuri) Beliefs
    and Customs*. Edited by John Parret.
    Imphal (India): Manipur State
    Archives.

Higham, Charles
1989   *The Archaeology of Mainland
    Southeast Asia*. Cambridge:
    Cambridge University Press.

Higham, Charles, and Tracey L.-D. Lu
1998   "The Origins and Dispersal of Rice
    Cultivation." *Antiquity* 72: 867–77.

Ho, Ping-ti
1975   *The Cradle of the East: An Inquiry
    into the Indigenous Origins of
    Techniques and Ideas of Neolithic and
    Early Historic China, 5000–1000 B.C.*
    Hong Kong: Chinese University
    Publications Office.

Hoff, Frank, trans.
1971   *The Genial Seed*. New York:
    Grossman Publishers.

Hollander, J. J. de
1852   *Manik Maja: Een Javaansch Gedicht
    (tembang)*. Batavia: Verhandelingen
    Bataviaasche Genootschap, no. 24.

Hooykaas, C.
1974   *Cosmogony and Creation in Balinese
    Tradition*. Bibliotheca Indonesica 9.
    The Hague: M. Nijhoff,

Hori, Ichiro
1968   *Folk Religion in Japan: Continuity
    and Change*. Chicago: University of
    Chicago Press.

Hosein Djajadiningrat
1955   "In Memorian Ir J. L. Moens."
    *Tijdschrift van Bataviaasche
    Genootschap* 85 (1952–57): i–iv.

Howard, David
2001   *The Last Filipino Head Hunters*. San
    Francisco: Last Gasp of San
    Francisco.

Howe, Leopold E. A.
1980   "Pujung: An Investigation into the
    Foundations of Balinese Culture."
    Ph.D. diss., University of Edinburgh.

Hung, Wu
1986   "Buddhist Elements in Early Chinese
    Art (Second and Third Centuries
    A.D.)." *Artibus Asiae* 47, no. 3–4:
    262–352.

Iijima, Shigeru
1985   "The Significance of Wet Rice
    Cultivation for Swidden Cultivators:
    The Case of the Karens." *East Asian
    Cultural Studies* 24: 93–98.

Imao Kazuo, ed.
1972a   *Imao-ke shozo Edo moyo hinakata-hon
    dai ikkai haihon* (Imao family collec-
    tion of sample design books of Edo
    patterns), vol. 1. Kyoto: Hakuousha.
1972b   *Imao-ke shozo Edo moyo hinakata-hon
    dai nikai Haihon* (Imao family col-
    lection of sample design books of
    Edo patterns), vol. 2. Kyoto:
    Hakuousha.

IRRI (International Rice Research Institute)
2000   *The Rewards of Rice Research*. Los
    Baños (Philippines): International
    Rice Research Institute. (Annual
    Report 1999–2000).
2001   *Rice Research: The Way Forward*. Los
    Baños (Philippines): International
    Rice Research Institute. (Annual
    Report 2000–2001).

Ito Takeshi
1996   *Dewa Sanzan* (Dewa's three moun-
    tains). N.P.: Michinoku-shobo.

Jacques, Claude
1997   *Angkor: Cities and Temples*. London:
    Thames and Hudson.

Jain, Jyotindra
1989   *National Handicrafts and Handlooms
    Museum*. Museums of India. New
    Delhi: National Handicrafts and
    Handlooms Museum.

Jay, Robert R.
1969   *Javanese Villagers: Social Relations in
    Rural Modjokuto*. Cambridge, Mass.:
    MIT Press.

Jenista, Frank Lawrence
1987   *The White Apos*. Quezon City
    (Philippines): New Day Publishers.

Jessup, Helen Ibbitson
1990   *Court Arts of Indonesia*. New York:
    The Asia Society Galleries.

Johnson, David, Andrew J. Nathan, and
    Evelyn S. Rawski, eds.
1985   *Popular Culture in Late Imperial
    China*. Berkeley, University of
    California Press.

Jones, Rex L., and Shirley Kurz Jones
1976   *The Himalayan Woman: A Study of
    Limbu Women in Marriage and
    Divorce*. Explorations in World
    Ethnology. Palo Alto, Calif.:
    Mayfield.

Jonsson, Hjorleifur
2000   "Yao Minority Identity and the
    Location of Difference in the South
    China Borderlands." *Ethnos* 65, no. 1:
    56–82.

Jordaan, Roy E., and Anke Niehof
1988   "Sirih Pinang and Symbolic Dualism
    in Indonesia." In *Time Past, Time
    Present, Time Future: Essays in
    Honour of P. E. de Josselin de Jong*,
    edited by David S. Moyer and Henri
    J. M. Claessen, 168–77. Dordrecht
    (The Netherlands): Foris
    Publications.

Kam, Garrett
1993   *Perceptions of Paradise: Images of Bali
    in the Arts*. Ubud (Bali): Yayasan
    Dharma Seni Museum Neka.

Keeler, Ward
1983   *Symbolic Dimensions of the Javanese
    House*. Centre of Southeast Asian
    Studies, Working Paper no. 29.
    Melbourne: Monash University.

Kerr, Rose, ed.
1991   *Chinese Art and Design: Art Objects
    in Ritual and Daily Life*. Woodstock,
    N.Y.: Overlook Press.

Kettes, Bunyong
1995   "Bun Beg Faa: Phuum panjaa kased-
    trakon Isan (Opening Sky
    Ceremony: Isan farmers' wisdom)."
    In *Chiiwid Thai chud buuchaa
    Phajaa Thaan* (Thai life: A series on
    worshiping Phajaa Thaan). Bangkok:
    Kurusaphaa Ladpraaw Press.

Keuning, J.
1839/1849 *De tweede Schipvaart*. 7 vols. The
    Hague: Linschoten Vereeniging.

Keyes, Roger
1984   *Surimono: Privately Published
    Japanese Prints in the Spencer
    Museum of Art*. New York: Kodansha
    International.

Keyes, Roger, and George Kuwayama
1980   *The Bizarre Imagery of Yoshitoshi: The
    Herbert R. Cole Collection*. Los
    Angeles: Los Angeles County
    Museum of Art.

Khemachaarii, Phraya Ariyanuwat
1983   *Prapheenii booraan Isaan baang ryan*
    (Some selected Isan ancient tradi-
    tions), vol. 2. Maha Sarakham
    (Thailand): Maha Sarakham
    Teachers' College.

Kim, Kumja Paik
1997   "Traditional Ceramics." In *An
    Introduction to Korean Culture*, edited
    by John H. Koo and Andrew C.
    Nahm, 387–98. Elizabeth, N.J.:
    Hollym Publishing Company.

Kirsch, A. Thomas
1977   "Complexity in the Thai Religious
    System: An Interpretation." *Journal
    of Asian Studies* 36, no. 2: 241–66.

Klopfer, Lisa
1994   "Confronting Modernity in a Rice-
    Producing Community:
    Contemporary Values and Identity
    among the Highland Minangkabau
    of West Sumatra." Ph.D. diss.,
    University of Pennsylvania.

Knez, Eugene I.
1997   *The Modernization of Three Korean
    Villages, 1951–1981: An Illustrated
    Study of a People and Their Material
    Culture*. Smithsonian Contributions
    to Anthropology, no 39.
    Washington, D.C.: Smithsonian
    Press.

Koentjaraningrat
1984   *Kebudayaan Jawa*. Jakarta: Balai
    Pustaka.

Kondō, Hiroshi
1984   *Saké: A Drinker's Guide*. Tokyo:
    Kodansha International Ltd.

Korean Overseas Culture and Information
    Service
1998   *Handbook of Korea*. 10th ed.
    Elizabeth, N.J.: Hollym Publishing
    Company.

Korvald, Tordis
1999   "Notes on Cultural Performances
    Potential of the Tharu of Far West
    Nepal." In *Nepal: Tharus and Their
    Neighbours*, edited by Harald O. Skar
    et al., 235–52. Kathmandu: EMR
    Publications.

Kotani, Heishichi, ed.
1942 *Mon zukushi* (Comprehensive set of crests). Kyoto: Geīsō-dō.

Krauskopff, Gisèle
1987 "La féminité des poissons: Un motif aquatique du mythe d'origine et des chants de mariage Tharu (Népal)." *Cahiers de littérature orale* 22: 13–28.
1989 *Maîtres et possédés: Les rites et l'ordre social chez les Tharu (Népal).* Paris: Editions du CNRS.
1996 "Emotions, mélodies saisonnières et rythmes de la nature: La littérature orale des Tharu de Dang (Népal)." In *Traditions orales dans le monde indien,* edited by Catherine Champion, 383–401. Paris: Editions de l'Ecole des Hautes Etudes des Sciences Sociales..
1999 "A Marshland Culture: Fishing and Trapping among a Farming People of the Tarai." *Himalayan Research Bulletin* 19, no. 2: 21–37.
2000 "From Jungle to Farm: A Look at Tharu History." In *The Kings of Nepal and the Tharu of the Terai,* edited by Gisèle Krauskopff and Pamela Deuel, 25–48. Kathmandu: CNAS et RUSCA Press.

Krauskopff, Gisèle, and Pamela Deuel Meyer, eds.
2000 *The Kings of Nepal and the Tharu of the Terai.* Los Angeles: Rusca Press.

Kruijt, Albert C.
1935 "De Rijstgodin op Midden-Celebes, en de Maangodin." *Mensch en Maatschappij, Journal of the Dutch Anthropological Society* 11: 109–22.

Kuhn, Dieter
1976 "Die Darstellung des *Keng Chih Tu* und ihre Wiedergabe in popular-enzyklopädischen Werken der Ming-Zeit." *Zeitschridt der deutschen morgenländischen Gesellschaft* 126, no. 2: 336–67.
1987 *Die Song-Dynastie (960 bis 1279): Eine neue Gesellschaft im Spiegel ihrer Kultur.* Weinheim: Acta Humaniorum VCH.

Kumar, Tuk-Tuk
1988 *History of Rice in India: Mythology, Culture, Agriculture.* Delhi: Gian Publishing House.

Lairenmayum, Ibohal, and Khelchandra Ningthoukhongjam
1997 *Cheitharol Kumbaba.* Imphal (India): Manipur Sahitya Parishad.

Lambrecht, Godfrey
1970 "Survivals of Gaddang Religion." *Journal of Northern Luzon* 1, no.1: 1–133.

Lansing, J. Stephen
1987 "Balinese 'Water Temples' and the Management of Irrigation." *American Anthropologist* 89, no. 2: 326–41.
1991 *Priests and Programmers: Technologies of Power in the Engineered Landscape of Bali.* Princeton, N.J.: Princeton University Press.

Lebar, Frank M., et al
1964 *Ethnic Groups of Mainland Southeast Asia.* New Haven. Conn.: Human Relations Area Files Press.

Ledderose, Lothar
2000 *Ten Thousand Things: Module and Mass Production in Chinese Art.* Princeton, N.J.: Princeton University Press.

Lee, Jung-Hee
1998 *Azaleas and Golden Bells—Korean Art in the Collection of the Portland Art Museum.* Portland, Ore.: The Portland Art Museum.

Lee, Oh-Young
1994 *Korea in Its Creations.* Seoul: Design House Publishers.

Lee, Young-Chul
1993 "Culture in the Periphery and Identity in Korean Art." In *Across the Pacific: Contemporary Korean and Korean American Art,* 10–17. New York: The Queens Museum of Art.

Lim, Young-Ju
1997 "Traditional Furnishings and Food Vessels." In *Korean Cultural Heritage—Traditional Lifestyles,* vol. 4, 208–13. Seoul: Korea Foundation.

Link, Howard A.
1979 *The Art of Shibata Zeshin.* Honolulu: Honolulu Academy of Arts.

Little, Stephen
2000 *Taoism and the Arts of China,* Chicago: Art Institute of Chicago.

Lolarga, Elizabeth
2001 "Reviving Old Rituals to Preserve Ifugao Terraces." *Philippine Daily Inquirer* (February 22).

Lombard, Denys
1990 *Le carrefour javanais: Essai d'histoire globale.* 3 vols. Paris: Editions de l'Ecole des Hautes Etudes des Sciences Sociales.

Los Angeles Times
2002 "Rice with Added Gene Withstands Drought." *Los Angeles Times,* (November 30): A16.

Lovelace, George W., et al., eds.
1988 *Rapid Rural Appraisal in Northeast Thailand: Case Studies.* Khon Kaen (Thailand): KKU-FORD Rural Systems Research Project, Khon Kaen University.

Luce, Gordon H.
1965 "Rice and Religion: A Study of Old Mon-Khmer Evolution and Culture." *Journal of the Siam Society* 53, no. 2: 139–52.

Madale, Nagasura T.
1974 "Kashawing: Rice Ritual of the Maranaos." *Mindanao Journal* 1, no. 1: 74–80.
2000 *Tales from Lake Lanao and Other Essays.* Manila: National Commission for Culture and the Arts

Maji, Baikunthanath
N.D. *Brihat Lakshmi Charitra.* Calcutta: Rajendra Library.

Majumdar, Ashutosh
1999 *Meyeder Bratokatha.* Calcutta: Deb Sahitya Kutir.

Manikmaya
1980 *Literature of Java,* vol. 4. Bibliotheca Univesitatis Leidensis. Leiden: Leiden University Press.

Manikmaya
1981 Vol. 1. Jakarta: Proyek Penerbitan buku Sastra Indonesia dan Daerah.

Maraini, Fosco.
1999 "*Ikupasuy*: It's Not a Mustache Lifter!" In *Ainu: Spirit of a Northern People,* edited by William W. Fitzhugh and Chisato O. Dubreuil, 327–34. Washington, D.C.: National Museum of Natural History.

Mayer, L. Th.
1897 *Een Blik in het Javaansche Volksleven.* 2 vols. Leiden: E. J. Brill.

McDonaugh, Christian
1984 "The Tharu of Dang: A Study of Social Organization, Myth, and Ritual in West Nepal." Ph.D. diss., Oxford University.
1989 "The Mythology of the Tharu: Aspects of Cultural Identity in Dang, West Nepal." *Kailash* 15, no. 3–4: 191–205.

Meyer, Kurt, and Pamela Deuel, eds.
1998 *Mahabharata: The Tharu Barka Naach.* Los Angeles: Deuel Purposes.

Modiano, G., et al.
1991 "Protection against Malaria Morbidity: Near Fixation of the Alpha-Thalassemia Gene in a Nepalese Population." *American Journal of Human Genetics* 48, no. 2: 390–97.

Moore, E.
1989 "Water management in Early Cambodia: Evidence from Arial Photography." *The Geographical Journal* 155, no. 2: 201–14.

Morse, Peter
1989 *Hokusai: One Hundred Poets.* New York: George Braziller, Inc.

Mutua Museum
1999 *Tribal Profile of Manipur.* Imphal (India): Mutua Museum.

Narayan, Kirin
1997 *Mondays on the Dark Night of the Moon.* Oxford: Oxford University Press.

Narayanan, Vasudha
1996 "Śrī: Giver of Fortune, Bestower of Grace." In *Devī: Goddesses of India,* edited by John Stratton Hawley and Donna Marie Wulff, 87–108. Berkeley: University of California Press.

National Folk Museum of Korea
1991 *Hanguk chip munwha: Uri salpwiburi* (Straw in Korean rural life). Seoul: National Folk Museum.

The National Library, Fine Arts Department
1990 *Lag silaacaaryg pho Khun Ramkhamhaeng Maharat* (The inscription of King Ramkhamhaeng the Great). Bangkok: The National Library.

National Museum of Korea
1993 Seoul: Tongchon Publishing Company

Needham, Joseph, and Wang Ling
1965 *Mechanical Engineering.* Vol. 4, pt. 2 of *Science and Civilization in China,* edited by Joseph Needham. Cambridge: Cambridge University Press.

Needham, Joseph, et al.
1971 *Civil Engineering and Nautics.* Vol. 4, pt. 3 of *Science and Civilization in China,* edited by Joseph Needham. Cambridge, Cambridge University Press.

Nelson, Sarah Milledge
1993   *The Archaeology of Korea*. New York: Cambridge University Press.

Ngariyanbam, Kulachandra
1979   *Phouoibi Khurumnaba*. Imphal (India): Ng. Kulachandra.
1998   *Meitei Lai-Haraoba*. Imphal (India): Ng. Kulachandra.

Ngô Quang Nam and Xuân Thiêm
1986   *Ðia chí Vĩnh Phú: Văn hoá dân gian vùng dất tổ* (Vĩnh Phú monography; Folklore in the ancestral land). Vĩnh Phú (Vietnam): Sở Văn hoá và Thông tin tỉnh Vĩnh Phú (Vĩnh Phú Department of Culture and Information).

Nguyễn Ngọc Binh
1985   "Vietnamese Poetry: The Classical Tradition." In *Vietnam: Essays on History, Culture, and Society*, edited by David P. Elliott, 79–98. New York: Asia Society.

Nguyễn Xuân Hiên
2001   *Glutinous-Rice-Eating Tradition in Vietnam and Elsewhere*. Bangkok: White Lotus Press.

Ningthoukhongjam, Khelchandra
1978   *Ariba Manipuri Longei*. Manipur (India): Manipur State Kala Academy

Nonghyob Agricultural Museum
1997   *Agricultural Artifacts of Korea*. Seoul: Nonghyob Agricultural Museum.

Nooy-Palm, Hetty
1986   *The Sa'dan Toraja: A Study of Their Social Life and Religion*. Vol. 2 of *Rituals of the East and West*. Verhandelingen van het Koninklijk Instituut voor Taal-, Land- en Volkenkunde, 118. Dordrecht (The Netherlands): Foris Publications.

Normile, Dennis, and Elizabeth Pennisi
2002   "Rice: Boiled Down to Bare Essentials." *Science* 296, no. 5565 (April 5): 32–36.

O'Connor, Richard
1995   "Agricultural Change and Ethnic Succession in Southeast Asian States: A Case for Regional Anthropology." *The Journal of Asian Studies* 54, no. 4: 968–96.
2000   "Who are the Tai?: A Discourse of Place, Activity, and Person." In *Dynamics of Ethnic Cultures across National Boundaries in Southwestern China and Mainland Southeast Asia*, edited by Y. Hayashi and G. Yang, 35–50. Kyoto: Center for Southeast Asian Studies.

Ohnuki-Tierney, Emiko
1993   *Rice as Self: Japanese Identities through Time*. Princeton: Princeton University Press.

Okada, Shigehiro, et al.
1989   *National Museum of Japanese History English Guide*. Sakura: National Museum of Japanese History.

Ottino, Arlette
2000   *The Universe Within: A Balinese Village through Its Ritual Practices*. Paris: Editions Karthala.

Paerels, J. J.
1913   *De Rijst: Onze koloniale Landbouw*, vol. 5. Haarlem: H. D. Tjeenk Willink & Zoon.

Park, Tae-Sun
1993   "Life's Milestones: Ceremonies and Food." *Koreana: Korean Art and Culture Journal* 7, no. 3: 20–27.

Paterno-Locsin, Maria Elena
1999   *Beyond Rice*. Manila: Centro Escolar University.

Pemberton, John
1994   *On the Subject of Java*. Ithaca, N.Y.: Cornell University Press.

Perryman, Jane
2000   *Traditional Pottery of India*. London: A. and C. Black.

Phayomyong, Mani
1994   *Prapheenii sib song dyan Lanna Thai* (The Lanna Thai twelve-month traditions). Chiang Mai (Thailand): So Sap Kan Pim.

Phimworamathakul, Bunkerd
1996   *Phayaa* (An anthology of Isan oral poetry). Khon Kaen (Thailand): Khon Kaen Publication.

Phinthong, Priichaa
1991   *Prapheenii boran Thai Isaan* (Isan's ancient traditions). 7th ed. Ubon Ratchathani (Thailand): Siritham Offset.

Pigeaud, Th.
1980   *Literature of Java*, vol. 4. Bibliotheca Univesitatis Leidensis. Leiden: Leiden University Press.
1994   *Javaans-Nederlands Woordenboek*. Leiden: KITLV.

Pingali, Prabhu L., Mahabub Hossain, and Roberta V. Gerpacio
1997   *Asian Rice Bowls: The Returning Crisis?* New York: CAB International.

Plutschow, Herbert
1996   *Matsuri: The Festivals of Japan*. Surrey: Japan Library.

Poeroebaja, B. P. H.
1939   "Rondom de Huwelijken in de Kraton te Jogyakarta." *Djawa* 6: 295–329.

Poerwardarminta, W. J. S
1930   *Baoesastra Djawa*, vol.1. Yogyakarta: Mardi Basa Triwikrama.
1939   *Baoesastra Djawa*. Groningen/Batavia: J. B. Wolter's.

Poor, Robert J.
1979   *Ancient Chinese Bronzes, Ceramics, and Jade in the Collection of the Honolulu Academy of Arts*. Honolulu: Honolulu Academy of Arts.

Powers, Martin J.
1991   *Art and Political Expression in Early China*. New Haven, Conn.: Yale University Press.

Prawiroatmojo
1981   *Bausastra Jawa-Indonesia*. Jakarta: Gunung Agung.

Premchit, Sommai, and Amphay Dore
1992   *The Lan Na Twelve-Month Traditions*. Chiang Mai (Thailand): So Sap Kan Pim.

Ramseyer, Urs
1977   *The Art and Culture of Bali*. Oxford: Oxford University Press.

Ranggawarsita, Raden Ngabei
1993–1994 *Serat Pustakaraja Purwa*. 3 vols. Edited by Kamajaya. Surakarta (Indonesia): Yayasan "Mangadeg"; Yoyakarta: Yayasan Centhini.

Rapier, David
1986   *Masks and Paradox*. Berkeley: University of California Press.

Rassers, W. H.
1959   *Pañji, the Culture Hero: A Structural Study of Religion in Java*. Koninklijk Instituut voor Taal-, Land- en Volkenkunde Translation Series 3. The Hague: M. Nijhoff.

Rawson, Jessica, et al., eds.
1996   *Mysteries of Ancient China: New Discoveries from Early Dynasties*. London: British Museum.

Reizei, Tamehito, et al.
1996   *Mizuho no kuni Nippon: Shiki-kosakuzu no sekai* (The land of vigorous rice plants, Japan: The world of four-season agricultural cultivation drawings). Kyoto: Tankosya.

"Rice Terraces"
2002   "Rice Terraces Task Force Abolition Deplored." *Manila Bulletin* (February): B-16.

Rosenfield, John M., ed.
1979   *Song of the Brush: Japanese Paintings from the Sansō Collection*. Seattle: Seattle Art Museum.

Santhi, G.
1994   *Folklore Survey of Thanjavur District*. Thanjavur (India): Tamil University.

Sarathasananan, San
1995   "Bun Khuun Laan." In *Chiiwid Thai chud buuchaa Phajaa Thaan* (Thai life: A series on worshiping Phajaa Thaan.), 1–5. Bangkok: Kurusaphaa Ladpraaw Press.

Sax, William S.
1994   "Who's Who in the Pandav Lila?" In *The Gods at Play: Lila in South Asia*, edited by William Sax, 131–55. New York: Oxford University Press.
1996   "Worshiping Epic Villains: A Kaurava Cult in the Central Himalayas." In *Epics and the Contemporary World*, edited by Margaret Beissinger et al. Berkeley: University of California Press.

Sayers, Robert, with Ralph Rinzler
1987   *The Korean Onggi Potter*. Smithsonian Folklife Studies, no. 5. Washington, D.C.: Smithsonian Institution Press

Schafer, Edward O.
1967   *The Vermilion Bird: T'ang Images of the South*. Berkeley: University of California Press.

Scubla, Lucien
1999   "'Ceci n'est pas un meutre'; ou, Comment le sacrifice contient la violence." In *De la violence: Séminaire de Françoise Héritier*, 135–70. Paris: Editions Odile Jacob.

Sen, Rai Saheb Dineshchandra
1920   *The Folk Literature of Bengal*. New Delhi: B. R. Publishing Corporation.

Setia, Putu
1978   "10 tahun sekali di desa Selat : Sang Bethara turun menginjak nasi sepanjang 1 km." *Bali Post* (March 13).

Shaughnessy, Edward L., ed.
2000   *China: Empire and Civilization*. Oxford: Oxford University Press.

Sibeth, Achim
1991 *The Batak: Peoples of the Island of Sumatra.* New York: Thames and Hudson.

Simoons, Frederick J.
1991 *Food in China: A Cultural and Historical Inquiry.* Boca Raton, Fla.: CRC Press.

Singh, P.
1994 *Excavations at Narhan, 1984–1989.* Varanasi (India): Benares Hindu University.

Sinha, Surajit
1985 "Social and Ecological Context of Rice Cultivation in India." *East Asian Cultural Studies* 24: 83–92.

Skeat, Walter William
1966 *Malay Magic.* New York: Barnes and Noble.

Smutkupt, Suriya, et al.
1991a *Bun Phawes of Isan: An Anthropological Interpretation* (in Thai). Khon Kaen (Thailand): Isan Anthropological Collection, Department of Sociology and Anthropology, Khon Kaen University.
1991b *Bun Khaw Pradabdin and Bun Khaw Sag: Ritual, Rice, and Man in the Sociocultural Context of Isan* (in Thai). Khon Kaen (Thailand): Isan Anthropological Collection, Department of Sociology and Anthropology, Khon Kaen University.
1993 *The Religious Rituals and Beliefs as Traditional Authority for the Indigenous System of Natural Resource Management in Isan Peasant Communities of the Chi River Basin* (in Thai). Nakhon Ratchasima (Thailand): Isan Anthropological Museum Project, School of General Education, Institute of Social Technology, Suranaree University of Technology.

Smutkupt, Suriya, and Pattana Kitiarsa
1998 *Dominant Symbols in Bun Bung Fai: An Anthropological Interpretation* (in Thai). 2d ed. Nakhon Ratchasima (Thailand): Thai Studies Anthropological Collection, Institute of Social Technology, Suranaree University of Technology.
2000a "Flying High: Kites, Place, and Imagination in Thai Popular Culture" (in Thai). In *Phaakhaawmaa, yaam, waaw: Khaamriang waaduay raangkaai attaluk lae phuenthii nai watthanatham Thai* (Loin cloths, shoulder bags, and kites: Essays on bodies, identities, and places in Thai culture), 83–154. Nakhon Ratchasima (Thailand): Thai Studies Anthropological Collection, Institute of Social Technology, Suranaree University of Technology.
2000b *Cultural Politics and the Secularization of the Bun Phawes in Roi-et Market Town* (in Thai). Nakhon Ratchasima (Thailand): Thai Studies Anthropological Collection, Institute of Social Technology, Suranaree University of Technology.

Smyers, Karen
1999 *The Fox and the Jewel: Shared and Private Meanings in Contemporary Japanese Inari Worship.* Honolulu: University of Hawaii Press.

Soekawati, Tjokorde Gde Rake
1926 "Legende over den Oorsprong van de Rijst en Godsdienstige gebruiken bij den Rijstbouw onder de Baliérs." *Tijdschrift voor Indische Taal-, Land- en Volkenkunde* 66: 423–34

Srisoong, Buasri
1992 *Hiid kong Isan* (Isan's folk traditions). Bangkok: The Phumpanya Foundation.

Srivastava, Surendra Kumar
1948–1949 "Spring Festival Among the Tharus." *Eastern Anthropologist* 2, no. 1: 27–33.
1958 *The Tharus: A Study in Culture Dynamics.* Agra: Agra University Press.

Stuart-Fox, David J.
1974 *The Art of the Balinese Offering.* Yogyakarta: Penerbitan Yayasan Kanisius.
1982 *Once a Century: Pura Besakih and the Eka Dasa Rudra Festival.* Jakarta: Penerbit Sinar Harapan and Citra Indonesia.
1996 *The Dictionary of Art,* edited by Jane Turner. S.v. "Besakih." London: Macmillan.
1998 "Cycles and Ritual Centres." In *Indonesian Heritage.* Vol. 9 of *Religion and Ritual,* ed. by James J. Fox, 118–19. Singapore: Archipelago Press.
2002 *Pura Besakih : Temple, Religion, and Society in Bali.* Leiden: KITLV Press.

Stutterheim, W. F.
1956 *Studies in Indonesian Archeology.* The Hague: M. Nijhoff

Supartha, I Gusti Ngurah Oka
1998 "Persembahan dan pemujaan karya patabuh gentuh dan pamijilan Ida Bhatara Ngerta Gumi, setiap 10 tahun di desa adat Selat." *Lontar* 9 (tahun 3): 51–55.

Tagore, Abanindranath
N.D. *Banglar brata-katha.* Unpublished manuscript.

Tagore, Rabindranath
1931 *Gitabitan.* Shantiniketan (India): Vishwabharati.

Tambiah, Stanley J.
1970 *Buddhism and the Spirit Cults in North-east Thailand.* Cambridge: Cambridge University Press.

Tang Shenwei, comp.
1600 *Chong kan Jing shi zheng lei da quan ben cao: San shi yi juan fu tu.* [China]: Ji Shan shu yuan chong qie, Ming Wanli geng zi. (First published 1108).

Tarcelo-Balmes, Florian
1999a "The Last Mumbaki." *Today* (Manila; April 23): 7.
1999b "Highland Erosion." *Today* (Manila; April 23): 7.

Thach Phuong and Le Trung Vu
1995 *Sixty Traditional Festivals of Vietnam* (in Vietnamese). Hanoi: Social Science Publishing House.

Thompson, Stuart E.
1988 "Death, Food and Fertility." In *Death Ritual in Late Imperial and Modern China,* edited by James L. Watson and Evelyn S. Rawski, 71–108. Berkeley: University of California Press.

Togawa Yasuyuki
1969 "Haguro-san no Toshiya Matsuri. (Toshiya Festival of Mount Haguro)." *Monthly Journal of Bunkazai* (December 1969).

Tran Quoc Vuong
1993 *Tracing the History Line* (in Vietnamese). Hanoi: Cultural Publishing House.

Trankell, Ing-Brit
1995 *Cooking, Care, and Domestication: A Culinary Ethnography of the Tai Yong, Northern Thailand.* Uppsala Studies in Cultural Anthropology, no. 21. Uppsala: Uppsala University.

Treu, Manfred G.
1993 "A Translation of Laksmiprasada Devkota's 'The Fifteen Days of Asar.'" *Contribution to Nepalese Studies* 20, no. 2: 149–64.

Tuân, Dao Thê
1985 "Types of Rice Cultivation and Its Related Civilizations in Vietnam." *East Asian Cultural Studies* 24: 41–56.

Turner, Victor
1967 *The Forrest of Symbols: Aspects of Ndembu Ritual.* Ithaca, N.Y.: Cornell University Press

Übelhör, Monika
1969 "Hsu Kuang-ch'i (1562–1633) und seine Einstellung zum Christentum." Ph.D. diss., University of Hamburg.

Ueno, Saeko
1974a *Review Essays on Sample Design Books* (in Japanese). Vol 1 of *Collections of Sample design books of Kosode Patterns,* edited by Yamanobe Tomoyuki. Tokyo: Gakusha-Kennkyusya.
1974b *Review Essays on Sample Design Books* (in Japanese). Vol 2 of *Collections of Sample design books of Kosode Patterns,* edited by Yamanobe Tomoyuki.. Tokyo: Gakusha-Kennkyusya.

Van der Goes, Beatriz
1997 "Beru Dayang: The Concept of Female Spirits and the Movement of Fertility in Karo Batak Culture." *Asian Folklore Studies* 56: 379–405.

Van der Weijden, Gera
1981 *Indonesische Reisrituale: Basler Beiträge zur Ethnologie 20.* Basel: Ethnologisches Seminar der Universität und Museum für Völkerkunde.

Van Setten van der Meer, Nancy Claire
1979 *Sawah Cultivation in Ancient Java.* Oriental Monograph Series, no. 22, Faculty of Asian Studies. Canberra: Australian National University Press.

Ventura, Bill, and Benjie de Yro
2002 "Arroyo Vows Preservation of the Banaue Rice Terraces." *Manila Bulletin,* (January 13): K2.

Visaya, Villamor
2002 "Ifugao Governor Hits Transfer of Rice Terraces' Supervision." *Philippine Daily Inquirer* (January 30): A13.

Visser, Leontine
1989   *My Rice Field Is My Child: Social and Territorial Aspects of Swidden Cultivation in Sahu, Eastern Indonesia.* Dordrecht (The Netherlands): Foris Publications.

Wagner, Donald B.
1996   *Iron and Steel in Ancient China.* Leiden: Brill.

Wang Chaosheng, ed.
1995   *Zhongguo gudai gengzhi tu* (Farming and weaving pictures in ancient China). Beijing: China Agriculture Press.

Wang Heyan
1744   *Shou shi tong kao.* [China]: Jiangxi shu ju.

Wang Shucun
1992   *Paper Joss: Deity Worship through Folk Prints.* Beijing: New World Press.

Wang Xiangkun and Sun Zhuanqing, eds.
1996   *Zhongguo zaipei dao qiyuan yu yanhua yanjiu zhuanji* (Collected papers on the origins and spread of rice cultivation in China). Beijing: Chinese Agricultural University Press.

Wang Zhen
1963   *Nong shu* (Agricultural treastise). Ju zhen ban cong shu [282–86]; Bai bu cong shu ji cheng, zhi 27. Taipei: Yee Wen Publishing Co., Ltd.

Wang Zhongshu
1982   *Han Civilization.* Translated by K. C. Chang, et al. New Haven, Conn.: Yale University Press.

Wangkhemcha, Tomba
1996   *Loutarol.* Imphal (India): Premier Publishers, Mayang Langjing.

Waterson, Roxana
1984   *Ritual and Belief of the Sa'dan Toraja.* Center for Southeast Asia Studies Occasional Paper. Canterbury: University of Kent.
1990   *The Living House: An Anthropology of Architeture in South-East Asia.* Singapore: Oxford University Press.

Webster's Ninth New Collegiate Dictionary
1988   Springfield, Mass.: Merriam-Webster Inc.

Wickman, Michael
1978   *Korean Chests: Treasures of the Yi Dynasty.* Seoul: International Tourist Publishing Company.

Wirz, Paul
1927   "Der Reisbau und die Reisbaukulte auf Bali und Lombok." *Tijdschrift voor Indische Taal-, Land- en Volkenkunde* 67: 216–345.

Wittfogel, Karl August
1957   *Oriental Despotism: A Comparative Study of Total Power.* New Haven, Conn.: Yale University Press.

Wolf, Arthur P.
1974   "Gods, Ghosts and Ancestors." In *Religion and Ritual in Chinese Society,* by Arthur P. Wolf and Robert J. Smith, 131–82. Stanford, Calif.: Stanford University Press.

Worcester, Dean C.
1912   "Head-Hunters of Northern Luzon." *National Geographic Magazine* 23, no. 9: 833–930.

Wright, Edward R., and Man-Sill Pai
2000   *Traditional Korean Furniture.* Tokyo: Kodansha International Ltd.

Wuhan Hydroelectric Institute
1979   *Zhongguo shuili shigao* (A brief history of water control in China). 2 vols. Beijing: Hydroelectric Press.

Yamanobe Tomoyuki, ed.
1974a   *Kosode moyo hinakata-bon shusei ichi* (Collections of sample design books of Kosode patterns), vol. 1. Tokyo: Gakusha-Kennkyusya.

1974b   *Kosode moyo hinakata-bon shusei ni* (Collections of sample design books of Kosode patterns), vol. 2. Tokyo: Gakusha-Kennkyusya.
1974c   *Kosode moyo hinakata-bon shusei san* (Collections of sample design books of Kosode patterns), vol. 3. Tokyo: Gakusha-Kennkyusya.
1974d   *Kosode moyo hinakata-bon shusei yon* (Collections of sample design books of Kosode patterns), vol. 4. Tokyo: Gakusha-Kennkyusya.

Yampolsky, Philip B., trans.
1967   *The Platform Sutra of the Sixth Patriarch: The Text of the Tun-Huang Manuscript.* New York: Columbia University Press.

Yee, Cordell D. K.
1996   *Space and Place: Map Making East and West.* Annapolis: St. John's College Press.

Yoo, Tae-Jong
1997   "Korea's Best Loved Wines." In *Korean Cultural Heritage: Traditional Lifestyles,* vol. 4, 228–37. Seoul: Korea Foundation.

Yoshida, Mitsukuni
1989   *Naorai: Communion of the Table.* Hiroshima: Mazda Motor Corporation.

Yule, Col. Henry, and A. C. Burnell
1903   *Hobson-Jobson: A Glossary of Colloquial Anglo-Indian Words and Phrases, and of Kindred Terms.* London: John Murray.

Zainal Kling
1985   "Rice Field Rituals in Malaysia." *East Asian Cultural Studies* 24: 131–54.

Contributors

Aurora Ammayao is a native of Banaue, Ifugao (Philippines), who has an intimate knowledge of the region's culture and rituals. **Gene Hettel** is head of Communication and Publications Services at the International Rice Research Institute (IRRI).

Mutua Bahadur is the founding director of the Mutua Museum, the primary museum in the state of Manipur, northeastern India. He is the author of many books including *Traditional Textiles of Manipur* and *Cane and Bamboo Crafts of Manipur*.

Dr. Francesca Bray is Professor of Anthropology at the University of California, Santa Barbara. Her publications include *Technology and Gender: Fabrics of Power in Late Imperial China*, *The Rice Economies: Technology and Development in Asian Societies*, and *Agriculture* (volume 6, part 2 of *Science and Civilization in China*).

Francine Brinkgreve is the author of *Offerings: The Ritual Art of Bali*. She currently works for the Foundation for the Oral History of Indonesia, Royal Institute of Linguistics and Anthropology, Leiden (Netherlands).

Dr. Eric Crystal is Emeritus Professor of Anthropology at the University of California, Berkeley, and former Coordinator for that university's Center for South and Southeast Asia Studies. In addition to his research experience in Indonesia and Vietnam, he is also an accomplished photographer whose work has been featured in numerous exhibitions.

Dr. Mariko Fujita is Professor of Cultural Anthropology at Hiroshima University, Japan. Her current research focuses on the economic history of the sake industry in Saijo, Hiroshima Prefecture, and she is also co-author of *Life in Riverfront: A Middle Western Town Seen through Japanese Eyes*.

Ruth Gerson is an independent scholar who has conducted many years of research on topics of Thai culture from her home base in Bangkok. She is the author of *Traditional Festivals in Thailand*.

Roy W. Hamilton is Curator of Asian and Pacific Collections at the UCLA Fowler Museum of Cultural History. His previous books include *From the Rainbow's Varied Hue: Textiles of the Southern Philippines* and *Gift of the Cotton Maiden: Textiles of Flores and the Solor Islands*.

Dr. Stephen C. Headley is an anthropologist specializing in Javanese studies at the Centre National de la Recherche Scientifique (CNRS), Paris. His publications include *From Cosmogony to Exorcism in a Javanese Genesis: The Spilt Seed*.

Rens Heringa's many publications based on her research in East Java have provided a valuable counterpoint to the court-centric focus of most Javanese cultural studies. She is author of *Spiegels van ruimte en tijd: Textiel uit Tuban*, and co-author of *Fabric of Enchantment: Batik from the North Coast of Java*.

Garrett Kam is an independent scholar and specialist in Balinese arts. He is the author of *Ramayana in the Arts of Asia* and *Perceptions of Paradise: Images of Bali in the Arts.*

Thitipol Kanteewong teaches in the Faculty of Thai Arts, Chiang Mai University, Thailand. He is an accomplished performer of northern Thai musical genres and has a master's degree in musicology from Nakornpathom University.

Dr. Pattana Kitiarsa is a lecturer at Suranaree Institute of Technology, Nakhon Ratchasima, Thailand. He has a doctorate in sociocultural anthropology from the University of Washington.

Dr. Gisèle Krauskopff is Chargée de Recherche at the Centre National de la Recherche Scientifique (CNRS), Paris. Most of her research deals with the Tharu of the Dang Valley, Nepal, and she is the author of *Maîtres et possédés: Les rites et l'ordre social chez les Tharu.*

Dr. Nanditha Krishna is Director of the C.P.R. Environmental Education Centre, Madras, India. Her numerous publications on the arts of southern India include *The Art and Iconography of Vishnu-Narayana* and *The Arts and Crafts of Tamilnadu.*

Dr. Kwang-Kyu Lee is a leading figure in Korean anthropology and Professor Emeritus at Seoul National University. Among his many publications are *Overseas Koreans* and *Social Anthropology of Korean Families.*

Kurt W. Meyer and **Pamela Deuel** have conducted research in the Tharu region of Nepal since 1993 and are the authors of *Mahabharata: The Tharu Barka Naach* and *The Kings of Nepal and the Tharu.* Kurt Meyer is also a Fellow of the American Institute of Architects.

Dr. Sohini Ray is Lecturer in Anthropology at the University of California, Irvine, and former fellow at the Center for the Study of World Religions, Harvard University. Her research subjects have included the dance of Manipur and the rice-related cultural traditions of her native West Bengal.

Dr. Michael Reinschmidt is Lecturer in Anthropology at California State University, Chico. He has previously organized exhibitions as Curator at the Korean American Museum and served as Associate Director at Academia Koreana, both in Los Angeles.

Dr. Toshiyuki Sano is Associate Professor of Anthropology at Nara Women's University. He has his doctorate from Stanford University and is co-author of *Life in Riverfront: A Middle Western Town Seen through Japanese Eyes.*

Suriya Smutkupt is Museum Director of the Thai Studies Anthropological Collection at Suranaree Institute of Technology, Nakhon Ratchasima, Thailand. His publications in Thai include *Liké in Thai Popular Culture* and *Cultural Constructs of Local Knowledge: Ethnic Thai-Lao Spirit-Medium Cults in Khorat Villages.*

Kik Soleh Adi Pramono is Artistic Director of the Mangun Dharma Arts Center (*Padepokan Seni Mangun Dhara*) in Tumpang, East Java. As a puppet master (*dhalang*), he works to maintain the vibrancy of traditional arts and local rituals.

David J. Stuart-Fox has published extensively on Balinese culture, including his most recent book, *Pura Besakih: Temple, Religion, and Society in Bali.* He is Librarian at the National Museum of Ethnology, Leiden (Netherlands).

Vi Văn An is Curator at the Vietnam Museum of Ethnology, Hanoi. A native Tai, he specializes in the culture of the Tai communities of northern Vietnam.

Vũ Hồng Thuật is Curator at the Vietnam Museum of Ethnology, Hanoi. His research focuses on the festivals and religious practices of the people of Vietnam's Red River Delta.

Index

544   INDEX